DIARY OF A FARMER AT THE FOOT OF MT. KANPŪ

DIARY OF A FARMER AT THE FOOT OF MT. KANPŪ

Living off the Land in Northern Japan, 1935–1936

Yoshida Saburō

Translated, Edited, and Annotated by

Donald C. Wood

University of Hawaiʻi Press

Honolulu

The publication of this book is supported by the Suntory Foundation Support for Overseas Publication program.

Originally published in 1938 in Japanese by the Attic Museum, Tokyo, as:
Oga Kanpūsanroku Nōmin Nichiroku (男鹿寒風山麓農民日録)

Printed in the United States of America

First printed, 2025

Library of Congress Cataloging-in-Publication Data

Names: Yoshida, Saburō, author. | Wood, Donald C.,
translator, editor, annotator.
Title: Diary of a farmer at the foot of Mt. Kanpū : living off the land in northern Japan, 1935-1936 / Saburō Yoshida ; translated, edited, and annotated by Donald C. Wood.
Other titles: Oga Kanpū sanroku nōmin nichiroku. English | Living off the land in northern Japan, 1935-1936
Description: Honolulu : University of Hawaiʻi Press, 2025. | Translation of: Oga Kanpū sanroku nōmin nichiroku. | Includes bibliographical references and index.
Identifiers: LCCN 2024049317 (print) | LCCN 2024049318 (ebook) | ISBN 9780824898021 (hardback) | ISBN 9798880701193 (epub) | ISBN 9798880701209 (kindle edition) | ISBN 9798880701186 (pdf)
Subjects: LCSH: Yoshida, Saburō, 1907—Diaries. | Farm life—Japan—Diaries. | Farmers—Japan—Social life and customs. | Farm life—Japan—Oga-shi. | Oga-shi (Japan)—Social life and customs. | LCGFT: Diaries.
Classification: LCC S522.J3 Y6713 2025 (print) | LCC S522.J3 (ebook) | DDC 630.92 [B]—dc23/eng/20250410
LC record available at https://lccn.loc.gov/2024049317
LC ebook record available at https://lccn.loc.gov/2024049318
Library of Congress Cataloging-in-Publication Data

Jacket art: *Background* Mt. Kanpū and the hamlet of Ōkura. *Portrait* Yoshida and his family. Photographs by Ogawa Tōru.

University of Hawaiʻi Press books are printed on acid-free paper and meet the guidelines for permanence and durability of the Council on Library Resources.

This book is dedicated to all of
the descendants and other relatives of Yoshida Saburō
who assisted me with this project, and especially to those
who departed this world before its completion.
Would that I had finished it while they were living.

Contents

List of Figures and Maps

Notes on Terms

With the exception of some names in the acknowledgments, Japanese names are written in the conventional style, with surnames first, except for the names of Japanese authors of works published in the reverse order in English. No names are used to refer to some people (e.g., most of the author's siblings and his elder brother's wife) mentioned in the original work. Annotations are given where necessary. Pseudonyms are used where the author used them, and also in one other case, which is explained.

Macrons are used to indicate long vowels. Exceptions include words like Tōkyō and tōfu, which are widely recognized in English, and the names of Japanese authors of works published in English.

Throughout this book, I refer to the original author as Yoshida and as Saburō interchangeably on a case-by-case basis, due to the large number of people with the Yoshida surname, and to my own feelings of affinity to him, gained by translating his 1938 book, through reading his other writings, and from my many conversations about him with his daughters and other close relatives.

Acknowledgments

Many people helped me bring Yoshida Saburō's 1938 masterpiece out into the world in English. First, I thank Itō Abito, my professor at the University of Tokyo, for introducing it to me, and Kamada Yukio of North Asia University for introducing me to Saburō's relatives and helping with the clarification of some terms. However, I am most grateful and indebted to Yoshida Fusae, Saburō's eldest daughter, for hosting me at her home for hours at a time on many occasions over more than a decade. From her I gained priceless insights into the village of Wakimoto in the 1930s, its people, the isolated life on the small mountain where her family's little farm was situated, and the time period. She also taught me about her parents and other relatives. This translation would have had many holes and been far less precise without her cooperation.

I also thank Fusae's son, Sōkō, for providing contextual information on his grandfather and the two letters in Appendices 6 and 7, and for giving me access to his grandfather's camera. I received invaluable assistance from Saburō's younger daughter, Yaeko, and her family. I spent many hours at their home during the first few years of the project, scanning photographs and other materials from Saburō's photo albums, files, and scrapbooks, while learning from her and her husband, Nakamichi Takurō, about Saburō, his life and times, and Oga and Akita in the 1930s. Takurō, a former physics professor with deep knowledge of the geographical area, taught me much. I appreciate his generosity and candor. It was his eldest son, Kō, who—having shouldered the responsibility of organizing and cataloging his grandfather's photographs and articles—kindly made these available to me. This allowed me to amass a large collection of images. I frequently took printouts of these to Fusae's home and asked her to tell me about them, which proved extremely fruitful. When I found that my image scans were inadequate for publication—a serious problem, as Nakamichi Takurō, Yaeko, and Kō had passed away by that time—Kō's brother Yō, and his sister Hino Chika, saved the day by allowing me to borrow several of their grandfather's photo albums. Thanks to their trust and generosity, I was able to create high-quality copies of nearly all of the photos in the original book, and more. I remain deeply grateful.

I also thank Yoshida Dōichi and his wife, Mutsuko, who taught me about Dōichi's father, Shirō, and Shirō's elder brother, Saburō. They also provided me with valuable printed materials and helped me solve some difficult puzzles. In addition, I thank the Institute for the Study of Japanese Folk Culture at Kanagawa University, successor to Shibusawa Keizō's Attic Museum, for granting me permission to publish my translation, for giving me access to certain materials, and for additional support.

A number of other people helped me, including Nagai Miho of the Shibusawa Eiichi Memorial Foundation, and Sensui Hidekazu, Sano Kenji, and Katō Tomoko of the Institute for the Study of Japanese Folk Culture, as well as former Institute member Kobayashi Kōichirō. Research librarians at the Akita Prefectural Library also assisted by unearthing useful information. My wife, Akiko, was instrumental in getting the project off the ground and in the early stages of translating. Her gracious assistance and support at the beginning allowed me to move forward on my own. I would not have been able to complete this project without her. I also thank Seiko for the typing, and for translation and other assistance.

I am grateful to the Japan Society for the Promotion of Science for the two research grants (JP21720317 and JP18KO1164) that made this project possible, and to the Donald Keene Center at Columbia University for selecting the first-draft manuscript of this book as a winner of the 2022–2023 Lindsley and Masao Miyoshi Translation Prize, which helped move things forward near the end. Specifically, Yoshiko Niiya's support is much appreciated. Furthermore, my gratitude goes out to the Suntory Foundation for the generous fiscal 2024 Support for Overseas Publication grant, which helped cover the costs of publishing this book: crucial subvention, due to the book's many illustrations and length.

I also thank the University of Hawaiʻi Press for publishing this book. I am especially grateful to Masako Ikeda for believing in this project from the time of my initial proposal, and for all of her support along the way, as well as to Malia Collins. Two anonymous readers whose critiques helped me make significant improvements also deserve acknowledgment, as do Jack Walsh, for creating the maps in Appendix 2, and Tom Mayers, for introducing me to Junichi Saga's *Memories of Silk and Straw.* The meticulous work of copy editor Wendy Bolton and proofreader Joyce Li is also deeply appreciated. Last, I am grateful to Robert J. Smith for his attention, bibliographic suggestions, and encouragement. Communicating with Bob about my research project and translating during the last few years of his life was an unforgettable and intellectually stimulating experience. Finally, all opinions expressed here (other than those in the original text and those attributed to others) are my own, and despite the support, cooperation, and collaboration of many people and organizations, I am solely responsible for any errors or shortcomings in this book.

Translator's Introduction

DONALD C. WOOD

> *Studies of economic and social change in Japan during the period of rapid modernization starting in the 1860s have tended to focus on the cities . . . where modernity was most visible. But if we are to understand the process of social transformation in Japan's modern era, we must look in the villages. It is, quite simply, where most Japanese were.*[1]

Japan in the 1930s was a land of rapid change and marked contrast. As reflected in the epigraph, major cities boomed and glittered with new things and opportunities. Movie theaters beckoned with colorful billboards. "Modern" boys and girls frequented cafés, from which there were at least 37,000 to choose nationwide by the middle of the decade.[2] Department stores employed thousands of people and raked in more profits than ever.[3] This urban culture existed in the smaller cities as well, to different degrees. In contrast to the lights and glitter, however, the ever-tightening grip of the state and its military apparatus cast a foreboding shadow. Moreover, the construction booms upon which the diffusion of modernity depended were fueled by the muscles of temporary laborers—mostly second, third, and subsequent sons of farmers—and factories powered by the toil of their sisters, some of whom were effectively sold into slavery. Furthermore, the homes of the urban well-to-do were cleaned, and the cooking and childcaring done, also by girls from the villages.

By the dawn of the decade, urban industry had switched places with agriculture, relegating the latter to occupying under 50 percent of the employed population and generating less than 20 percent of the nation's economic output.[4] Yet, also reflected in the epigraph, at least two-thirds of Japan's nearly 70 million residents were living in rural places; most people were in the villages during this final decade of "normal" life before all was upended by war.[5] As in the cities, modernity was evident in market towns and villages, albeit less noticeably. Popular magazines, boater hats and newsboy caps, three-wheeled motorized delivery vehicles and motorcars: all had become ubiquitous across the land.

Nevertheless, the rural/urban gap was growing, and life for most farmers was getting worse. The Wall Street crash and Great Depression had reduced demand for Japanese silk, and rising national farm productivity[6] led to price collapses that especially hurt the small-scale growers numerous times in the early to mid-1930s.[7] Liberal political trends were quenched, and administrations toppled, by problems relating to the deepening rural poverty and generally poor economic conditions.[8] Although recovery from the financial crisis and military spending–driven inflation began to help the rural economy from around the midpoint of the decade,[9] agriculture had already come to be a drag on the rapidly advancing, and mostly industrial, national economy.[10] Significant paddy land reorganization and the introduction of new techniques and tools aside, farming and village life remained largely unchanged—at a fundamental level—from the situation nearly 70 years earlier when the Meiji Restoration replaced the final feudal era with a new, modern, order that included a land tax system that subjected the former peasants to a fickle and unyielding cash-based economy.[11]

In 1880, nearly 30 million people farmed 4.4 million hectares (ha) of land for an average of about seven individuals per hectare, and in 1930 slightly over 32 million farmed nearly six million ha—roughly six per ha.[12] Farming remained a labor-intensive small-scale enterprise, performed manually, often by tenants. By 1930 the national figure for average individual farm size had hovered at approximately one ha for decades, with 90 percent in the 1.5–2.0 ha range, and nearly 50 percent under tenancy cultivation (up from roughly 30 percent in the 1870s).[13] Overall agricultural productivity also remained relatively stagnant, due mainly to lagging technical development. During the 1880–1940 period, labor productivity per workhour and land productivity roughly doubled, whereas the former figure rose by a factor of five, and the latter by about half of that, from 1955 to 1985.[14] A long-term case study illustrates the relative stasis in Japan's prewar agrarian situation: a stable rice farming family in Niigata Prefecture spent about the same amount of time performing roughly the same tasks in the 1930s as it had three decades earlier, with mounting debt to show for its hard work—a problem that plagued nearly all farmers in the first half of the 1930s.[15]

Tenant anger had been simmering for years before it began to explode in the 1920s in the form of open disputes, and the establishment of over 4,000 tenant unions by 1930.[16] This "tenant farmer problem" (*kosaku mondai*) was rooted in endemic rural poverty and in the relationship between small-scale tenant (*kosaku*) labor and typical paddy geographical and ownership arrangements. On the one hand, although single-heir impartible inheritance was a common ideal among agrarian households in much of the country, there was a tendency for larger landed farmers—particularly in the northeast and southwest—to found tenant households by granting small parcels of land to second, third, and successive sons, thereby increasing the number of small-scale, poor, farmers.[17] On the other hand, it was common for households with comfortable holdings to farm larger, contiguous paddies as family units and to rent out smaller, inconveniently located paddies to non-family.[18]

Although it was not always easy for a landlord to remove a tenant from such land without warning or compensation, *kosaku* were at the mercy of the landlords, with whom local authorities usually sided in disputes.[19] The plight of poor farmers was a major focal point for writers of the proletarian literature movement. Novels such as Nagatsuka Takashi's *The Soil* (1912) and Kobayashi Takiji's *The Absentee Landlord* (1929) well illustrate the hardships faced by small-scale farmers, and their dependence on the benevolence—and suffering at the hands—of landlords.[20] The situation gave rise to a nationwide farmer rights movement (*undō*), inspired by Marxism and powered by thousands of tenant farmer labor disputes; in 1934 nearly 500 such protest gatherings occurred in the geographically isolated northeastern province of Akita Prefecture, which had emerged as one of the top three regional centers of such activity by the end of 1928.[21] It is no surprise that Akita became a major proletarian literature hub.

It was in Akita Prefecture, in this severe environment, that an impoverished 30-year-old farmer named Yoshida Saburō put pen to paper on March 13, 1935, and began writing a detailed one-year account of his life—focusing on work—and of agricultural matters and other affairs of his natal household and of his community, the tiny hamlet of Ōkura, and the larger village of Wakimoto, nestled in the foothills of Mt. Kanpū on the Oga Peninsula (see map, Appendix 2). When Yoshida finished his task on March 12 of the following year, he had written 366 daily entries, with the help of his wife, who wrote some entries in his stead.[22] His valuable record presents a vivid picture of daily life in a mid-1930s Japanese farming village, bluntly testifying to, and dramatically illuminating, the situation described above. The relative stagnation of agriculture is evident; yet so are social, political, and technical changes over time. Yoshida also frequently contrasts the farming methods of others in his community with his own, which he generally views as superior. Yet, he honestly reports his failings. Tenant-landlord conflicts, and their outcomes, also receive attention.

The story behind the manuscript points to some of its most exceptional attributes.[23] First, although Yoshida kept a diary for most of his life, he would not have methodologically recorded each day's weather, meals (including sources of food items and cooking methods), expenditures and income, happenings at his natal home and in the village, farming methods and labor performed, cultural and technological change and more, and in ways understandable to outsiders, in magnificent detail, complete with dozens of diagrams, had he not been assigned to do so by his patron in Tokyo: Shibusawa Keizō (1896–1963), grandson and heir of Meiji-era magnate and "father of Japanese capitalism" Shibusawa Eiichi (1840–1931). Having published Yoshida's first book[24]—an ethnographic and historical account of his community—Shibusawa asked him to produce a precise and comprehensive record. This he did, as a graduate student would write a dissertation to please a professor, or as a feudal retainer would execute a command for his lord.[25] Second, Shibusawa armed Yoshida with a camera—a rarity in the village—and gave permission to use it freely. Third, Yoshida was well-read. He regularly perused newspapers, and had been consuming critical political commentary in magazines for years. He was also familiar with the works of Karl Marx and with proletarian literature. Furthermore, he had studied the teachings of numerous agrarian leaders. Therefore, Yoshida had a well-informed goal: to create a clear and objective record that would not only testify to the harsh conditions faced by poor farmers but might also prove valuable as a catalyst for greater change, or at least as a starting point.

During the writing, Yoshida supported his family of four by growing vegetables together with his wife on about one-quarter of a hectare (2,430 m^2) of dry, uneven, farmland on the far side of a small mountain in his hamlet, which he rented for 20 yen annually. He thus ranked among the poorest farmers, and would have even if he had owned his fields. Their tiny house, which overlooked the farm, had cost him 30 yen to build in 1930. Yoshida had no tenant rights to paddy land, and so therefore grew no rice, although he and his wife helped others with such work, to supplement their income. His natal household, beneath the mountain on the other side, was also poor, albeit less so than many others in the community. It had access to roughly the national average of 1.1 ha of farmland—some dry fields (*hatake*) but mostly scattered paddies (*tanbo* or *suiden*)—yet owned only about one-quarter of this. The remainder it farmed under tenant arrangements.[26] If in 1935 the family had owned all 1.1 ha, it would have ranked among the 22 percent of the 94,900 Akita Prefecture farming households with holdings the size of the national average; instead, it aligned with the 47 percent that possessed one-half of that amount or less, and with the 42 percent that grew crops on both owned and rented land.[27]

Yoshida's account testifies to the fact that, although agriculture *had* by the 1930s become a problematic sector of the national economy, it had not been abandoned by the government or concerned individuals with the knowledge and power to make a difference. One such individual was Yoshida's good friend Ōnishi Goichi (1898–1992), a national leader in the farmer education and agricultural ecologism movements who strongly influenced and encouraged him. Furthermore, it shows that dereliction had not necessarily happened in the northeastern (Tōhoku) region, comprised of the six prefectures of Aomori, Akita, Iwate, Yamagata, Miyagi, and Fukushima (see map, Appendix 2), which has often been overlooked or seen as backward and underdeveloped.[28]

For one thing, overshadowing Yoshida's work—his entire mindset—was the influence of Akita's most famous farmer, Ishikawa Rikinosuke (1845–1915), who advocated industriousness, frugality, and scholarship, and pursued the lifestyle with devotion.[29] Yoshida held Ishikawa and his examples in high regard, and admired his emphasis on the importance of villages functioning as single, cohesive units, and of farmers seeking knowledge and embracing change to improve their productivity and their lots.

Ishikawa was one of many *rōnō* (elder/senior/experienced—literally, "old"—farmers) who acted as rural leaders and encouraged and inspired their fellow agriculturalists, often with the support of regional administrations, and were selected and upheld as examples. Indeed, Ishikawa traveled widely as an educator on modern and efficient farming methods on behalf of Akita's prefectural government, and was instrumental in national-level efforts to unite all farming households of the country under the Zenkoku Nōjikai agricultural umbrella association, through Nōkai association branches—an arrangement that helped preserve local landowner-led development and counteract the power of absentee landlords.[30] Ishikawa's legacy remains alive and well in Akita today; his name is known widely and children learn about him in school. *Rōnō* were of the same general category of regional leaders known as *chihō meibōka,* often wealthy landowners, who tended to play pivotal roles in regional and local development projects beginning in the late nineteenth century, and who often came to hold public offices.[31] These important people, usually crucial in the introduction of new techniques and technologies, were integral to the sociopolitical system that had been established during the earliest decades following the Meiji Restoration in 1868.[32]

The significance of Ishikawa and the general *rōnō* ideal is unmistakable in Yoshida's lifestyle, revealed in his chronicle. It can be seen in the way he constantly seeks information about new farming methods and techniques, and applies these experimentally. The effects of the message and spirit of agricultural modernization and improvement, to which the *rōnō* were central, are also evident in the numerous mentions in Yoshida's chronicle of visitors coming to see his tiny farm. We can see that he was functioning as a conduit of the *rōnō* message and spirit in his own little corner of Akita Prefecture, yet a very political one, in contrast to the *rōnō* themselves, who—like *meibōka* in general—were expected by the state and by regional and municipal administrations to be action-oriented but essentially apolitical in their approaches to solving rural problems and bettering farmers' lives.[33]

Although Yoshida was born too late to have learned from Ishikawa personally, he sought similar firsthand tutelage whenever possible. For example, in 1929 he spent some months apprenticed to an experienced farmer—who may have been local *rōnō* of sorts—in the southwestern Akita county of Hiraka, where he learned vegetable farming and hotbox techniques.[34] Prior to that, he spent a winter at a young farmers' cooperative agricultural training camp in Shizuoka Prefecture. This had been founded by social activist and philosopher Okamoto Rikichi (1885–1963), who professed the small-scale, compact, efficient, self-sustaining "Denmark Model" farming philosophy popular in Japan in the 1920s. The experience heavily guided Yoshida's future, probably including his choice to grow vegetables in his home village while doing daily work locally rather than engage in seasonal labor away from home (*dekasegi*).[35] He was, in other words, a product of his home environment and of the larger sociopolitical agrarian landscape, across which new ideas and practices were being introduced, energetically yet unevenly, and with varying degrees of reception and adoption.

The uniqueness of Yoshida's chronicle was mentioned above, yet not sufficiently explored. For one thing, it stands out among Japan's firsthand historical records. There is an abundance of such material offering insight into daily life in agrarian households and communities from the Edo (1603–1868) and Meiji (1868–1912) eras. Much has been published in readable, modern Japanese, with annotations and analyses.[36] However, such chronicles were generally produced by relatively, if not very, wealthy households. Moreover, although some of these are quite detailed and cover many months or years, none approaches the precision of Saburō's 1935–1936 record. Some sources[37] make excellent use of such original materials to present fine overviews and analyses of historical situations and trends over time, but again, the limitations of the original materials prevent deep understanding of the annual (much less daily) lives of poor farmers, including women and children of the villages. Finally, very little of this material is available in English, although some has been used well by researchers writing in English.[38]

As suggested above, male elites have been the dominant producers of firsthand historical chronicles.[39] A well-known early example is the captivating manuscript authored by the pseudonymous Edo-era samurai Buyō Inshi, a new edited and abridged translation of which was published in 2017.[40] It is fortunate that Buyō bequeathed his marvelous record to later generations, in which he comes across as somewhat of an early nineteenth-century Adam Smith, lamenting the failure of his society to conform to an ideology he treasures and believes to be correct. However, as an elite male Edo (now Tokyo) resident and member of the warrior class, he was not interested in the particulars of peasants' lives, although it cannot be said that he was wholly unsympathetic toward those who worked hard, avoided temptations, and stayed in their proper positions. Notably, Yoshida and Buyō both express frustration over their powerlessness to change their worlds and their places in them, and complain about ebbing morals and growing selfishness, but the former—writing only 12 decades after his warrior forebear—was looking upward from the bottom, believed in the power of hard work coupled with study, and possessed a different knowledge base and a more flexible mindset.

A memoir written by an elite man that resonates more in tune with—yet from the opposite perspective of—Yoshida's report is the translated and edited partial diary of Aizawa Kikutarō (1866–1963), a wealthy landowner in what is now part of the city of Sagamihara, Kanagawa Prefecture, who eventually became mayor of the amalgamated village of Aihara. The original diary was not meant for public consumption. Nevertheless, selected translated sections are deftly used by author Simon Partner to tell a compelling longitudinal story of development in a place that was originally as rural as Saburō's home village of Wakimoto, from the perspective of an instrumental *chihō meibōka*. Landlord-tenant relations and interactions are seen from the former's perspective, but Aizawa and Yoshida occupy common ground in their shared grievance over growing covetousness.[41] However, small-scale and tenant farmers—and other marginalized, poor, and politically weak people—are seen but not much heard in Aizawa's diary. Furthermore, although it is about village life, it is not about farming life.

Some chroniclers and researchers have given multiple people opportunities to share autobiographical stories that relate to Yoshida's 1930s life and environment, and presented them in single collections. Tenacious traveler and prolific ethnographer Miyamoto Tsuneichi does this expertly in his famous book, *The Forgotten Japanese*—hence its endurance.[42] Miyamoto, the most famous of Shibusawa Keizō's many disciples, was acquainted with Yoshida, yet the two do not seem to have been especially close. Another monograph that deserves mention is Dr. Junichi Saga's delightful 1987 collection of personal stories he collected of people living in his hometown of Tsuchiura in Ibaraki Prefecture, north of Tokyo.[43] Most of the storytellers were born around the turn of the nineteenth century, and some reflect not only on their own lives but on those of their parents or grandparents, making for a very informative and entertaining study of life in that place—where Nagatsuka Takashi's novel, *The Soil,* was set—during the late nineteenth and early twentieth centuries. Nevertheless, it lacks the detail, depth, and focus of Yoshida's account.

Other publications present relevant case studies of individuals or families over time. In *Toshie,* Simon Partner traces the story of a Niigata Prefecture woman and her natal family over five decades beginning in the 1920s, based partly on her personal recollections.[44] As her parents were poor tenant farmers at the time of her 1925 birth, the section on the family's prewar situation is especially pertinent, providing a broader picture of what Yoshida experienced in the 1930s. The author's focus on tracing the consequences of individual choice also resonates with the revelations of Saburō's record, which does this to different degrees, both short- and long-term. A major difference between the situation Partner explores and that of Yoshida, who lived only about 240 km to the north, is that the family in *Toshie* grew rice and raised silkworms, and Saburō did neither.

In a chapter in an edited volume, Nishida Yoshiaki relies partly on personal diaries to tell a story of a farming family's situation and experiences through the twentieth century.[45] Many conditions, situations, and incidents highlighted in the text could have been lifted from Yoshida's record, including the general poverty, the weak position of tenant farmers vis-à-vis the landlords, and disputes, the formation of tenant unions, the economic downturn of the 1930s and mounting farmer debt, and the effects of militarization, national mobilization, and war. Yoshida's account differs, though, in its intensely personal and subjective handling and view of these matters, and its presentation of multiple cases of tenant-landlord problems with which he was intimately familiar, in addition to its fine detail and focus on daily life.[46] Finally, Gail Bernstein's *Isami's House,* which follows a Fukushima Prefecture family over several centuries with a focus on the early modern period and the twentieth century, offers an excellent overview of the general prewar environment, but as it was a very wealthy landowning family, the situation was quite different from Yoshida's.

In sum, Yoshida's monograph is unique not only among Japan's firsthand memoirs and other records and chronicles but also amidst biographical studies in general in that it was written by a poor, small-scale, tenant farmer about his own daily life, and at a level of detail surpassing that which would normally be found in a private diary. And as an intricate microhistory of the minutiae of everyday life in an ordinary 1930s village—though written by a fairly uncommon person—it has not gone unnoticed by researchers over the years. Since its publication by Keizō Shibusawa's Attic Museum ethnological association in 1938, Yoshida's record has been utilized on many occasions to support academic arguments, even though it has never been republished or digitized.[47] It also inspired film producer and cinematographer Miki Shigeru (1905–1978) and famed ethnologist Yanagita Kunio (1875–1962) to embark on a joint project in the 1940s to document rural life in the same geographical area—an initiative that resulted in a motion picture, most of which was later lost, and in a book of valuable photos that were taken during the filming.[48]

Yoshida's record is special in other ways. One is its focus on daily meals and the annual diet, which is unparalleled among translated accounts of the period, and probably also among those in the original Japanese. Another is the monograph's relative lack of bias. To be sure, his sympathies lie with his fellow tenant farmers, and his distrust of wealthy landlords—and contempt for some—is almost palpable at times.[49] But what bias exists comes from the low end of the social hierarchy: a welcome bias, as such people are underrepresented in the general body of autobiographical accounts, and in that of studies relying on diaries, memoirs, and other similar documents, due to a focus on elites and to a lack of material. Opportunities for someone like Saburō to write and publish a document like his chronicle, full of raw, candid opinions, were few and far between. As Partner notes, impoverished tenant farmers left few records of their lives.[50] Yoshida is a rare exception. The inclusion of dozens of photographs taken by the author with the camera provided by Shibusawa is also extraordinary. It was a very rare thing at the time for someone in Yoshida's position to be able to photographically document his life, family, and community as he pleased,[51] much less for illustrating his own written account. Here, Yoshida has produced a visual ethnographic record that is somewhat more "natural" than those created by expert "outsider" academics, including Attic Museum photographic collections.[52]

Furthermore, Yoshida's chronicle illustrates the confluence of crucial elements in Japan's course of agricultural modernization and evolution into the modern state it is today. The critical role that the larger landlords came to play, as seen in the abovementioned figure of Partner's mayor of Aihara, is hinted at.[53] These were often elected to public offices, and frequently became important figures in the new agricultural associations.[54] The influence of *rōnō* in change and agricultural development at the time is also evident throughout the record, as mentioned above, and also in the spread and permeation of the Nōkai association network, which was led largely by *rōnō*. This generally concentrated on practical goals such as farmland consolidation, technique modernization, and output-boosting, while helping to extend the hand of the central government into the lives of the nation's villagers—directly into their productive activities.[55] Nōkai influence in Yoshida's 1935–1936 life and that of his community is evidenced numerous times in the pages of the present book.[56] Along these lines, another notable attribute of the record is how it reveals other ways in which the national government reached into the daily lives of people as far away as distant Wakimoto in Akita Prefecture, such as in the form of directives regarding morals, social behaviors, and money, and in the form of required attendance for military reserve drills.[57] Yoshida's chronicle also shows that his tiny village was connected to the outer world in the names of major national figures, including a government official, a suffragette and future Diet member, and a high-ranking military colonel, who stopped there to deliver lectures while he was writing.[58]

There is also much in Yoshida's record about economic policy and daily activity. A policy of particular interest is the Economic Rehabilitation Movement (Nōsangyōson Keizaikōsei Undō), a government initiative launched in 1932 to improve farmers' living conditions that offered debt repayment extensions and financial assistance for buying fertilizer.[59] Some of its local effects—not all positive for small-scale farmers—under the direction of a wealthy and powerful mayor, can be seen. The viselike effects of the generally poor economic conditions are always evident, but Yoshida highlights unfair advantages enjoyed by certain villagers. Economic activities, aside from tending to vegetable and rice crops, include making a myriad of items, such as ropes, shoes, and mantles, for home use—rope-weaving having been a major source of extra cash, especially for young boys. Saburō also provides explanations of labor exchange[60] between households, particularly regarding rice paddy work, as well as his and his wife's seasonal paddy work for wages. Rotating credit associations (*tanomoshikō* or *mujinkō*), which had largely disappeared from villages by 1950, also receive mention.[61]

It should be apparent to readers versed in the anthropology of twentieth-century Japan that Yoshida's record meshes with John Embree's classic 1939 ethnographic study of the Kumamoto village of Suye Mura, as well as Smith and Wisell's 1982 companion book (Embree's wife's experience of the same fieldwork), part of which was undertaken when Yoshida was writing his manuscript, over 1,000 km to the northeast.[62] Such readers will also notice that Yoshida's record stands between the firsthand personal accounts discussed above and Embree's and Smith and Wiswell's books, in its very personal and sometimes subjective yet highly structured, methodological, and often analytical, approach. Even though, unlike Embree, Yoshida was not expressly interested in outlining social structures, he does offer insight.[63] Although he was not necessarily focused on illuminating gender roles or highlighting the position of religion in local society and culture, Yoshida provides considerable elucidation.[64] And despite the priority he had to place on his family's survival, he delivers a sharp picture of the operation of 1930s rural economy, characterized in part by a clash of feudalistic and modern (capitalistic, cash economy) forces, through which Yoshida and other poor farmers struggled to navigate daily.[65] It *was* a struggle; Saburō did not break even financially from 1930 to 1936, despite his, and his wife's, extremely hard work.[66]

Finally, Yoshida's manuscript also stands apart from other records and studies of the prewar period because he did not grow rice—which, along with sericulture, has received an inordinate amount of attention both in government reports and in academic studies.[67] Although Akita Prefecture has long been known for its high production levels of quality rice, farming there has been extremely varied and did not necessarily always center on the crop, and growing vegetables has always been an important side activity, even for rice farmers.[68] Data from the 1878–1987 period reveal that non-rice crops accounted for 30–35 percent of national agricultural production in terms of 1934–1936 prices during that entire time span, but that rice lost over half of its initial 64 percent share to a sharp post-1950 rise in livestock production.[69] Yoshida was therefore not a demographic outlier even though he grew no rice of his own; he can be seen as representative of not only a large section of very poor farmers but also of a majority who spent considerable time, energy, and land area on vegetable farming. Yet, his intimate familiarity with rice agriculture is also reflected in his record.

Yoshida's chronicle may in fact be the most detailed account of one person's farming practices and their results of not only the 1930s but also the entire 1868–1945 period. The record of his carefully measured, expert-informed, scientific, trial-and-error approach to farming is a rare gem. It helps give a face to the late prewar period farmer—mixing and applying chemical fertilizers while slopping about in night soil,[70] performing intensive manual labor with both newer and older tools, and scrambling to obtain cash in a rapidly changing economy where payment for labor was still often handled with rice—on the cusp of rural Japan's transformation into the more egalitarian and rapidly changing post–Land Reformation world described by researchers who began their investigations in the late 1940s and 1950s,[71] and connecting the Meiji-era villager to those of the current century.

No introduction to Yoshida's chronicle would be complete without a clearer explanation of its author's background and of exactly how he came to find himself in the company of viscount Shibusawa Keizō, and to have two of his books published by the heir to the Shibusawa financial empire and future Bank of Japan governor and finance minister. According to the records, Yoshida Saburō was born on February 2, 1905. Officially, he was the fourth child, and the third son—a girl and two boys had preceded him—but the second-born son had died at the age of four, while Saburō was still in the womb, making him a prenatal "murderer." It was said in his village that if a child in the womb was especially wild and difficult, it might take too much of the mother's energy and attention, causing her youngest child to perish of *odomimake*. Saburō was often accused of this as a child, when scolded for misbehavior: "You terrible little brat—you even killed Keiji!"[72]

It seems that non-conformity was always part of Yoshida's personality, but he managed to survive childhood, and to complete his compulsory education by the age of sixteen,[73] when he began seriously helping on the family farm.[74] He loved farmwork, and took it seriously, but had already begun to feel a need for farmers to not only work diligently but also to ground their efforts in knowledge. This led him to enroll in a local night school, probably in 1920, bringing him into contact with new teachers and ideas, and facilitating his introduction to the teachings of *rōnō* Ishikawa Rikinosuke. Yoshida used his savings—earnings from rope-weaving and day labor—to buy books and magazines, and read them voraciously (see Fig. 0.59). "Plow when it's sunny, read when it's rainy": the proverb had become his way of life by the time he left his teens.

Saburō's big turning point came toward the end of this five-year period of study at the night school in the form of an encounter with a new magazine called Aozora (Blue Sky),[75] which must have resonated with him due to its proletarian leanings, agrarian concerns, and educational focus. Aozora was the brainchild of Ōnishi Goichi and some of his associates; the copy Yoshida acquired had apparently been sent by Ōnishi to one of Yoshida's teachers, with whom Ōnishi had connections through membership in the Keimeikai, Japan's first teachers' union, founded by Shimonaka Yasaburō. Yoshida's senior by a mere seven years, Ōnishi—a former schoolteacher—was already on his way to becoming a national leader in ecologism and farmer education. Aozora contained an invitation for readers to join Ōnishi's fledgling farmer education and empowerment association, Aozora. Saburō did this immediately, thereby starting a chain of events that would change his life.

In 1926 Yoshida got his first look at Tokyo. Having been invited by Ōnishi to attend a meeting of his new association at his house in Kamitakaido, Saburō packed his things in a cardboard Morinaga caramel box, tied everything else up in a *furoshiki,* sneaked out of the house under cover of darkness—as he had not obtained his father's permission—and boarded a train. The study session brought Yoshida into contact with a stimulating new range of people and ideas, just at the end of the so-called era of Taishō Democracy.[76] Back home again, Saburō was wed to a woman of the same hamlet, Yoshida Katsue, in an arranged union. In Ōkura in the 1930s, about 80 percent of the approximately 80 households—divided into lineage groups (*maki*)[77] corresponding to the *dōzoku* in much of the anthropological literature—shared the Yoshida surname.[78] Such families of different *maki* did not necessarily consider themselves kin, but all knew the story of the formation of their hamlet: a single family brought the surname to the spot in the distant past, settling and subsequently splitting multiple times, thus creating branches that also split over generations.[79]

Despite his marriage, and the devastating loss of his father in May or June of 1927, Yoshida headed off to Shizuoka at the end of that year to participate in Okamoto Rikichi's agricultural training camp (mentioned above). This time, instead of gaining a wife upon returning, Yoshida lost her, albeit temporarily. She had been taken home by her family due to her being left behind for the winter and to a miscarriage. Yoshida returned to helping on his family's farm, and managed to reclaim his wife. During the summer of 1928, he became acquainted with Izumi Kanjirō, a regional tenant farmer movement (*nōmin undō*)[80] leader, who taught him about Marxism. Yoshida became involved to a degree in the farmer protest movement in the area during this 1928–1929 period through his association with Izumi, but probably not deeply, and although he was influenced by Marxism, he did not take it much to heart.[81]

The next few years were busy ones for Yoshida. Over four or five months in the spring of 1929, as mentioned above, he learned about growing vegetables as an apprentice to an expert vegetable farmer named Itō Chōjirō some 85 km to the southwest in Hiraka County.[82] Subsequently, Yoshida founded his own farm up on the mountain, which included having the house built, in the spring of 1930—a year that also marked the birth of his eldest daughter, Fusae. Throughout this period, Yoshida kept in touch with Ōnishi Goishi by way of letters and postcards. Fascinated by tales of Wakimoto Ōnishi heard from Yoshida during a 1932 trip to Akita, Ōnishi encouraged Yoshida to write about his home village, based on his own knowledge and on what he could learn from local elders, to create a record. Saburō set about this "native ethnography"[83] project almost immediately, researching and writing whenever he could manage it.

Yoshida mailed his completed manuscript to Ōnishi in April of 1933.[84] By this time, he had become the father of two girls; his youngest, Yaeko, had been born. Therefore, although he might have stopped to visit his younger sister and her family at their house near the train station on his way home from the post office, he probably hurried home to tend to his farm; it was the start of another year of growing vegetables on the mountain, and he had plenty to do. Meanwhile in Tokyo, Ōnishi found an opportunity to share the document with Yanagita Kunio, who reacted encouragingly. Next, Ōnishi presented it to Shibusawa Keizō, whose self-funded Attic Museum ethnological research association had begun publishing monographs. Shibusawa highly valued accounts of rural residents he referred to as the *seikatsu-sha,* or the people who live (in the ethnographic "field").[85] Soon Yoshida's manuscript was in the Attic Museum's publication pipeline, and his remote hamlet of Ōkura was on Shibusawa's travel agenda.

In late August or early September of 1934, the people of Ōkura were surprised by the appearance of an Attic Museum entourage that included Shibusawa, future Minister of Agriculture Ishiguro Tadaatsu,[86] Miyamoto Keitarō,[87] Takahashi Buntarō,[88] and likely also Ogawa Tōru.[89] Even more surprising: the group lodged at Yoshida's natal household, where they enjoyed local food and saké and bunked in the house's *tatami* mat living room. The team's stay in Ōkura, though brief, was long enough to observe Yoshida's farming and living situation, and to take many photographs for inclusion in his first book—*Notes by a Farmer at the Foothills of Mt. Kampū in Oga*—which was published the following year by the Attic Museum, as number four in its monograph series. To celebrate the book's publication, a party was held at Shibusawa's Mita, Tokyo, estate in early March of 1935 (see Fig. 0.60). At the event, Shibusawa asked Yoshida to produce a new manuscript: a detailed annual chronicle. Saburō was apprehensive. He knew it would be a gargantuan undertaking, dwarfing his first writing project. But he was in no position to refuse.[90]

Determined to succeed despite his concerns, Yoshida headed home and set about the task, making his first entry on the thirteenth day of the month. Shibusawa did not only send Saburō home with an assignment; he gave him two luxurious gifts—a phonograph, and the abovementioned German Contessa Nettel camera, complete with film and instructions to send all exposed film to him for processing (see Fig. 0.61).[91] Yoshida's uneasiness about his new project is reflected in a letter from Shibusawa dated March 27 (see Appendix 6), a response to a March 18 letter he had received from Yoshida.[92] With the arrival of Shibusawa's letter, his worries were solved; he was ready to push ahead with his greatest writing project and bring it to fruition. The following March, as the original author preface to this book attests, Yoshida headed to Tokyo to submit his manuscript to Shibusawa in person. He had completed his task; his second book was published as number 16 in the Attic Museum's monograph series in 1938.

Sixty-one years later, in the summer of 1999, I encountered Yoshida's chronicle for the first time, in the library of the cultural anthropology department of Tokyo University, as a temporary visiting post-graduate student. My host for that summer, Professor Abito Itō, introduced the book to me because I had previously lived and conducted research in Akita Prefecture. "Look at this detail," said my host as he held the delicate copy of Yoshida's 1938 book in his hands. "He was just a poor farmer. Wouldn't it be wonderful if this were translated into English?" I agreed, and took his words as a challenge, but did not think I would ever be able to translate the book. I did make a complete photocopy, in the hope of eventually finding a way to make use of it.[93] As it went, I entered the department's doctoral course the following spring, as one of Professor Itō's students, and later returned to Akita Prefecture, where I began conducting my dissertation fieldwork.

The turning point came in 2003, when my wife went to the office of a member of the teaching faculty of North Asia University to ask him about the Namahage of the Oga Peninsula.[94] A native of Oga and expert on all things ethnological about Akita, Professor Kamada Yukio provided her with all she needed to know about the Namahage in order to write an article for a magazine published by the company who employed her at the time. At some point that day, Professor Kamada mentioned Yoshida Saburō and showed my wife a copy of his 1938 book, which she immediately recognized as the book I had encountered in Tokyo four years earlier. After hearing about this encounter, I dug out my photocopies of Yoshida's prewar books and began to seriously consider Professor Itō's 1999 challenge. Six years later—a full decade after my initial encounter with the two books—I finally managed to start a research project on Yoshida's life and works. Professor Kamada kindly introduced me to Yoshida's family members, and I began the process of translating the 1935–1936 chronicle.

This translation of Saburō's 1938 book represents an attempt to remain faithful to the original in as many ways as possible, with some additions and other changes that were either unavoidable or were intended to help better convey the author's messages, facilitate a deeper understanding of the issues and events of the period, and otherwise enrich the reader's experience. First, there is the handling of the photographs—most of which were taken by Yoshida and used in the original book. The set of 62 numbered photographs presented in segments in this book is identical to the original regarding content, except for one substitution and one omission—I was unable to find adequate copies of the originals—and the addition of the last 11, all but one of which came from Yoshida's personal collection.[95] (The original set appeared as a single unbroken series in the front of the 1938 book.) The arrangement and handling of the photographs and diagrams (figures) appearing in the daily entries of this book reflects an attempt to remain faithful to the original, but numbers have been assigned to all, and captions added to those that had none before. Their positioning remains close to that of the original, yet some have been moved to new locations. One photograph was substituted for a nearly identical one that was in the 1938 book but has since apparently been lost. Photos that were not in the original book are so labeled. Saburō's style of writing the daily entries was also preserved, but some small adjustments were made to their arrangement, for continuity and uniformity. A decision was also made to preserve the old-style measurements in the original book and provide metric equivalents, as these were integral to Yoshida's ways of perceiving and ordering his world. This is also the case regarding cash, as the sen (¥ 0.01) was more commonly used in daily life than the yen at the time.

Major differences between this translation and the original book include the addition of annotations, mainly in the form of endnotes. In some cases, I have embedded small comments or clarifications in angular brackets inside the text, and in some figure captions. These are my words. In contrast, text inside parentheses represent Yoshida's original or implied words. The way in which this translation differs most from the original book is in the content and handling of the appendix material, although Appendix 1 is essentially identical to the 1938 book. The map with inset in Appendix 2 does not appear in the original publication, which contains no such map, probably because it was published for a domestic audience and because Saburō's 1935 book does contain a similar map. Additionally, the house diagram in Appendix 2 has been moved from its original position at the front of the book, and placed together with pre-existing field diagrams. The tables in Appendices 3, 4, and 5 are also extra backmatter content in the original book. Appendices 6, 7, and 8 are new; they do not appear in the prewar publication.[96] One part of the backmatter in the 1938 version—a table of Yoshida's family's meals for the entire 366-day period that covers 26 pages—was omitted from this translation due mainly to space concerns. I hope Saburō would be pleased with this translation, and would appreciate my attempt to remain largely faithful to his original.[97]

Finally, as a coda to this introduction, it seems prudent to briefly explain what happened to Yoshida and his family after he delivered his second manuscript to Shibusawa in March of 1936. First, of course, he set to work preparing his farm for another year. As the months passed, though, he worried about the future because he knew that he would not be able to provide adequately for his family—especially for his daughters—without making a change. He had previously considered relocating to the Imperial colony of Manchuria. There, according to a recruiter, he would mainly have overseen farm laborers from atop a horse. But he had put the idea on hold to concentrate instead on what he was truly good at: being a "real" farmer.[98]

The following April, a letter from Shibusawa solved Yoshida's problem. It was an invitation to serve as the salaried caretaker of a new Japanese Society of Ethnology facility in the Tokyo village of Hōya (see Appendix 7). The entire family was invited to live in the building, at no charge. Yoshida accepted the offer without hesitation. On the seventh of July he and his wife and daughters boarded a train and headed to Tokyo, and to a new life. He then took up a spot in the heart of ethnological research in Japan, and at the center of an organization that would evolve into one of the largest anthropological associations in the world—the Japanese Society of Cultural Anthropology.[99] In Tokyo, Yoshida and his wife and daughters lived on the ground floor of the Hōya building (see Fig. 0.62). While he dealt with artifacts, looked after the facility, contributed articles to the Attic Museum's newsletter, joined occasional research sojourns, and grew vegetables—including pumpkins and watermelons, for which he became widely known. His wife, Katsue, served tea and snacks to the researchers and learned how to cook new dishes from some of their wives, and his daughters went to school.[100]

Life in Tokyo was good for the family, despite the effects of the war, and the bombings toward the end. Yoshida, although fundamentally different from most of the Attic Museum researchers, appears to have had good relations with most of them. However, while there he formulated a new plan for his future. This time, instead of waiting for an offer to come from on high, he made use of his connections with Shibusawa and with Ishiguro Tadaatsu—who by that time had risen to a position of considerable political influence—and arranged to receive the rights to clear a parcel of land north of Akita City, along the coastline, and found a new group farm there, which he did in 1945. In a sense, he had come full circle and taken on the role of a prewar quasi-*rōnō* through his own connections to people in high places. Although Saburō's original plan to gather a cohort of young, eager people interested in becoming farmers did not pan out quite as he had hoped, he was able to find many motivated people, including his younger brother Shirō, and build an operation that sustained them well for several decades. But that is another story.

DIARY OF A FARMER AT THE FOOT OF MT. KANPŪ

Author's Preface

It's the middle of March 1936. I've made it through my sixth winter of shoveling heaps of snow to survive on the mountain—and managed to write this manuscript about my farming life, even under such harsh conditions—and now I'm off to Tokyo.

I'd like to record a few paltry thoughts, for posterity.

As I stare suspiciously at the stack of paper that comprises the manuscript sitting on the desk in a corner of my 2 *ken* (3.6 m) by 2.5 *ken* (4.5 m) shack, I don't know what words might help me explain my feelings in a way that everyone can understand.

First, I can say that I've been relieved of a heavy burden. Truly, I feel happy from the bottom of my heart.

But that's all—maybe it's just my natural disposition, but even though I've felt such unbounded joy since finishing, I have not been able to simply revel in it with no thoughts of my initial motivations for starting, and of the entire process.

Shibusawa-sensei,[1] Ōnishi-sensei,[2] my colleagues at the Attic Museum, all the various people around me [who taught and helped me], and my wife and the others—thinking about everyone who has assisted and supported me, I can see all their faces spinning around like images projected from a revolving lantern.

I have not done this all on my own power, and in fact, from the start I didn't think I was qualified to go through the very rough process of completing this manuscript. Indeed, when I think about it again quietly and carefully, I know that I could not have written this without being mindful of some sort of driving force that was flowing through my heart the entire time.

I have been keeping a diary daily, without missing a single day, since I was eighteen years old. This is my fifteenth year. This manuscript is but one part of that entire record—the part that covers the period from March 13, 1935, to March 12, 1936, making for 366 days—and I wrote it as the record of my own farmwork and that of my natal household ("the main house"), including exactly what they did, and who did what, every single day. Furthermore, I not only wrote about what people did, but also explained in correct detail exactly how the things they did were done, step by step, so that even someone with no farming experience would be able to understand what is on the pages.

In addition, I also recorded what we ate, three times a day, every day, and how our meals changed over time, how the food was prepared, how we obtained food items, and also where they came from.

And beyond simply being a chronicle of farming life, the manuscript is also a record of the place where we live—the village itself—and all its happenings, which I have repeatedly tried to report truly and clearly, facts as facts, along with photographs and written notes.

Moreover, I have included six years' worth of maps of my fields[3] to show how I've used the land over time, and tables showing our farm income and our living expenditures.[4]

So now, please permit me to explain in this preface—at least in part—the painful struggle of completing this manuscript.

Just looking back now and thinking about the very tough period of time I spent writing it makes me feel like a lifetime's worth of fatigue and weariness is about to swallow me up.

Simply to survive financially and to provide adequate education for the kids in the amount of time we had each day, it was necessary to complete the basic daily morning work by seven o'clock from spring until winter. After that, breakfast had to be prepared: cooking rice, making soup, getting out the tea, and washing my own face and nostrils. The rice had to be taken out of the pot it had been cooked in. Then, the kids had to be awakened and dressed, and the floor had to be cleaned up—the bedding put away—and then we'd have breakfast. After eating, the bowls and dishes and pots and pans and utensils for eating and for cooking had to be washed and put away. The whole breakfast making, eating, and cleanup process took two hours. Afterward I'd rest up a bit, and then get back to work. Well, I suppose this might have seemed like an easy life when viewed from the outside.

But, just to get by, my wife had to head out well before sunrise every day to sell vegetables. And there was the May rice planting season, and the harvest season, when my wife and I had to get up and go work for other people for day wages—we'd have to wake the kids in the early morning, and drag them along with us and head out of the village, while they rubbed their sleepy eyes with both hands. Just thinking of those faces now brings tears to my eyes. During such times of the year, it would be after eight o'clock by the time we'd finished supper. And I would write a record of that day after that. From autumn to spring, at least, there were no mosquitoes or fleas, but from around June 20 or so the mosquitoes would multiply rapidly, and there would be thousands, or tens of thousands, of them and they'd just fill the little shack. I'd sit at my desk and the mosquito squadron would aim at me from head to toe and swarm forth, blowing their blood-seeker attack force trumpets. If I swatted at mosquitoes on my leg, others would get my back, and if I went after the ones on my back, others would get my head, and if tried to get those with my right hand, my left hand would get ravaged. That's how it was. And even if I burned some cedar leaves and green plants and such to keep the mosquitoes away, an hour later a massive horde would reappear and advance in a great buzzing clamor. It really was just as if a gigantic swarm of honeybees had flown from their boxes all at the same time, or like the drone of dozens of airplanes flying high overhead. The kids would quickly take refuge under the mosquito nets, but I had to sit at the desk and write. There were times when I just couldn't take it anymore, and I'd remorsefully get my wife to help me by waving a fan around. Even if I burned two or three mosquito repellent sticks, it was like driving nails into a glob of fermented rice bran—no effect whatsoever. And once I'd finally gotten the day's entry written down, I'd relax for a moment and crawl into my own *futon* with a feeling of relief, and then the flea generals would commence their biting attack. Even though I'd catch and kill 40 or so when putting away the *futon* every morning, night after night they'd emerge from somewhere and gather up and come—it was really weird. No matter how exhausted I was, I never slept all night undisturbed until morning.

Moreover, this winter we've had snowfall the likes of which is rarely seen here, but despite the harshest snowstorms, I've gone into the village every day, without fail, to see—and hear about—the situation and activities at the main house, and things happening in the village. I mean, I had to investigate and such. And, I had enough of my own worries to make me feel like I was going nuts, and despite my struggles I had to make things with straw during wintertime. For me—a man definitely lacking in innate literary talent—writing about this work and all the processes clearly enough for anyone to understand was an extremely difficult task. I have no idea how many times I tossed my pen aside with a sigh, like someone suffering from delirium. And even when I tried to explain by providing diagrams, I couldn't draw them very well. I marveled at how difficult it was to explain in words the processes of making things that I can easily make very quickly. So, I think that there must be many unclear points in this manuscript, given the conditions under which it was written. Considering this, all I can do is to hope that anyone who kindly takes the time to read this will exercise wise judgment. And in addition, although I've used many people's true names in reporting about happenings in the village, in some cases I was compelled to use pseudonyms for people due to certain circumstances.

The thing about the decision to use pseudonyms is that in the future, when someone looks at this document as one piece of recorded history, they might feel that there are some big contradictions, and I really have worried about this.

Furthermore, regarding my intention to make the contents of this diary public, I have had no specific extra desire in mind. I never doubted my ability to complete it—although I did have a feeling of proud contentment—I just accepted Shibusawa-sensei's command and followed. However, now that I've completed it, I feel a need to try to write some sort of critique or appraisal of my work.

That is to say, I hope that even if a variety of people with different circumstances and points of view read this, it will serve as a useful reference source. It seems to me when I think about it carefully, that we farmers have been unconsciously shedding our old skins—sprouts coming out of eggplant seeds and then shedding the shell, and then growing and casting off the old husk along the way while growing more. The eggplant seedling that sprouts becomes an adult plant, and flowers, and then bears fruit—like that. And just as the farmer checks the plant's health, and pays close attention to its growth pattern, and is then able to tell how it's going to produce its fruit, we farmers of today—our agrarian lifestyles central to our survival—must learn about how we used to live, and at the same time see how our lives are going to change. From this standpoint, I like to believe that this diary will be useful for solving major problems faced by farmers in the future.

But these are just my personal feelings, after finishing the manuscript. The only reason why I undertook this arduous task was to keep the vow I made to Shibusawa-sensei. I promised unequivocally to write it, so I was bound to complete my mighty mission. Every time I thought about that, I felt the weight of my responsibility grow. Therefore, I decided to aim for as much precision as possible, which drove me to include many photographs of village life, complete with detailed explanations.

Finally, once again, I'd like to express my deepest gratitude to Shibusawa-sensei, Ōnishi-sensei, all my peers at the Attic Museum, my *senpais* and other brothers, and my wife and everyone else.

March 15, 1936

As I prepare to leave for Tokyo, I stare at this completed manuscript on my desk, and think wistfully of my long-lost father.

Yoshida Saburō

Chapter 1: March 1935

March 13, Wednesday

(February 9 according to the old calendar)

Partly cloudy
Strong west wind
2°C at 1:00 p.m.

Breakfast:	rice, miso soup,[1] pickled daikon[2]
Lunch:	rice, pickled daikon, salted[3] *hatahata*[4]
Supper:	rice, pickled daikon, miso soup
Rice consumed today:	1 *shō* (1.8 L) (daikon: ½ of one)

Miso soup: 4 *gō* (721.6 ml) of water and about 50 *momme* (187.5 g) of miso. We put a bit of taro (*sato imo*) stems in the soup. This was the amount for breakfast. For supper, the amounts of water and miso were equal, but the stock was made from roasted dried anchovies, and then some chopped and dried daikon was added. This was the amount for all four of us—me, my wife, and the two kids—for each meal. The taro stems were harvested last autumn. We removed the outer skin and dried the stems in the sunshine. They were boiled once and added to the miso soup.

I went to my wife's natal household to talk about my (recent) trip to Tokyo. My crazy neighbor[5] came over to visit. I heard doves calling for the first time this year.

My work for today: I transported a bunch of straw I bought this morning. Soon I'll use it in a hotbox.[6] In the afternoon I built a new cover for the concrete basin that holds the excrement from the toilet.

My wife did laundry, cooked, and went over to her natal home on errands.

> At the main house: Kanezō, the *wakaze*,[7] had already collected firewood (when I arrived)—three bundles in the morning. He took the afternoon off. My brother had gone out and his wife was cooking. My mother has been helping with cooking and such at the house my younger sister married into—in a different hamlet—because she's pregnant.

Since this is the monthly anniversary of my father's passing (*tsuki meinichi*),[8] the temple priest visited the main house to chant sutras.

On February ninth every year (according to the old calendar) everyone takes off from work and eats *Ogano mochi.*[9]

March 14, Thursday

(February 10 according to the old calendar)

Sunny, cloudy later
Fairly strong west wind
6°C at 1:00 p.m.

Breakfast:	rice and pickled daikon with *wakame*[10] soup flavored with *shōyu no mi*[11]
Lunch:	yesterday's leftover rice—poured boiling water over it and ate it in bowls—also, pickled daikon and salted salmon
Supper:	rice with daikon soup and *hatahata* and pickled daikon (We ate six *hatahata.*)
Rice for today:	8 *gō* (1.4 L)

I went to my wife's younger brother's house to deliver a gift from Tokyo. My aunt came over to my house. In the morning I fixed up a pit outside, to build a hotbox.

In the afternoon I went over to get night soil[12] from my younger sister's house near the train station. My wife was preparing dried daikon and cooking and sewing and such today.

At the main house: Kanezō took the horse up into the mountains to cut brushwood. He returned around two o'clock with six bundles, and he made *habaki*[13] for himself. My brother went out on some business. His wife[14] sewed and cooked.

March 15, Friday

(February 11 according to the old calendar)

Cloudy
Strong west wind
8°C at 1:00 p.m.

Breakfast:	rice, daikon and potato miso soup, pickled daikon, and *shōyu no mi*
Lunch:	rice, *hatahata,* salted krill,[15] and pickled daikon
Supper:	rice with taro stem and whale miso soup[16] and pickled daikon
Rice for today:	one *shō* (1.8 L)

The master of Kōshin-en[17] [Yoshida Kōtarō], Yoshida Tokiji,[18] and some others came over.

Today everyone in my household helped to stamp down the straw in my hotbox. In the afternoon my wife went to the post office, in the main part of the village,[19] where she ordered seeds and sent a copy of my book to an acquaintance of mine. For a while, I'll leave the cooking to her.

I think my way of making hotboxes is different from most others, so I'll explain.[20]

The pit width is 4 *shaku* (1.2 m) and the length is 12 *shaku* (3.6 m). The depth is between 1.5 *shaku* (45.5 cm) and 2 *shaku* (60.6 cm). I dug a pit in a spot where the water drains well and where the sun hits, running from east to west lengthwise. Once I'd dug the pit, I lined the inside walls with rice straws, standing on end—just enough so the dirt was hidden. After that, I laid straws down flat across the floor of the pit, and then stamped them down. After stomping once, I put down another layer at a 90-degree angle to the first layer, then wet them with about 1 *ka*[21] of water and stepped on them again. I kept going like this, putting down layers (at right angles to one another), and added about 2 *ka* of water or urine as I went, all the way to the top of the pit, stamping each layer down hard as I worked my way up. Once done, I set up a wooden frame and put [wood-framed covers with] glass panels on top of that, and that was it for today (see Fig. 1.01). About 250 bundles of straw are needed, but this figure is for straws that have had all the stamens and pistils removed. If I work with straws that still have all that on them, I'll need about 1,000 bundles.

Figure 1.01. The author with a hotbox. The sections of the wood-framed cover with glass panels are on the left, beneath his elbow. From Yoshida's personal collection: not included in 1938 book.

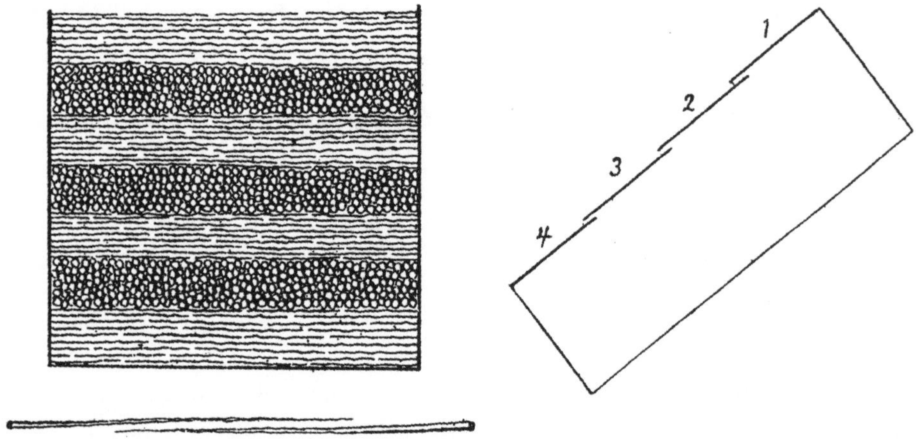

Figure 1.02. Hotbox diagrams. Left: a cross-section view of a hotbox, showing the layers of straw going in different directions (only the ends are visible in the second, fourth, and sixth layers from bottom). Right: the overlapping glass panels as seen from the side; Bottom: straws as they were placed in the hotbox, overlapping and pointing in opposite directions.

Here we call rice straws with the stamens and pistils (*shibe*) removed *shinagodawara.* Those that still have them are called *shibetorazu.* If you take five bundles of *shibetorazu* and strip off the stamens and pistils then you'll have one bundle of *shinagodawara,* or so people say. One of these days I'm going to count the number of straws in a bundle of *shinagodawara* and write about it. The wooden frames (I use) are about 1 *sun* (3 cm) thick and are made of pine. I set them up around the pits, with the south side a bit low and the north side a bit higher, at an angle of about 35–40 degrees. I set the glass panels in a way so that if rain falls it won't go into the pit. I'll include a drawing to show that I mean (see Fig. 1.02).

But it's important to make sure that the straws are laid down with the tips overlapping one another, with about 1 *shaku* (30.3 cm) of the bases sticking out on each side.

Today's expenditures:

stamps	30 sen
seeds	1 yen 64 sen
postal fee for book	40 sen

At the main house: Kanezō cut straw to feed the horse early in the morning. After that, he went out to cut firewood. He took the horse up into the fields[22] owned by the household and cut bamboo and brushwood and brought back 6 bundles. In the afternoon he went into some forest land owned by another household and came back with 8 bundles of pine and cedar branches. A workhorse can carry about 6 bundles of brushwood in a day. Two horses could carry about 16 bundles of bamboo or pine and cedar branches in a day, and that's about it.[23] The master of the house[24] is serving as an officer of the village household head association,[25] which keeps him busy, so he was out dealing with related matters. His wife sewed and cooked.

March 16, Saturday

Rain, then snow, and then rain again
Southeast wind, later out of the north
6°C at 1:00 p.m.

Breakfast:	rice, *wakame* soup, pickled daikon and *shōyu no mi*, and we also ate some miso-preserved vegetables (*misozuke*), and daikon
Lunch:	rice and salted salmon and pickled daikon
Supper:	rice and daikon miso soup (stock made from dried roasted anchovies); also: *takuan*[26] and daikon tips and salted krill
Rice for today:	9 *gō* (1.6 L)

My friend Yoshida Shūzō came over.

Today I went over to my mother's natal home to buy straw. Due to the bad economy and such, the government is encouraging people to make things with straw on the side, so many people are coming to the village to buy straw these days. In addition, because of last year's poor harvest, we're about 30–40 percent short on straw as compared to normal. This has caused a big jump in the straw price. Last year I had to pay 5–6 rin for one bundle of *shinagodawara,* and only 1 rin for a bundle of *shibetorazu.*[27] But this year a bundle of *shibetorazu* costs about 2 rin 3 bu, and a bundle of *shinagodawara* is about 9 rin or 1 sen—real explosive inflation. But even at these prices I couldn't get enough (from my mother's home) so I had to scramble around all day looking for straw, but was only able to get about a hundred (bundles).

My wife was at home sewing all day. She also cooked and did general house chores.

Today's expenditures:

kerosene (4 *gō* (721.6 ml))	12 sen
a new glass cover for the lantern	10 sen
ink from Maruzen	30 sen
straw	1 yen

At the main house: Kanezō wove rope all day, from the morning. He used eight bundles of straw (*shinagodawara*) to make six bundles of rope. In this case, one bundle of rope is about 40 *hiro* (72 m), with 1 *hiro* (about 1.8 m) being considered equal to 5 *shaku.*[28]

The master of the house was home today. He manned the counter of the little shop.[29] His wife cooked and sewed.

March 17, Sunday

Partly cloudy
Very strong west wind
6°C at 1:00 p.m.

All three meals were nearly the same as yesterday.

Katō Kentarō's wife gave birth to a baby girl.

Today my wife and I started warming the hotbox, and I prepared topsoil for growing plants in it. My wife borrowed a big pot that holds about 8 *shō* (14.4 L) from the main house and boiled water in it. I took a piece of bamboo that I'd hollowed out like a pipe and stuck it down into the straw inside the hotbox, to a depth of about 1 *shaku* (30.3 cm). We then poured a potful of hot water down the pipe. The pit is 2 *ma* (12 *shaku* (3.9 m^2)) in size, so it needed two full pots (1 *to* 6 *shō* (28.9 L)). We poured in that amount while changing the position of the pipe several times (see Fig. 0.23). After pouring in all the water, we stamped the straw down further. Then, I prepared the hotbox topsoil. For planting seeds, it should only be about 4 *sun* (12.2 cm) deep. If you want to use it to raise plants (bought as seedlings) for a while until you put them out in the field, you can make it up to about 5 *sun* (15.5 cm) deep.

For hotbox soil, you need a blend of 60 percent plain soil and 40 percent soil with plenty of nutrients mixed in. The latter is called *baiyōdo,* and I always make it the previous autumn. I take some well-composted dung and mix a bit of plain soil in with that, and add 10 parts nitrogen, 6 parts phosphoric acid, and 4 parts potassium (potash) and mix everything well. The amount of soil and ratios of different ingredients should be adjusted by the person on a case-by-case basis. This *baiyōdo* and plain soil must be sifted with a 3 *bu* (9 mm) sieve before being used in a hotbox.

Regarding the soil, I first make a layer of small bits of hard clay. On top of that, I spread the *baiyōdo* I mixed, and then sprinkle some ash on the surface. I then spread a layer of sand about 5 *bu* (1.5 cm) thick across the top. It's fine if the soil part is about 4 *sun* (12.1 cm) deep. Finally, I use a big wooden spatula to smooth it all out nice and flat. I'll draw the tools used and a cross section of the topsoil for a hotbox (see Fig. 1.03).

My wife and I finished this job during the morning, so she went to the main house in the afternoon with 2 *shō* (3.6 L) of polished rice and borrowed a mortar and pestle and ground the rice into

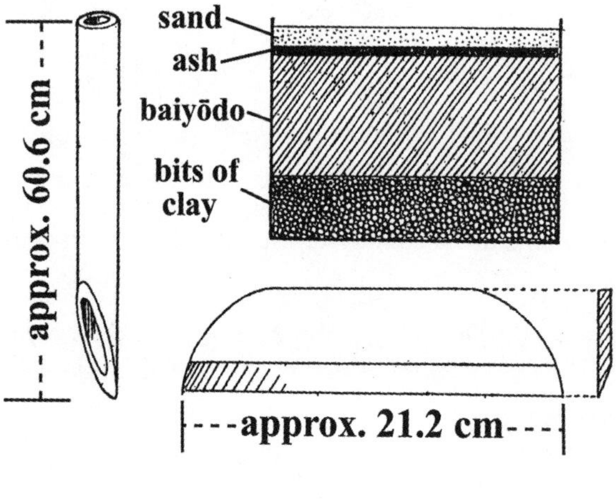

Figure 1.03. Tools and hotbox planting surface. Left: the bamboo pipe for pouring hot water into a hotbox; Upper right: a cross-section view of the planting surface; Bottom right: the wooden spatula for smoothing the sandy surface.

powder. It's time for the spring *higan*,[30] so this will be used for making *dango*.[31] In the afternoon I went around the village looking for straw to buy. I was finally able to buy some from Sangobē's household.[32] I also bought some from the big main house.[33]

Today's expenditures:

straw	2 yen
candies[34]	5 sen

At the main house: Kanezō was out all day, cutting pine and cedar branches. He collected 16 bundles—two horse loads—returning around two o'clock. Later, he made a scythe handle. The master manned the shop counter. His wife went out to get firewood. This is called *shiogari* in local speech. In a day, a person can gather perhaps six bundles of brush or about 12 bundles of sticks and bamboo, and that's about it. Today she collected sticks and bamboo. She said she wouldn't grind rice powder for making *dango* at home for Nirvana Day[35] and *higan* but rather buy some leftover powder[36] from a saké brewery and make *dango* with that instead.

March 18, Monday

Sunny
Strong south wind
12°C at 1:00 p.m.

This morning I went to get the straw I bought yesterday. I borrowed the horse from the main house and brought back 150 bundles of *shinagodawara*. I divided them into six *maro*, each equal to 25 bundles (*maro* means *taba*, and *maroku* means *tabaneru*). I put two bundles together at the thin ends of the straw (of each) to make 25, then I put one more in the center and tied them in two places in the middle. I didn't use rope; I used *tsunagi*—pieces of straw tied together. Some might want to make larger *maro;* they'll need to make the *tsunagi* longer. But today I did *yōba-tsunagi* to tie up all 25 bundles. *Yōba-tsunagi* means to pull straws out four times. Later in the morning,[37] I went out to collect brushwood.

My wife went out to collect bamboo (in the morning). She finished that pretty early, and then she made *dango*. We rested in the afternoon. It's a village tradition to do that for spring *higan*.

Breakfast:	rice, *miso* soup with daikon in it, *takuan*, boiled cabbage mixed with slices of Chinese yam (*nagaimo*),[38] which we ate with soy sauce
Lunch:	We made *dango* with the powdered rice from yesterday.

Yesterday, my wife ground 2 *shō* (3.6 L) of rice into powder, but that's for eating on the first day of *higan*, on the middle day, and on the final day—three times. Today we put one-third of the powder in the large mortar, added a little water and kneaded it into a firm paste, and then made *dango* with the paste and dropped them into a pot of boiling water. They must boil for a little while.

After they're done you take them out of the pot and put them on a bamboo strainer. When we eat *dango* we mix sugar with mashed red beans and eat them with that, or with sugar and dried soybean powder (*kinako*) or with ground roasted sesame seeds mixed with sugar and water, etc. Today my wife served them with (sweet) red bean paste. We ate these *dango* with (boiled) cabbage flavored with soy sauce, and *takuan,* etc.

Today, from my wife's natal household, we received a dish of potatoes, carrots, and *konnyaku*[39] and such cooked with soy sauce, one dish of sliced boiled daikon and cabbage, etc., flavored with miso and ground sesame seeds, and one dish of cucumbers that must have been preserved since last year mixed with chrysanthemum petals and flavored with *keshimiso*[40] and such, and we ate these dishes.

Also, I gave eight *dango* to the crazy neighbor (see Fig. 1.04), so in return he gave a dish of small shrimp and minnows that he'd scooped up from a catchment pond with a bamboo strainer and cooked with salt. We ate that, too. Everything was quite tasty. For supper we ate *dango* from lunch and leftover rice from breakfast, miso soup with daikon in it, *takuan,* cabbage flavored with soy sauce, and salted roasted *hatahata.*

Rice for today: 5 *gō* (902 ml), *hatahata:* eight fish.

Today's expenditures:

soy sauce (3 *gō* (541.2 ml))	8 sen
sugar	10 sen

At the main house: Kanezō made rope early in the morning. Later during the morning, he cut a horse load of pine and cedar branches and brought them back. It was still early so he went to cut more and returned with three bundles of bamboo. He took off in the afternoon. My elder brother manned the shop counter. His wife was busy making *dango* for lunch from the (early) morning. She also went out to cut some bamboo during the morning, and she rested in the afternoon. Even though I say "rested," in general all women are still busy sewing and cooking and such when they "rest" and are not exactly just playing around.

When you carry firewood on a horse, you point the roots down in the case of bamboo and brushwood, and for pine, cedar, and other branches you point the thicker ends upward. In other words, bamboo and brushwood go the opposite direction.

Today is the first day of *higan,* and the evening before the day when Shaka-sama[41] attained nirvana (according to the lunar calendar), so villagers cook various dishes and feast. *Higan* starts today and goes on for a week. If the *meinichi* of any deceased family member falls during this week, household members take saké and food to Enmeiji[42] and feed the old women of the village. This goes both for the "true" [yearly] *meinichi* as well as the monthly *meinichi.*

Figure 1.04. The man with a mental problem who lives next door (March 22).

March 19, Tuesday

(February 15 according to the old calendar)

Cloudy
Weak south wind
12°C at 1:00 p.m.

Breakfast:	miso soup with daikon and salted eggplant and such in it, pickled daikon, cabbage and grated daikon, etc., with soy sauce
Lunch:	salted *hatahata* soup with green onions, pickled daikon and such
Supper:	daikon miso soup, salted krill, pickled daikon, *takuan*, etc. We finished the salted *hatahata* today, but we started eating the *takuan* today as well.[43] This kind of pickled daikon is also known as *ipponzuke*[44] in local speech.
Rice for today:	1 *shō* (1.8 L)

I split firewood and cleaned some fields to prepare for planting green onions this morning. The Chinese yams and potatoes I buried last autumn hadn't rotted—they looked great—and some had even started to sprout. These will be seed potatoes this time. Until last year I kept them in various places, such as above the ceiling, instead of burying them. But they always froze and got ruined. This year I tried a new way of storing them and the result was very good, as I explained.

It's a very easy method—no trouble at all. You dig a hole in a sunny spot where the water drains well. It should be from about 1 *shaku* (30.3 cm) to about 1.5 *shaku* (45.5 cm) deep. You line the bottom of the hole with some straw or rice grain chaff (*momigara*) and then put the potatoes in, while sprinkling some wood ash over them as you go. Finally, you put more straw or *momigara* on top and then at least 5 *sun* (15 cm), or maybe 6 or 7 *sun* (20 cm), of soil on top of that and you're done.

In the afternoon I rested. My wife was out gathering firewood all day.

Miura Kanezō came over to see my book.

It's the day that Shaka-sama attained nirvana so all the villagers rest in the afternoon.

At the main house: The master[45] manned the shop in the morning, but he went to the temple [Enmeiji] in the afternoon. He was summoned because it's Nirvana Day. His wife cut firewood in the morning and watched the shop in the afternoon, and cooked. Kanezō cut hay for the horse in the (early) morning, then dug a small pond for soaking rice seeds in water. After that he took down the (miscanthus) thatching on the *fuyu kakoi*[46] around the house. In the afternoon he rested.

This year, wheat bran (*fusuma*) for feeding horses costs more than 2 yen 30 sen for 8 *kan* (30 kg), so all the farmers are complaining. At the main house, they bought 10 *kan* (37.5 kg) of rice bran for 1 yen 70 sen instead, and have been feeding that to the horse—about 4 *shō* (7.2 L) per day. They take okay care of the horse, but farmers usually feed a horse about 1 *to* (18 L) per day. And the manure you get from a well-fed horse is very good stuff—full of nutrients and great for growing crops in both dry fields and rice paddies.

March 20, Wednesday

(February 16 according to the old calendar)

Cloudy
North wind
4°C at 1:00 p.m.

Two young village fellows came over to see my hotbox.

Breakfast: miso soup with salted eggplant and butterbur (*fuki*) we received from my wife's elder brother's household,[47] cabbage with soy sauce, pickled daikon, *shōyu no mi,* etc.
Lunch: pickled daikon and salted salmon
Supper: miso soup flavored with whale and with taro stems in it, pickled daikon, salted krill with grated daikon mixed in
Rice for today: 1 *shō* 2 *gō* (2.2 L)

Today my wife gathered firewood and cooked all day. I went out and planted seeds for eggplants and tomatoes and such in the hotbox during the morning. In the afternoon I chopped logs.

Planting seeds in a hotbox is different from person to person. I'll write about my way.

I'd already prepared everything for planting seeds, so all I had to do today was decide on the spacing between the seeds. For planting I use the area within a hotbox going up to about 2 *sun* (6.1 cm) away from the south side of the frame, and about 2 *sun* away from the east and west sides, and about 1 *sun* (3 cm) away from the north side. I use a ruler to draw thin lines in the dirt—about 2 *sun* apart for eggplants and about 2.5 *sun* (7.6 cm) apart for tomatoes. Once that's done, it's time to plant the seeds. Some people plant only a few seeds at once and some plant a lot at once. I like to plant a lot. Along one line I plant about 150 seeds in the case of eggplants and for tomatoes about 100 seeds per line. Then I sprinkle fine sand over the top—about 1 *bu* 5 *rin* (4.5 mm) to 2 *bu* (6 mm) deep—and put a thin layer of straw over the whole thing and that's the end of my seed planting.

Today I planted three different kinds of eggplant seeds: (1) Akita Marunasu (eighteen lines), (2) Kawabe Naganasu (eleven lines), and (3) regular long eggplants (six lines). That made for a total of thirty-five lines. I planted them in the center of the hotbox. And I planted three different types of tomatoes, on both sides of the eggplants: Early Jewel along eleven lines on the west side and seven lines of Hinode and four lines of Best of All. This made for fifty-seven total lines of seeds in the hotbox.

At the main house: The master manned the shop counter during the day. His wife collected firewood and cooked. Kanezō took the horse out to get branches in the morning, and went out for brushwood in the afternoon. In the early morning he sharpened the axe blade.

March 21, Thursday

Sunny
Northwest wind
7°C at 1:00 p.m.

Breakfast: *wakame* soup, pickled daikon, cabbage with soy sauce
Lunch: *dango,* pickled daikon, cabbage with soy sauce

Supper:	I went to my younger sister's house (near the train station) and ate horsemeat. My wife and daughters went to her natal house and ate *dango* and such.
Rice for today:	5 *gō* (902 ml)

Satō Kameji's wife gave birth to a baby boy.

I chopped firewood early, and then prepared rabbit pelts till about three o'clock. My younger sister's family asked me to do this. It's something that any farmer should be able to do, and my wish to learn how came true when I took a course last autumn. It was sponsored by the Akita Prefecture Agricultural Association (Akita-ken Nōkai) and the teacher was Professor Sawayama Satoshi of Hokkaidō Imperial University.[48] I'll write about what I did.

Three days ago, I brought four pelts that had already been stretched out and nailed to a board to dry home, and put them right away into some clean, cold water to soak. Today is the third day, so I took them out and squeezed them as hard as I could and then got ready to strip the inside part of the skin by scraping it from top to bottom with a piece of split bamboo (to do *urauchi*). Fresh, undried pelts only need two days in water. But it's very difficult to tan pelts that are more than a year old. Also, pelts obtained between October and March are said to be the best because they have a lot of fur.

To do *urauchi,* you need a couple of tools—a *kamabokodai*[49] and a curved scraper with dull blades generally known as a *sentō.*[50] These are the tools that the teacher said to use. But I'm not doing this for fun or entertainment—this is important for me to live. You need a carpenter to build a proper *kamabokodai,* and a blacksmith to make a *sentō.* Both cost money, so I made my own tools.

First, I went to the lumberyard and bought a cedar plank—about 3 *sun* (9.1 cm) thick and 4 *shaku* (1.2 m) long—that was left over after sawing the sides off a big log, so it was naturally shaped like *kamaboko.* I shaved off the rough, uneven parts with a knife. I also got another board—about 6 *sun* (18.2 cm) wide, 5 *shaku* (1.5 m) long, and 1 *sun* (3 cm) thick—and (laid it down and) attached one end of the *kamabokodai* to the second board by cutting a notch in the latter, with about 3 *sun* (9.1 cm) of it remaining at the end. To make it so that a person can sit on the lower part and work on the upper part at about stomach level, I held the *kamabokodai* up at the angle I wanted and then went down about 5 *sun* (15.2 cm) from its high end and cut a notch in the long board directly below that. I made the notch about 5 *bu* (1.5 cm) wide—adequate for another board—and deep enough to sink a board into it. I also fastened the low end of the *kamabokodai* to the long board on the ground. Finally, I prepared another board long enough to raise the *kamabokodai* to about stomach level of a man standing and stuck it into the notch I'd cut—standing it vertically—and then attached the top of that to the underside of the top board and in this way, I made a great *kamabokodai* (see Fig. 1.05). Next, I used a piece of bamboo instead of a *sentō.* I took a piece about 5 *sun* (15.2 cm) in diameter, and about 1.5 *shaku* (45.5 cm) long, and split it down the center lengthwise and removed all the knots and such inside and shaved the edges where I'd split it nice and smooth, and in this way I fashioned my own handmade *sentō.* I was satisfied with the result (see Fig. 1.06).

Figure 1.05. Tools for cleaning rabbit pelts: The *kamabokodai* (April 7).

Figure 1.06. Tools for cleaning rabbit skins: The bamboo blade resting on a wooden box (April 7).

To get started, I attached one pelt to the *kamabokodai* with the fur side up and the lie of the fur going downward. Then I put another skin on top of that one with the head end pointing downward and the fur side down, so that all the fat and muscle tissue still clinging to the skin was facing up. The purpose of *urauchi* is to remove all the fat and muscle and tissues from the skin. First, I scraped the anal area with my fingernails. Then, I started scraping downward with the bamboo blade I'd made, going toward the head part.

After cleaning each pelt, it needs to be washed off with water once. After that, you need to soak it in detergent to remove oils. How to make detergent (for ten pelts):

laundry soap	30 *momme* (112.5 g)
hot water (40°C)	6 *shō* (10.8 L)

Put each one in and wash thoroughly. After washing, rinse each one well with water, and then they must be soaked in alum. Alum solution for ten pelts is made as follows:

alum	200 *momme* (750 g)
salt	150 *momme* (562.5 g)
water	1 *to* (18 L)

If you use aluminum potassium sulfate, you only need one-half of this amount of alum. Raw alum doesn't easily dissolve in water, so I used hot water and that worked pretty well. I plan to let the pelts soak in this solution for four or five days, stirring from time to time. This was my work for today. I rested some in the afternoon.

My wife went out to collect firewood in the morning, and cooked and rested a bit in the afternoon, and went to the temple for *hyakumanben*.[51] I don't know if it's because of the bad economy or what, but people don't come for this much anymore.

> At the main house: Kanezō cut hay for the horse in the early morning and then later in the morning he went out to gather firewood. The master was watching the shop, and his wife and my mother cooked some dishes for *hyakumanben*. Everyone rested in the afternoon. My mother just returned from my younger sister's house.

In the old days, people used a big rope for *hyakumanben*, but they didn't do that this year. I heard that this time each person had to bring a piece of thin rope to the temple, just for tying around their own head.[52] Even the religious rituals change like this—little by little.

March 22, Friday

Cloudy
Light snow, sunny later
Northwest wind
8°C at 1:00 p.m.

Figure 1.07. An old woven-mat house that belongs to an Ōkura family that has lived in poverty since long ago (March 21).

Breakfast:	rice, miso soup with salted butterbur stems, pickled daikon, cabbage with soy sauce
Lunch:	leftover *dango* from yesterday, rice, salted salmon, pickled daikon, etc.
Supper:	rice, miso soup with taro stems flavored with whale, pickled daikon, salted krill
Rice for today:	1 *shō* (1.8 L)

Yoshida Yasuke asked an *edako*[53] to perform a *nanakurayose*[54] for his child. All women of the village gathered for that (at his house).[55]

In the morning, I went to my sister's house near the station with the horse and brought back about 3 *to* (54 L) of night soil in one trip. Also, the fertilizer I ordered had arrived, so I brought that back as well (another trip). I plowed my fields all day long—in the morning and in the afternoon—with the help of the horse. I borrowed the horse all day for that. In return for borrowing the horse frequently like this, I often go and do jobs at the main house. Because my fields are in a bad location, even working all day with a horse I can only plow about 3 *tanbu* (0.3 ha). I could probably plow 5 *tanbu* (0.5 ha) in a day if the fields were flat. I'll explain about the tools I use for plowing later.

My wife collected firewood in the morning, and she used a hoe to turn over the soil in the parts of the fields that couldn't be plowed with the horse in the afternoon.

Today's expenditures:

fertilizer	12 yen 50 sen

Figure 1.08. Rabbits I'm raising in front of my house (March 22).

itemized:

sulfuric acid ammonia, 1 *kamasu*[56] (10 *kan* (37.5 kg))	3 yen 80 sen
(strong) superphosphate of lime, 1 *kamasu* (8 *kan* (30 kg))	1 yen 60 sen
dried powdered anchovy, 1 *kamasu* (10 *kan*)	3 yen 80 sen
oil cake (*aburakasu*), 1 *kamasu* (10 *kan*)	3 yen 30 sen
brown sugar	10 sen
candies	6 sen
alum	4 sen

At the main house: The master watched the shop but also went out on village business. His wife cooked and went out for some firewood. My mother went to the *nanakurayose*. Kanezō repaired the trough for the horse's hay in the early morning. He went out to cut branches (cedar and pine) both in the morning and in the afternoon. He finished that pretty early, so he cut more hay for the horse.

March 23, Saturday

Clouds and sun
West wind
8°C at 1:00 p.m.

Breakfast: miso soup with salted eggplant and butterbur stems that we got from my wife's elder brother's household, pickled daikon, *misozuke*

	(miso-preserved vegetables), *shōyu no mi*, etc. Today we finished the *shōyu no mi* we got from my wife's natal home.
Lunch:	I skinned and butchered the rabbit that died yesterday, and that made one plate of meat, so we cooked it in a pot with green onions and soy sauce for flavor. We also ate pickled daikon, etc.
Supper:	miso soup with dried taro stems and flavored with whale, pickled daikon, *takuan*. Also, we got a plate of steamed krill flavored with soy sauce from my wife's brother and that was really good.
Rice for today:	1 *shō* 2 *gō* (2.2 L)

I went to the forest land of another household early and dug up cedar stumps.[57]

Today my wife and I planted green peas (*endō mame*) all day. First, I leveled out one of the fields I plowed yesterday. Then, I made rows about 2 *shaku* 5 *sun* (75.8 cm) apart. In local speech, people say *ni shaku go sun ni hatake o ebaru*.[58]

I always make rows in fields using a ruler, based on careful calculations, for efficiency. But most people simply make rows without tools—just by estimating. I take two thick poles about 3 *shaku* and 4 or 5 *sun* (1.1 m) long and make little marks on them at 5 *sun* (15.2 cm) intervals to the 3 *shaku* (90.9 cm) point. Then I tie ropes to them and measure with them. It's a very convenient way because you can easily measure distances of 2 *shaku* (60.6 cm) to 2.5 *shaku* or 3 *shaku* between rows, depending on the type of plants you want to grow. Also, if the land you're planting is long from north to south and narrow from east to west and flat, then it doesn't matter which direction you start making rows from. A person can start from the north or the south. But if the field is on a slope, you need to work from the top down or else it won't go very well. Everyone here in the village does it this way.

Also, making rows is different for *hidarikko* and *migikko:* left-handed and right-handed people.[59] For example, when making rows along an east-west line, a right-handed person faces east and pulls soil back to the west while moving eastward. A left-handed person does the opposite.

After the rows have been made, for long north-south plots with rows running in the same direction, I always dig a hole with a single chop of the hoe on the north side of each spot where I'm going to plant something. I put fertilizer in these. Other villagers don't do this—they just spread fertilizer between the rows and then plant seeds and cover them with dirt. I always do it as I wrote above. I put some fertilizer in the holes I dug, and then some night soil, and then some of the superphosphate of lime, spread a thin layer of soil over that, then plant peas, then sprinkle some ash over the top of the peas, and then spread some soil over everything, and that's it. For covering seeds with soil, using east-west rows as an example, a right-handed person faces west and pulls back to the east, but a left-handed person does the opposite.

I planted about 40 *tsubo* (132 m^2) of green peas today. There are about 32 rows.

Fertilizer:	
composted straw	about 30 *kan* (112 kg)

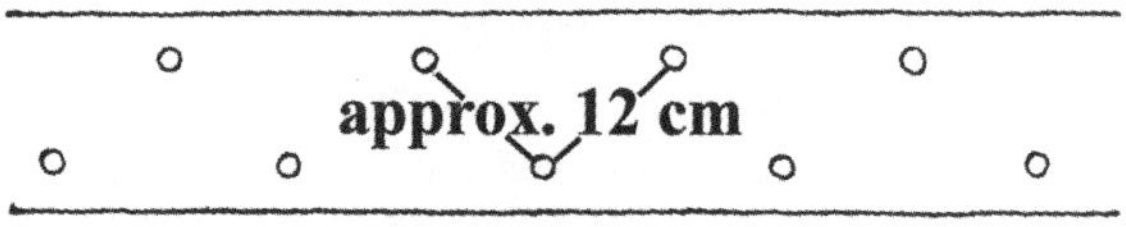

Figure 1.09. Green pea seed-planting pattern.

Figure 1.10. A woman and a girl carrying woven straw mats on their backs to sell at the agriculture warehouse, in front of Yoshida Tetsuya's house (March 22).

night soil	2 *ka* (with about 2 *ka* of water mixed in)
superphosphate of lime	2 *shō* 2 *gō* (4 L)
ash (grass and wood)	1 *to* 2 *shō* (21.6 L)
seed peas	about 4 *gō* (721.6 ml)

I planted the seeds about 4 *sun* (12.1 cm) apart, as shown in the diagram (see Fig. 1.09).

Four of the rabbits that were born about twenty days ago are still alive, but when I looked carefully this morning, I saw that the eyes of one hadn't opened, and it seemed to be blind. While I was wondering about this, the crazy neighbor came over and told me that I should wet the rabbit's eyes with water and then after a while try to open them with my fingers, and then spray some water on them again. I tried that, and sure enough its eyes opened.

At the main house: My brother's wife went out to collect brushwood and cooked. Kanezō cut hay for the horse in the early morning, and he collected firewood the rest of the day. The master of the main house doesn't do any farmwork, so I won't write about him anymore.[60] I'll only write about him if he does something relating to farmwork. That goes for my mother as well.

March 24, Sunday

Cloudy
Very strong southeast wind
8°C at 1:00 p.m.

Breakfast:	miso soup with daikon in it, *takuan,* pickled daikon
Lunch:	*dango* with crushed sesame seed paste, cabbage with soy sauce, pickled daikon, *takuan*
Supper:	leftover *dango,* miso soup with daikon in it, pickled daikon, cabbage with soy sauce
Rice for today:	5 *gō* (902 ml)

I split cedar stumps early in the morning. And later I cut brushwood and split more cedar stumps. My wife cut bamboo and cooked. Both of us rested in the afternoon. It's the last day of *higan,* so everyone in the village rested in the afternoon.

At the main house: My mother cooked. Kanezō and my brother's wife collected firewood and then rested in the afternoon.

Over in the next village—Tennō Mura—the school caught fire, so many people went to help.[61] The school totally burned down, and many houses also burned as well. There appears to have been a lot of damage.

The Wakimoto Shinyō Hankō Kumiai[62] built a big warehouse[63] in front of Wakimoto Station, so to make it easier to transport goods, the little road of Ōkura has been designated a village road. This means that money is now coming from the prefecture government. It's going to be a great village road. The local roads were generally in bad shape, so the village received money from the prefecture and repaired the prefectural roads. The stones for fixing the roads came from Mt. Kanpū (see Fig. 1.11). The soil came from privately owned mountain land in Ōkura, and it was taken from right behind the houses, so those properties naturally became larger. The property owners are Yoshida Chūzō, Yoshida Kinjirō, and Yoshida Kasen.[64]

March 25, Monday

Rain
Light northwest wind, later strong out of the southwest
8°C at 12:30 p.m.

Breakfast:	miso soup with daikon in it, pickled daikon, cabbage with soy sauce
Lunch:	salted salmon, pickled daikon, etc.
Supper:	miso soup with taro stems in it (stock made from roasted dried sardines), pickled daikon, *takuan,* salted krill, etc.

Figure 1.11. Hauling stones from Mt. Kanpū by horse-drawn cart. From Yoshida's personal collection: not included in 1938 book.

I slept until breakfast time this morning, and I made rope for myself to use the rest of the day. I used eight bundles of straw and made seven of rope. It took seven hours.

My wife cooked and helped make a kimono for her brother's family.

Today's expenditures:

postcards (10)	15 sen
soy sauce (2 *gō* (361 ml))	6 sen
candy (30 pieces)	10 sen

At the main house: Kanezō cut hay for the horse in the early morning. Later in the morning, and in the afternoon, he made rope. He used ten bundles of straw to make seven bundles of rope. He made his rope for a different purpose than me, so it needed to be thicker, and that's why he used more straw to produce about the same amount of rope as I did. It took him eight and a half hours. My brother's wife had to go see her grandmother because she's been sick, so she didn't do any work.

I heard that a land dispute between Izaemon, a landlord in Tayazawa,[65] and Kakichi, a farmer in Ōkura, finally reached a satisfactory settlement. Izaemon bought five *tanbu* (approx. 0.5 ha, or 4,959 m^2) of paddy land from its previous owner—land that Kakichi had been farming—and declared that he was going to grow rice on the land himself. But Kakichi only had farming rights to one *chōbu* (approx. 1 ha, or 9,917 m^2) of land in the first place; he insisted that losing half wouldn't just make it difficult to feed his family of seven or eight people but would effectively condemn them to death. This sparked the dispute. A mediator stepped in, and it was finally resolved.[66] The points of the agreement are:

1. Kakichi can continue to farm the land for five years.
2. Of the 15 *hyō*[67] (approx. 900 kg) of rice grown on the land last year (the 1934 harvest), Kakichi keeps ten, and five go to Izaemon.
3. Izaemon will pay Kakichi ten yen to cover the fees relating to the dispute settlement.
4. One *tanbu* (991.7 m^2) of the land will be handed over to Izaemon in April of 1935.

March 26, Tuesday

Cloudy
Strong northwest wind
2°C at 2:00 p.m.

Breakfast:	rice, miso soup with daikon in it, pickled daikon, *takuan*, etc.
Lunch:	My wife ate pickled daikon, *takuan*, and leftover soup from morning by herself at home. Fusae[68] went to my wife's elder brother's house on an errand and ate lunch there. I took Yaeko[69] to the main house and we ate lunch there. We ate cabbage cooked with salted krill and pickled daikon, *takuan*, etc.
Supper:	miso soup with daikon in it, and we put some chopped spring *arame* (sea oak) in it. When you put this in hot miso soup it becomes nice and gooey and it's very delicious. We also ate pickled daikon, *takuan*, and salted krill.
Rice for today:	8 *gō* (1.4 L)

Today my wife went to the shore to collect seaweed, but she was only able to get a bit of *arame*. In the afternoon she sewed and cooked, etc. The place she went for seaweed was the shore of Wakimoto.

Early this morning I made two pairs of *ashinaka*[70] and later in the morning I made *mina*, until about three o'clock. *Mina* is a kind of rope for carrying stuff on your back. I took three bundles of *shinagodawara* (stripped straw) and removed all the remaining stamens and pistils from the straws and combined them to make just two bundles and pounded them until the straws became like strings of linen. Straw in that state is called *yakowara*.

I made two pairs of *ashinaka* and three *mina*. There are *mina* of different sizes—long and short—depending on the need. A regular one is about 5 *hiro* (7.6 m). The *mina* I made today are for my wife to use to carry a basket on her back, so I made them all 3 *hiro* (4.5 m) long.

When you make *mina,* you don't start weaving the rope from the end—you start from the middle. First, you grip one bunch of straw and divide it into two smaller bunches and start weaving them together from the center and work toward the ends, and you make the center part fairly thick.

Then, you go back and weave more straw into the rope you've made, to make it stronger. So, in the end, you have a piece of rope that was made from three small bunches of straw. This is called *mikko kaketa nawa.*

After finishing my straw work, I took the rabbit skins out of the alum mixture since it's the fifth day. I squeezed out all the liquid that I could and then stretched the skins out and coated the inner sides with a 4:2:1 mixture of water, starch, and egg yolk. You mix this thoroughly and apply it with a brush (like those used for writing). Then, you fold the skins over on themselves once—or perhaps twice—and press them down and leave them for one night like that. (Note: It's best to use one egg yolk for two skins, but areas where the fur on the other side is especially thick should be gone over again with the brush. And, you have to keep stirring the mixture constantly while painting it on the skins.)

At the main house: My brother's wife cooked and sewed. Kanezō made small *tawara*[71] for holding rice seeds (to be soaked). You need about 6 *hiro* (9.8 m) of rope to make one of these *tanedawara.* And as for straw, about two bundles are needed. But you have to remove any excess stamens and pistils from the *shinagodawara.* In the afternoon he took the horse out and brought back a load of wood, and after that he cleaned the dung (and straw) out of the horse's stall and threw it on the pile outside.

March 27, Wednesday

Cloudy, sometimes sunny, occasional snowflakes
Strong northwest wind
A little over 4°C at 2:00 p.m.

Breakfast:	miso soup with daikon and potato in it, pickled daikon, *takuan,* etc.
Lunch:	cabbage cooked with salted krill, pickled daikon
Supper:	miso soup with taro stems in it (stock made from roasted dried sardines), pickled daikon, salted krill
Rice for today:	1 *shō* (1.8 L)
Snack:	dried New Year *mochi*—roasted

My wife cooked and sewed, and she stayed home with Fusae because she got sick.

I split logs and collected firewood from the morning. I also hanged the rabbit skins from yesterday—which I had coated with egg yolk and starch and left overnight—in the sun (to dry). And the hotbox tomatoes started to sprout, so I removed the top layer of straw. The thermometer in the soil was reading 28°C, so I opened the glass panel above the tomato plants about three *sun*

(9.1 cm) and let in some air, adjusting the temperature a bit. This was for preventing the temperature inside the hotbox from getting too high. I forgot to write this before, but in the evening, I always lay thin *noma* mats made of straw or miscanthus (*kaya*) down inside my hotboxes. And, until the seeds start to sprout, I remove the mats and let in a bit of sunlight on sunny days, making sure not to allow the temperature inside to get too high. On cold days I leave the mats in place.

Today's expenditures:

candies	5 sen

At the main house: Kanezō cut hay for the horse in the early morning, and then took the horse out to get firewood both in the morning and in the afternoon. He brought back two horse loads for the day, and he also went out once by himself to collect firewood (branches and sticks). My brother's wife stayed in to take care of her younger son [Tsugio][72] because he got sick, and she sewed while keeping her eyes on him.

March 28, Thursday

Cloudy, sunny later
Light northwest wind
8°C at 1:00 p.m.

Breakfast:	miso soup with daikon in it—put *arame* in and ate it—and pickled daikon, *takuan*, etc.
Lunch:	My wife and the kids stayed home and ate salted krill, pickled daikon, *takuan*, etc. I ate lunch at the main house. We had salted herring—the first time for that this year. We also ate pickled daikon, *takuan*, etc.
Supper:	My wife ate last night's leftover miso soup, pickled daikon, *takuan*, salted krill, etc. Fusae got sick, so she hasn't been eating anything but *okayu*[73] since yesterday. Today I ate supper at the main house again. We had miso soup with daikon and potato in it, and pickled daikon, *takuan*, etc.
Rice for today:	5 *gō* (902 ml); miso: 20 *momme* (75 g)

Today's expenditures:

cold medicine (antipyrine)	20 sen
apples (2)	5 sen

In the early morning I kept my eyes on Fusae. After breakfast, I removed the straw *komo* mats from the hotbox surface and heated some water to about 20°C and sprinkled about 8 *shō* (14.4 L) into the hotbox. After that, I went to buy medicine for Fusae. In the afternoon I went over to the main house to help stack firewood. Once that was done, I dealt with the rabbit skins again. I took the skins—since they'd been out in the sun for two days (yesterday and today)—and sprayed some water on the inner sides and folded them over and pressed them down lightly. I'll leave them like that for one more night.

My wife cooked, sewed, looked after Fusae, and covered the hotbox surface with mats.

> At the main house: Kanezō cut hay for the horse early, and then took the horse out and brought back a load of sticks after breakfast. In the afternoon we stacked firewood together.

My elder brother's wife had to go to her grandmother's funeral, so my mother did the cooking.

March 29, Friday

Sunny, no clouds
No wind today
12°C at 2:00 p.m.

Breakfast:	miso soup with *wakame* in it, pickled daikon, *takuan,* miso with ground roasted sesame seeds mixed in,[74] etc.
Lunch:	Soy sauce–based soup with *shiroyo*[75] and tofu in it, dried strips of daikon and carp[76] cooked together with sugar and soy sauce, cabbage with soy sauce, *kata-uri*[77] preserved in miso, pickled daikon, *takuan,* etc.
Supper:	rice with hot water poured over it (*yuzuke*), salted herring from last year, pickled daikon, *takuan,* etc.
Rice for today:	1 *shō,* 2 *gō* (2.2 L), miso: 20 *momme* (75 g)

Five villagers came over to see my hotbox.

Someone I'd never met came from Tokyo because he wanted to learn more about farming villages.

My work for today: In the early morning I worked with the rabbit skins again. I took the skins that I'd wetted last night and scraped the inner sides against a blade-shaped piece of wood about 4 *sun* (12.1 cm) wide fixed to a box.[78] In other words, I held them (with both hands) and pulled one side and then the other—back and forth—scraping them with the piece of wood in the center. When you do this, you can put a lot of effort into it if you're holding the skin vertically (with the head part up and the fur pointing down) but you must be careful if you are holding the skins horizontally because they may rip if you scrape them too roughly.

After doing this long enough to make the skins fairly soft, you have to put them out in the sun with the fur side up for a day.

This morning, and this afternoon, I was busy chatting with visitors. My wife cooked and read a book while looking after Fusae.

Today's expenditures:

shiroyo (1 *gō* (180.4 ml))	10 sen
tofu (1 block)	6 sen

white sugar	5 sen
green tea	5 sen
soy sauce (2 *gō* (360.8 ml))	10 sen

At the main house: My brother's wife still hasn't returned. My mother cooked. My brother and Kanezō prepared rice seeds for planting during the morning. Kanezō went out to collect wood in the afternoon.

In the old days, people selected rice seeds for the following year by type—examining individual plants in the field one by one by hand, every autumn. Also, I've heard that since they used to plant so many seeds at once, they'd cram anywhere from 1 to 5 *shō* (1.8 L to 9 L) of seeds into each *tawara*. People could plant enough rice to get about 100 swipes with a scythe (at harvest time), or about 360 *tsubo* (1,190.2 m^2) with one such *tawara* of rice seeds. For 2,000 strokes you'd need 20 *tawara* full of seeds. You have to start soaking them around the time of the spring equinox.

But today our knowledge and skills are better, so nobody selects rice plants one by one anymore (well, some people might do that). Instead, they usually soak the seeds in salt water to find the best ones. You fill a big container with water and add salt and stir it up well. But you only need enough salt to make an egg float in the water. If you put rice seeds in that water, the bad ones will float up to the surface and you can easily discard them (see Fig. 0.13). Then you collect the ones that sank to the bottom and pack them into *tawara* and soak them in a pond.

Unlike the olden days, since people don't have to plant so many seeds per unit of land, we only pack about 1 *to* (18 L) of seeds into one *tawara*. You can raise healthier plants by spreading seeds more thinly. Today, you can get about 100 swipes with a scythe (at harvest time) with one *tawara* of seeds. In the old days, folks had to soak seeds in water for three or even four weeks, but now you only need to do it for a week to ten days. Most have a small pond for this, and people who don't usually ask someone who has one to let them soak their seeds in it. You lay a thick plank down across the pond and tie the *tawara* to it (in the water) and cover everything with reeds (*kaya*) (see Figs. 0.14 and 0.15).

March 30, Saturday

Cloudy, sunny later
Weak west wind
8°C at 2:00 p.m.

Breakfast:	miso soup with sea lettuce (*aosa*) and tofu in it, pickled daikon, *kata-uri* preserved in miso, *takuan*, miso with sesame seeds from yesterday, etc.
Lunch:	*shiroyo* soup left over from yesterday, salted herring, pickled daikon, *takuan*, yesterday's leftover food, etc.

Supper: miso soup (stock flavored with whale) with sun-dried taro stems in it, *takuan,* salted salmon, etc.

The sea lettuce came from the shore of Hirasawa Village, Minami Iso.[79] My wife's elder brother received it from someone and shared some with us. We also got the salted herring from them.

Rice for today: 1 *shō* (1.8 L); miso: 30 *momme* (112.5 g)

Today's expenditures:
nails 5 sen

This morning I worked on the rabbit pelts again like yesterday. If you do this twice before they fully dry you can get a better result, it seems. Then, I nailed each one to a board, to make them nice and flat. When you do this, you must nail the top part first, and then the sides, but not pull too hard on the skin while you do it.

I also collected firewood, and I added some fertilizer to the summer cabbages I planted last autumn—the first time to do that this year. I dissolved 200 *momme* (750 g) of ammonium sulfate in about 2 *to* (36.1 L) of water and sprinkled it on the field with a watering can. The area is about 3 *tsubo* (10 m^2).

The hotbox temperature isn't very good—it's uneven across the surface. There are warmer areas and cooler areas, so some plants are growing faster than others. The tomato plants were like that, so I thinned them out for the first time this year. You must do this to give them enough room to grow, relieving the clustering. I removed the mat covers at eight-thirty this morning and opened the glass panels about 3 *sun* (9.1 cm) at ten o'clock. At three o'clock I closed the glass panels, and I put the mats back in place at four-thirty.

My wife watched over Fusae and sewed and, of course, cooked.

At the main house: My brother's wife went out to get brushwood. Kanezō cut hay for the horse early, and went out for firewood later in the morning and again in the afternoon. My mother cooked and cut daikon into strips for drying in the sun, etc.

March 31, Sunday

Cloudy, sunny later
North wind
0°C in the morning, 8°C at 2:00 p.m.

Breakfast: miso soup with sea lettuce and potatoes in it, pickled daikon, *takuan,* etc.

Lunch:	*soba* (buckwheat) flour kneaded with a bit of hot water. This is called *nereke.* First, you boil water in a pot and then add *soba* flour and mix it up with a spatula.[80] We ate it with some soy sauce for flavor. In the old days people ate it with *shottsuru* made from *hatahata,*[81] I'm told. We also had pickled daikon, *takuan,* salted salmon, etc.
Supper:	miso soup (stock made from dried roasted sardines) with daikon in it, pickled daikon, *takuan,* salted krill, etc.
Rice:	1 *shō* (1.8 L)

I forgot to note exactly when the first sprouts appeared in the hotbox, so I'll do that now.

tomatoes	March 27
eggplants	March 30

I removed the hotbox mats at eight o'clock this morning and opened the glass panels at eleven. I closed them again at three o'clock and then put the mat back in place at four. Today I thinned out the eggplants. As I wrote yesterday, the hotbox temperature is uneven, so the seeds aren't all sprouting at the same time. According to what I've heard, not stomping all the straw down evenly (at the beginning) can cause this.

I also prepared a field for planting summer daikon. And I finally finished up my work with the rabbit pelts. I took them down from the board I nailed them to yesterday and laid them flat on a board and gently scrubbed them with a pumice stone to clean them, and the result was four quite excellent skins.

My wife cut daikon into strips for drying and did other tasks during the morning, and, of course, she cooked. In the afternoon she cut down some bamboo around the fields.

At the main house: The master went out on some business, so his wife watched the shop. My mother cooked. Kanezō cut hay for the horse in the morning, and he spent the rest of the day cutting firewood. He got two horse loads.

Chapter 2: April 1935

April 1, Monday

Sunny
Northwest wind from the afternoon
–2°C at 7:00 a.m., 8°C at 2:00 p.m.

Breakfast:	miso soup with daikon and tofu in it, pickled daikon, *takuan,* etc. (The tofu wasn't homemade—we bought it.)
Lunch:	I got some horsemeat from my sister's family (they'd bought it) and we cooked that with green onions that I grew, and with miso, and we ate that. The side dishes were the same as breakfast.
Supper:	miso soup with dried taro stems, roasted salted herring, and side dishes same as morning. This was what my wife and the kids ate. I had to go out on an errand, so I ate supper at my sister's house. We had soy sauce–based soup with carp (*funa*) caught in Lake Hachirō and green onions (the fish and onions had been purchased) and pickled daikon. There was also sole (*karei*) caught at Minami Iso cooked with potatoes (*nishime*). The fish was purchased, and the potatoes had been given to us by relatives.
Rice:	1 *shō* (1.8 L) miso: 20 *momme* (75 g)

Today's income:

my tanning work on four rabbit pelts	50 sen

Today's expenditures:

candies	3 sen
fee for mailing photo negatives	10 sen

In the morning I made a new handle for my hoe. I also made a new dipper (*hishaku*) for scooping night soil. Actually, I modified an old kitchen ladle to make it. In the afternoon I went to cast my vote for our representative in the Agricultural Association[1] election. My wife was out cutting bamboo leaves (*sasa*) all day. Today I did just about the same things as yesterday to the hotbox.

At the main house: My brother's wife watched the shop and my mother cooked. Kanezō was sick in bed all day. Today the Wakimoto Nōkai representative election was held at my old school.[2] The village is divided into areas, and we had to elect representatives for each. There were no contests in the other hamlets but in our hamlet, Tominaga,[3] we had a lot of trouble because someone entered the race. We had already decided how to divide our hamlet votes, in a meeting, when the Socialist Masses Party[4] put up a candidate. So, we had a contest. But the Proletarian Party[5] lost by two points in the end. Because the race was contested, according to what I heard, votes were bought for 30 sen. Apparently, there were a lot of other secret deals. My brother entered the race too, and he won!

April 2, Tuesday

Snow, then cloudy, later sunny
Weak northwest wind
0°C in the morning, 4°C at 2:00 p.m.
Show accumulation: about 4 *sun* (12.1 cm)

All three meals were basically the same as yesterday.

For the first time ever, at the main house, they bought fish called *gangari,* or *yudoyo*[6]—I don't know the scientific name—caught in Lake Hachirō and ate it for lunch. I ate some, too.

I made rope in the morning and in the afternoon. I used six bundles (*wa*) of straw but could only make five bundles of rope with that. My wife was busy sewing and cooking. It was really cold, so I didn't remove the (glass panels) of the hotbox at all.

At the main house: Everything was about the same as yesterday.

For winter season *dekasegi*[7] work this year, forty men and fifteen women from Ōkura went away. Most of the men went to Hokkaidō and most women went to the Tokyo area. Men usually fish and do construction work and the like. Women generally work as maids (*jochū*) or labor in factories, etc.

April 3, Wednesday

Snow, then sunshine
Northwest wind
2°C in the morning, 6°C at 6:00 p.m.
Snow accumulation: 1 *sun* (3 cm)

Breakfast:	miso soup with daikon, side dishes: same as yesterday
Lunch:	dried strips of daikon fried in Shirashimeyu[8] and flavored with soy sauce, side dishes: same as yesterday
Supper:	miso soup with dried taro stems, salted salmon, side dishes: same

Today I made things with bamboo. I made toy pistols for the kids and an ink container for myself. My wife was sewing all day.

I lifted the straw mats from the hotbox surface at eight o'clock and opened the glass panels at noon. I closed these at three o'clock and put the mats back in place at four.

At the main house: My brother's wife sewed, and my mother cooked. Kanezō got up today and worked. He made rope all day. He also made a *modotsu*.[9] This is a kind of rope device you attach to a horse's saddle for carrying loads of stuff.

April 4, Thursday

Sunny, cloudy later
South wind
–1°C in the morning, 11°C in the afternoon

Breakfast:	miso soup with daikon in it, pickled daikon, *takuan,* miso-preserved vegetables
Lunch:	salted krill, side dishes: same as morning
Supper:	a bit of *mochi* we got from the main house, side dishes: same as morning. I was invited over to the main house in the evening, so I ate *mochi* there before coming back home. *Mochi* for spring (March) festival time[10] would normally be pounded from *mochi* rice but the main house had a poor rice harvest last year so there isn't much *mochi* rice now. Therefore, they made *mochi* with regular rice—known as *yama mochi.* There were two kinds: one covered with sweet, cooked red *azuki* beans (*azuki-an*), and another in soup flavored with dried sardine (*niboshi*) stock (*ozōni*). Side dishes: boiled napa cabbage flavored with soy sauce, flounder from Minami Iso simmered in soy sauce, and herring roe from Hokkaido mixed with ground daikon and vinegar. In addition, we ate pickled daikon, *takuan,* etc.
Rice:	5 *gō* (902 ml), miso = 20 *momme* (11.25 g)

I made hanging screens (*sudare*)[11] from the morning—for protecting the fields from strong winds. I lifted the mats off the hotbox surface at eight o'clock this morning, and heated some water and poured it over the surface. I opened the glass panel covers at eleven, but mainly made *sudare.* In the afternoon I went to the main house to help collect two bundles of firewood and then closed the hotbox panels at two-thirty. I put the mats back in place at four-thirty. My wife went out on errands in the morning and chopped firewood in the afternoon.

At the main house: My brother's wife sewed and helped my mother pound *mochi.* Kanezō took the horse out to collect branches in the morning and did a whole day's work by lunchtime. In the afternoon he helped me gather those two bundles of firewood. He also helped pound *mochi.*

April 5, Friday

Rain, later just cloudy, then more rain
Weak northwest wind
6°C in the morning, 10°C in the afternoon

Breakfast:	*wakame* miso soup, chopped Chinese yam (*nagaimo*) flavored with soy sauce
Lunch:	same as morning
Supper:	miso soup with taro stems in it, salted herring from Hokkaidō, side dishes: same as morning. We put rich miso liquid (*miso no tamari*) on the roasted herring and ate it that way.

I wove *sudare* this morning and rested in the afternoon. I removed the straw mats from the hotbox surface at nine o'clock and put them back in place at four. My wife sewed all day.

Today's expenditures:

10 postcards	15 sen
6 candies for the kids	2 sen
4 herring from Hokkaidō	10 sen

At the main house: My brother's wife sewed, and cooked. Kanezō went out early to cut hay for the horse and then he wove rope later in the morning. He rested in the afternoon. Today is March 3 according to the old calendar, so it's festival day.

April 6, Saturday

Sunny, with a strong northwest wind
3°C in the morning, 8°C in the afternoon

Breakfast:	*wakame* miso soup, side dishes: same as every day
Lunch:	roasted salted herring flavored with strong miso liquid (*miso no tamari*), side dishes: same as breakfast
Supper:	miso soup with pieces of taro in it, salted krill, side dishes: same as morning
Snacks:	We got some leftover festival *mochi* from the main house, so we ate that.

Hige no bikki[12] cried for the first time this year today.

I thinned out the eggplant, tomato, and other seedlings in the hotbox—the second time.

Figure 2.1. A village girl searches for young shoots in a small water channel below my house (April 5).

I lifted the hotbox mats at eight o'clock and opened the glass panel covers about 2 *sun* (6.1 cm) on each side at nine-thirty and then to about 4 *sun* (12.1 cm) on each side at ten-thirty. I closed them at three o'clock and put the mats back in place at four-thirty.

I also made a basket for carrying things on my back with bamboo strips. If you buy one at a store, you'll pay 1 yen 50 sen. But I don't have much money, so I just got four or five bamboo stalks from the main house's land and made one myself: self-sufficiency! My wife collected firewood.

At the main house: My mother cooked. My brother's wife sewed while watching the shop. Kanezō went to Maeda[13] in the morning to do *tazukiri*.[14] This is done to the rice paddies with a long tool that resembles a halberd (*naginata*). You cut about 5 or 6 *sun* (15.2–18.2 cm) deep with the tool along both sides of the mud ridges (*kuro*) running along the edges of the paddies. This is done in advance, because it makes the soil on the edges of the paddies come apart from the ridges more easily. The hand plowing goes more smoothly.

In the old days people used to make the ridges by spreading mud from the paddies on each side and packing it down, so *tazukiri* wasn't necessary.

Since the old days (but less so today), there have been two types of these soil ridges between paddies: *yazukuro* and *nurikuro. Yazukuro* are lower than *nurikuro,* and they are full of weeds with strong root systems, so even if someone walks on them, they won't be damaged. Therefore, there has never been any need to cover them with mud from inside the paddies.

Nurikuro, on the other hand, don't contain weeds with strong roots, so they get damaged if walked on. And field mice can easily make holes in them, which prevents them from retaining water, so mud must constantly be spread over them and packed down. But these days, you never see anyone doing this. Instead of making *nurikuro,* people do *tazukiri* around the edges of the paddies, and they always lift some soil up onto the ridges when they turn it over with a hoe. It naturally gets patted down as people walk, so there's no need to pack and smooth mud ridges by hand. This makes for perfectly good ridges between the fields—people really thought up something good. Nobody really knows who first got the idea. It's a new development. Up until I was about twenty-four, everyone always worked hard to make *nurikuro* (to do *kuronuri*). We had to let water into the paddies as soon as the snow had disappeared and soften up the mud and then pile it up bit by bit to form soft ridges, scoop by scoop, with a hoe. Not only me, but lots of people would complain about lower back pain—especially women. We had to break through the ice in the early morning, or on especially cold days. Even now I shiver when I think about it. It was so cold that everyone wore straw boots (*waragutsu*) into the paddies, but even that didn't help so much; they always had to climb out and warm their feet on top of the ridges many times, even just in the morning. Today we wear rubber boots into icy paddies. And so, the old ways of making and reinforcing ridges from both sides gradually evolved into what everyone calls *katakuro* [or *kata-aze*] today.

Katakuro are made by packing mud onto only one side, so it takes only half the work. Once people started making these it only took about two or three years for everyone to quit the other method (making *nurikuro*). It was very fast.

In the afternoon, Kanezō prepared a rice paddy for planting seeds to raise seedlings. In Oga, we call that *nashiro o kohiru*.[15] First, you have to form the ridges to prepare for spreading mud on the sides. When you do this, you find a lot of dead leaves, roots, and other stuff [twigs, etc.] in the paddy so you must dredge it thoroughly and pull all that out using a hoe or a *komajari*[16] and put all the stuff (*godomori:* "trash") into a hole.

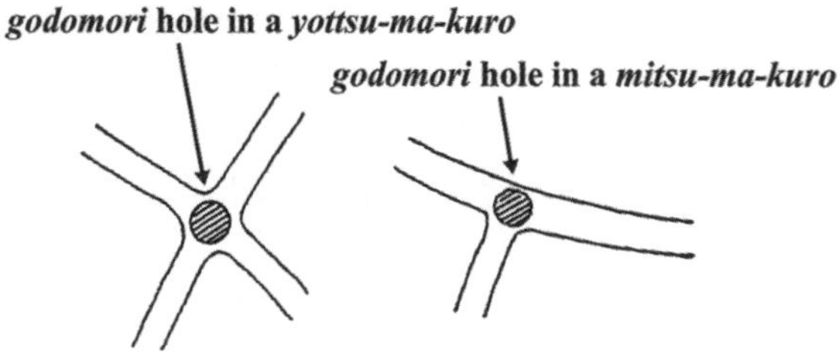

Figure 2.2. Four-way and three-way paddy ridge intersections, with *godomori* holes.

Komajari is a local word. I think it probably comes from *konoha-sarai* [dredging leaves]. You just dig a small hole at a ridge intersection and stuff in the piles of junk (*godomori*)—all the stuff you pulled from the paddy when you cleaned it out. Then, during winter when everything is covered with snow you use a sled to bring it to the house and make fertilizer with it. I suspect that *godomori* comes from *gomimori* [garbage pile]. Four-way ridge intersections, where this is often buried, are called *yottsu-ma-kuro,* and three-way intersections are known as *mitsu-ma-kuro* (see Fig. 2.2).

Kanezō finished this job pretty quickly, so next he hand-plowed the *nashiro* with a three-pronged hoe (*sanbon makka*).[17]

April 7, Sunday

Sunny, no clouds
–4°C in the morning, 9°C in the afternoon

Breakfast:	miso soup with some cabbage in it, side dishes: same as always
Lunch:	salted herring, side dishes: same
Supper:	miso soup with taro stems in it, salted krill
Snack:	roasted dried *mochi* (*hoshi mochi*), in the morning

I cut bamboo all morning. I removed the straw mats from the hotbox at eight o'clock and sprinkled warm water over them at eight-thirty. I opened the glass covers 5 *sun* (15.2 cm) on each side at nine o'clock and closed everything at three. I put the straw mats back in place at four-thirty.

I planted summer daikon seeds today. My wife helped me make rows in the field—tall rows. After they were ready, I dug holes (along the tops of the rows) about 6 *sun* (18.2 cm) apart, so my rows are about 2 *shaku* 5 *sun* (75.8 cm) wide, with 6 *sun* between each daikon. Then I put night soil diluted 1:1 with water into each hole I'd dug. The holes are about 2 *sun* (6.1 cm) in diameter. I used a tool like the one in the illustration to dig the holes (see Fig. 2.3).

I just grabbed the handle of the tool with both hands and stuck it into the soil on the top of the rows as I walked along and that made great holes. Then I made a fist with my hand and stuck my middle finger out just enough for the knuckle to protrude beyond the others and went along the rows again, pressing my knuckle into the tops of the rows just beside each hole into which I'd already poured night soil. I put daikon seeds into each shallow depression I'd made and covered them up and that was the end of my daikon planting (see Fig. 2.4).

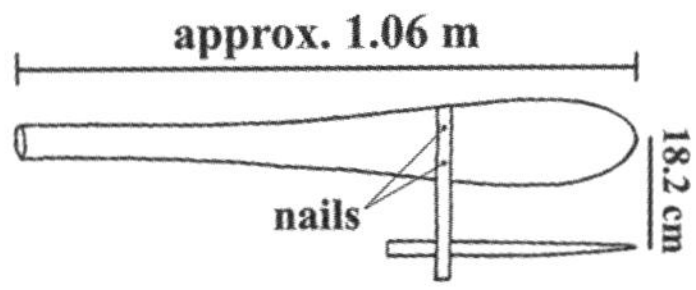

Figure 2.3. Tool for making seed-planting holes.

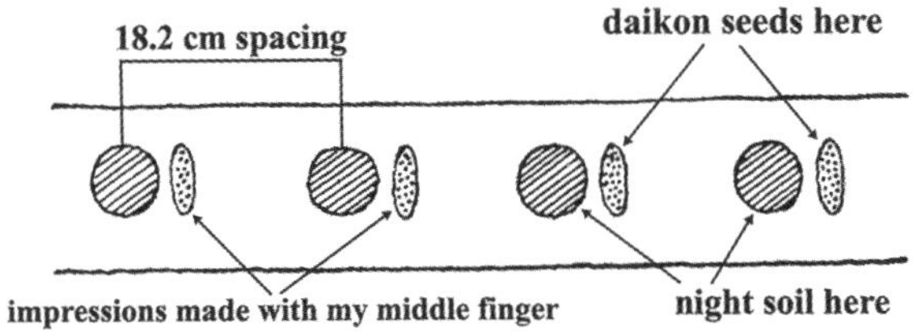

Figure 2.4. Daikon seed-planting and night-soil fertilizing pattern.

My daikon field measures about 70 *tsubo* (231 m^2). I used 3 *ka* of night soil and the same amount of water. I put about ten daikon seeds into each impression. The type of daikon I planted is Hayabutori Hanashirazu Tokinashi daikon.[18]

At the main house: My brother's wife cleaned up outside and my mother cooked. Kanezō cut hay for the horse early, and he plowed the *nashiro* later in the morning. In the afternoon he did *kuronuri.* Since he cleaned out the *nashiro* and formed ridges yesterday, today he just had to move along the ridges and pull some mud up onto them with a hoe as he went. When you do this, you have to start where the water enters the paddy and work from there. After you've done both sides of a single ridge, you go back and pour water over it using the hoe and then smooth it out nicely. If you do this, you're done. Today I took two photos of Kanezō's work—one of him putting mud on the ridges and another of him smoothing them out (see Fig. 0.11).

Regarding this, there are *migikko* [right-handed people] and *hidarikko* [left-handed people]. The former work with the right leg up close to the ridge and the latter work with the left leg up close to the ridge.

Today was the first time this year to use a three-pronged hoe, or any hoe, in the paddies (at the main house) so all family members got together and roasted and ate *kuwa no mochi* that had been made for the New Year.[19]

Today's expenditures:

Lion tooth (brushing) powder (1 bag)	9 sen
wire	7 sen
alum	5 sen
4 candies for the kids	2 sen

Figure 2.5. [My wife] doing some hand-plowing (*ibaru*) in our fields (April 7).

April 8, Monday

Sunny, no clouds
West wind
0°C in the morning, 14°C in the afternoon

Breakfast:	miso soup with daikon in it, side dishes: same as every day
Lunch:	salted krill, side dishes: same
Supper:	miso soup with taro stems in it, salted herring, side dishes: same

I basically did the same things as yesterday for my hotbox, but no watering today.

Today I studied about tanning rabbit pelts, which I was asked to do by someone a few days ago. After I finished that, I worked on basket weaving. My wife hand-plowed some of our fields—the parts I couldn't do with the horse.

Today's expenditures:
4 salted herrings from Hokkaidō 10 sen

At the main house: Kanezō cleaned up outside early and my mother cooked. My brother's wife and Kanezō plowed the fields this morning. Kanezō plowed with the horse, and she did the rest with a hoe.

Before the invention of the horse-drawn plow, people had to plow all their fields—no matter how big—entirely by hand (with hoes). Even now, of course, folks with no horse must do it all by hand. The first horse-drawn plows that came into use were called *kakai-shiki* [*kakai* style]. When pulled by a horse, they'd part the soil in a way that caused it to flow to both sides (of the plow). After that, a bunch of different kinds of plows came out. The one we got when I was 18—exactly 13 years ago—is still being used at the main house. It's called *mukai-shiki*. Today's popular plows are much easier to use, and far more efficient. The one I used this time, which is pretty popular now, is designed to move the soil only to one side of the plow when pulled by a horse.

In the afternoon, my brother's wife and Kanezō worked on the *nashiro* again. Kanezō pushed the plow, and she did *sahitori*.[20] Horse-drawn plows used to have six or seven iron bars on them. The number of bars then increased to nine. The one at the main house is a nine-bar plow, and it's still in use. New plows today are even bigger—twelve bars. And there are even huge ones with sixteen bars.

I'll show with a diagram the proper way to plow a *nashiro* (see Fig. 2.6).

You guide the horse through the paddy as shown in the diagram. It's not very efficient to cover the same ground over and over again. I used to do that all the time when I was a kid, and get scolded and smacked by the adults. It's boring work—you even get sleepy while doing it. You just go across the paddy once, and then over, then back across, and then over again. When you go lengthwise it's called *tatehaga*, and when you go crossways it's called *yokohaga*. The first plowing is known as *hitokuwa*, the second *futakuwa*, and the third *mikuwa*.

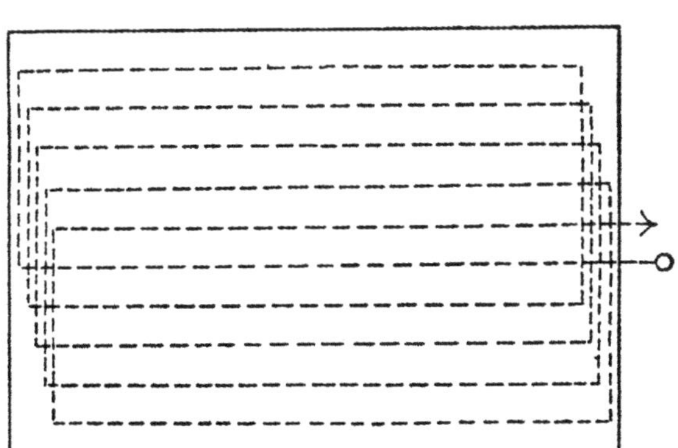

Figure 2.6. A representation of a seedling nursery paddy (*nashiro*), showing the proper plowing pattern.

I'll explain what was used today to fertilize the *nashiro*.[21] They used 5 *shō* (9 L), or 1 *kan* and 600 *momme* (6 kg), of lime nitrogen—known as *juppyō-maki* (enough to produce ten *tawara* full of rice grains).

Since it was the first time this year to use the horse-drawn plow, we ate *maguwa no mochi*. These are also known as *shōgatsu no medama mochi*.[22]

April 9, Tuesday

Sunny
0°C in the morning, 18°C in the afternoon

Breakfast:	*wakame* miso soup, side dishes: same as every day
Lunch:	various fish braised with salt, side dishes: same
Supper:	cabbage and potato miso soup, roasted salted herring, side dishes: same

I heard bush warblers (*uguisu*) singing for the first time this year today.

The fish we cooked and ate for lunch were caught by the crazy neighbor. He brought them over. They included loach, bitterlings, shrimp, killifish, *natogori*,[23] and others. He caught them in a marsh called Ōzutsumi, using only a bamboo strainer (*takezaru*).

Today I worked more on making a basket for carrying stuff on my back. While doing that I sprinkled water inside my hotbox, doing the same things as yesterday.

My wife leveled some fields for planting burdock and potatoes in the morning, and cleaned up in the afternoon—removing all the dried leaves and such.

Today's expenditures:
sweets[24] for the kids 3 sen

At the main house: My brother's wife sewed while watching the shop. My mother mainly cooked. Kanezō went out to cut hay for the horse early and made rows for planting potatoes in the fields later in the morning. In the afternoon he bundled *kyūhi*[25] for fertilizing the [dry] fields. How to bundle *kyūhi:* you separate one straw cord into four parts and lay the *kyūhi* straw across the center, over the knot, and then pull the four ends up and tie it in two places.

Today I received the following notice from the village office:
Lifestyle Improvement Articles (enacted 1932)
Regarding festivals:

1. Villagers shall cease the practice of inviting neighbors to their homes.
2. Villagers shall cease the practice of giving food and treats to one another.

April 1935, Wakimoto Village

In the old days, people used to eat at each other's homes all the time on festival days, but now we get a lot of nagging orders like this from "above," so even festival days, which come but once a year each, are kind of lonely and unsatisfying. The appearance of this sort of proclamation is just another sign of the bad economy.

I heard something today. I'll change the names of the people involved. "Sankichi" (50), the owner of the largest amount of mountain land in the village, spends each day patrolling his property with a hatchet in his belt. Recently, Sankichi discovered a poor woman, "Masako" (35), collecting branches (pine and cedar) on his land. He gave her a choice: going with him to the police or exchanging her own body for the branches right then and there. She decided that it would be better to solve the problem with what she could offer immediately than to disgrace herself and her family by allowing herself to be reported to the police, so she exchanged herself for three bundles of branches on the spot. This is actually the "profession" of this landowner. People say that there are many more women who have traded "favors" with him like that.

April 10, Wednesday

Sunny, later cloudy, a light rain late
South wind, dying later
11°C in the morning, 18°C in the afternoon

Breakfast:	*wakame* soup, side dishes: same as always
Lunch:	salted herring, side dishes: same
Supper:	miso soup with taro stems in it, salted krill, side dishes: same
Snack:	*medama mochi* (received from the main house)

I worked on my basket making more from the morning today, and by about nine o'clock I'd completed a rather fine basket, if I do say so myself. It's much better than the one I made last year—even a professional would admit defeat upon seeing this one . . . although it took me about twenty hours to make it!

I closed the covers of the hotbox early because the weather changed suddenly around noon. I put the straw mats back in place around four o'clock.

My wife helped me plant burdock in the morning, and I continued alone until around three. She cleaned out the fields in the afternoon. After I finished, I went out to scrape up some money, as I'd run out of cash.

The rows I made are fairly tall—2 *shaku* 5 *sun* (75.8 cm) wide—and the space between plants is about 4 *sun* (12.2 cm) on all sides.

I planted 24 *tsubo* (79 m^2) with burdock. For fertilizer: 2 *ka* of night soil diluted with the same amount of water. I used 1 *gō* 5 *shaku* (270.6 ml) of seeds.

I planted the burdock in a *nijōmaki* [double-row] pattern—1,080 little clusters (*kabu*) in total—and scattered the seeds over the remaining two rows.

As for night soil, I completely covered all rows with it, and I covered all the seeds I'd planted with about 1 *sun* (3 cm) of dirt.

At the main house: In the morning my brother's wife went out to do some hoe chopping and plowing (*ibaru*) in the potato fields. Kanezō went out early to do *tazukiri* [*tachikiri*] in the paddies, and then later in the morning he loaded the horse with bundled compost straw (*kyūhi*) and took it out to the dry fields and spread it around. The amount of fertilizer needed per field (the number of passes required) is pre-decided, and goes back a long time. The main house's fields are in Kogasaki,[26] so they need seven passes in the morning and in the evening.

My mother cooked today, and in the afternoon my brother's wife and Kanezō did *nakakagi* to the *nashiro*. You must plow the *nashiro* using the horse three times before it can be ready for planting seeds, and today was the second time so it's called *naka* (middle) *kaki* (plowing or turning the soil). You go up lengthwise and then across once, and then do it again, called *futa-kuwa*. After finishing, they cleaned the *nashiro* again (removing *goto*, or garbage). In the old days, removing this junk required all members of the household—picking everything up by hand—but then people started using *konoha-sarai*, or *komajari* instead of all that hard labor. Now people use something that's even easier: a tool called *gotosarai*, which was recently invented by modifying a *komajari*. [My brother's wife and Kanezō] put all the [compostable] junk they got out of the *nashiro* into *goto-ana* holes.

A man named Yoshida Denjirō bought paddy land from Yoshida Yasuke. He bought a 2 *tan*, 6 *bu* (606 *tsubo*, or 2,003 m^2) paddy for 840 yen. This paddy had already been fertilized (so it was ready for planting), and it was right next to Denjirō's other paddies, and his house is there, too. It's a very convenient arrangement for Denjirō, especially since it's getting so hard these days to buy and sell paddy land for the purpose of growing rice on it yourself, what with the *kosaku mondai* situation.[27] The paddy he bought is ready to use, hence the high price. Normally, such a plot would cost 200 or 300 yen.

And the main house bought a 1 *tan* 8 *se* (540 *tsubo*, or 1,785 m^2) field (*hatake*) from the household into which my aunt married, which is headed by Yoshida Kiyozō. The price was 140 yen—also pretty high, but it's because the field isn't very far from the village.

April 11, Thursday

Rain
South wind, dying later
11°C in the morning, 14°C in the afternoon

Today is *kigitsuke*, so we ate *yama mochi*.[28]

Breakfast:	miso soup with dried taro stems, side dishes: same as every day, plus some miso-preserved vegetables, and daikon
Lunch:	My wife and the kids ate the same as always. I had to go to another village on an errand, so I ate at someone else's house. I'll explain what I ate. You put raw krill into a mortar and add some flour and grind it up, and then you cook it in water with some soy sauce (to make it firm). I forgot—you also add some tofu. This is called *ishaja no kamaboko*. I also ate salted roasted *idoyo*.[29] There was some saké, too. The krill and *idoyo* were all from Lake Hachirō.
Supper:	We pounded regular rice to make *mochi* (*yama mochi*) and ate that. We ate them with *azuki* [red bean] paste and put them into soup broth made from roasted sardines as well (the soup was flavored with miso). Side dishes: same as morning

It was rainy today, so I just lifted the straw hotbox mats at eight o'clock and put them back in place at four. I didn't open the (glass) covers. I was out all day (from morning) on that errand. My wife mainly sewed.

Lately I've been hearing a lot of different birds that I don't even know the names of.

Today's expenditures:

brown sugar	10 sen
sweets	5 sen

At the main house: My brother's wife sewed and cooked with my mother. Kanezō went out early to cut hay for the horse, and the rest of the day he wove *tsukenawa* rope. He used eight bundles of straw but could only make five bundles of rope. In the evening he pounded *mochi*.

I heard that my crazy neighbor broke a sacred object in the Mishima Shrine yesterday afternoon. The god of this shrine has suffered many ordeals. Three years ago, the god's house burned, and now the god's body has been damaged by the crazy neighbor. As a matter of fact, this "crazy" man lives next door[30] to me, and even though I call him "crazy" he cooks and takes care of himself. His name is Yoshida Yūjirō,[31] and he receives food from his elder brother. I have no idea why he became crazy.

I think I'll write a little about the lifestyle of this crazy man.

He's kind of strange, even as a "crazy man." Most such "crazy" guys devote themselves to believing in some god, but this "crazy" man is different. He mentally divides gods into two types—good and evil.

"The good gods are in heaven. The evil ones are far more cunning and wily than the good gods, and know magic that the good ones don't know, and are always causing trouble for the good gods and for humans as well. So, from way back, some people have gotten poorer and poorer even though all they ever did was just work hard, save money, and live efficiently—this is the work of these evil gods. So, the people complained to the good gods about this, and the village gods appealed to the good gods of the prefecture, and these in turn appealed to the good gods of the entire country. But still the identity of the bad gods remained unclear. So then, the good gods of all the countries of the world got together and decided that Japan's local village gods, the prefecture-level gods, and even the national-level gods were incapable of preventing the bad gods from causing trouble and declared that they would get rid of the bad gods themselves. Then, these gods secretly ordered me, Yoshida Yūjirō, Yoshida Yūjirō of Akita—*Akita's Yoshida Yūjirō*—to invent an airplane that makes no sound. I worked hard, and finally finished it seven years and eight months ago, and humbly offered it to the gods of the world up in heaven.[32] I hear that right now they're flying around up there exterminating all those bad gods. So, once they've finished their job, the good gods will return all the wealth that was lost in the past to those of us who don't have anything anymore. Also, they've ordered the Home Ministry[33] to give me—Yoshida Yūjirō of Wakimoto Village in Akita—a reward of 10,000 yen for inventing that airplane. But even though I've been waiting for seven years and eight months, nothing has arrived yet."

He goes on like this day after day with his voice raised. Apparently, he really has been waiting for seven years and eight months. Every time he gets a bit of change from a sibling or other relative, he posts a letter to the Home Ministry or the emperor or some other high office or person, pressing them to hurry up with his reward money. But the post office just keeps all the letters without sending them. And so, the crazy guy just keeps saying:

"My money hasn't arrived despite all the letters I've sent. This is certainly the work of the evil gods. There's no way my letters could fail to reach the Home Ministry. I've prayed to the local village gods, but still—nothing!"

And finally, he declared, "This village god is no good!" So yesterday, April tenth, at about three-thirty in the afternoon, he finally damaged one of the sacred objects[34] in the shrine. They say he broke off the head and hands, among other parts. Then, to the villagers who gathered in response he said, "You people say I'm dumb, but you're dumber than me—because you worship this statue made of wood, saying it's a god. You're all just stupid!" Then he just walked away. He really is a strange, crazy guy.

A village meeting was held in response, and I heard that it was decided that the village would pay a bit toward having a cell built to hold him in solitary confinement.[35]

April 12, Friday

Sunny
Northwest wind
4°C in the morning, a little under 10°C in the afternoon

Breakfast, lunch, and supper:	essentially the same as always
Snacks:	festival food from a neighboring village (*sekihan,* tempura, *mugi-magi,*[36] etc.—received from my mother)

I cared for my hotbox basically the same way as every sunny day. If the sunshine lasts for some days, I'll sprinkle water only every other day.

I worked all day on a *tekago* handbasket (see Fig. 0.24). My wife collected firewood and made planting rows in the potato fields.

At the main house: My brother's wife went to her natal village for festival day and my mother went, too. Kanezō bundled compost (*marogi-goe*) early, and later in the morning he took it out to the fields. Of course, this is fertilizer for potatoes. In the afternoon he went out to turn over the soil in the rice paddies. He used a three-pronged hoe to do the work by hand today.

April 13, Saturday

Sunny
Northwest wind
2°C in the morning, 12°C in the afternoon

All three meals: same as yesterday

I did nothing unusual to the hotbox today.

In the morning, I went to the market at Nagane[37] to buy Chinese yam seeds and some other things. My wife spread fertilizer in the potato fields. In the afternoon we planted potatoes together, and also summer cabbages. The way we planted potatoes was basically the same as for the peas, except that we used some *mamekasu*[38] for extra fertilizer.

field area	about 18 *tsubo* (60 m^2)
fertilizer	about 30 *kan* (112.5 kg)
night soil	1 *ka* diluted 1:1 with water
superphosphate of lime	2 *shō* (3.6 L)
mamekasu	3 *shō* (5.4 L)
ash[39]	6 *shō* (10.8 L)
potatoes	100 (But those weighing over 20 *momme* (75 g) I chopped and cut off the parts with many eyes and covered them with ash and planted them, making for a total of 200 planted.)

The row width is 2 *shaku* 5 *sun* (75.7 cm), and the space between the potatoes measures 1 *shaku* 5 *sun* (45.5 cm).

The planting area for the summer cabbages amounts to 1 *tsubo* (approx. 3 m^2).

Fertilizer: night soil—4 *shō* (7.2 L), mixed with the same amount of water and 2 *shō* (3.6 L) of ash, etc., added when planting summer cabbages.

Seed amount: approx. ½ *shaku* (9 ml)

In general, planting seeds in a bed to grow seedlings that will be moved later is called *tane o shiiru,* or *tane o hiiru* in local speech.

Today's expenditures:

soy sauce (3 *gō* (541.2 ml))	8 sen
seeds	50 sen
Chinese yam seeds, 1 *kan* (3.75 g)	40 sen
green onion seeds, 1 *shaku* (18 ml)	10 sen
11 herrings	20 sen
rubber boots for the kids (2 pairs)	60 sen
10 stamps	30 sen
postage for mailing a copy of my book	10 sen
presents for the kids (3 apples, 6 candies)	12 sen

At the main house: My brother's wife watched the shop in the morning, and my mother went out to the fields alone. Both went out to the fields in the afternoon. In the early morning, Kanezō pulled all the rice seed–packed *tawara* out of the pond and spread them out to prepare for drying them in the sun. He carried *marogi-goe* out to the fields in the morning, and he carried night soil out there with the horse and planted potatoes in the afternoon. At the main house, they still plant potatoes in the old way. Basically, you dig holes about 6 or 7 *sun* (20 cm) in diameter about 2 *shaku* (6.06 cm) apart between the rows with a hoe, put fertilizer in the holes, pour night soil over them, place a very small potato right on top of each, and cover everything with soil.

field area	80 *tsubo* (265 m^2)
compost (*kyūhi*)	200 *kan* (750 kg)
night soil	40 *kan* (150 kg)
potatoes	2 *to* 7 *shō* (48.7 L) (One *shō* amounts to about 32 potatoes on average.)

About the *tawara* full of rice seeds from this morning, people used to always pull them from the water at the time of *doyō,*[40] but these days people usually plant them in the *nashiro* on *doyō,* so it's done earlier than before.

April 14, Sunday

Sunny, later cloudy
Wind out of the south
6°C in the morning, 16°C in the afternoon

Breakfast:	miso soup with *wakame* and potato, pickled (*mizuzuke*) daikon, *takuan, misozuke* (miso-preserved vegetables), etc.
Lunch:	The crazy neighbor brought over a potful of various small fish he'd caught in an irrigation ditch near the paddies, so we cooked them in soy sauce and ate them. Side dishes: same as morning.
Supper:	My wife and one of my daughters ate miso soup with dried taro stems and potatoes in it, roasted salted herring, sides: same as morning. I took my other daughter with me to the main house and there we ate *tororo meshi.*[41] Side dishes included pickled daikon and *takuan,* etc. There was also burdock root and dried daikon strips cooked with miso for flavor. To make *tororo jiru,* you make broth with burdock root and add miso to make miso soup with it. Then you grate some Chinese yam and once you have enough you add the miso soup slowly to that while stirring and then when it's nice and ready you pour it over a bowl full of rice. Then you eat it.

Hotbox care today: I sprinkled some warm water over it, but otherwise I just did what I normally do on sunny days.

My wife prepared for starting a new hotbox in the [early] morning. Of course, I did that, too. Then I went to the main house to help them finish preparing their nursery paddies. In the afternoon, my wife and I started making the new hotbox. This one will be for starting tomato seedlings to be transplanted later. We finished pretty early, so I returned to making the handbasket (*tekago*) I wrote about before.

Today's income:
profits from selling daikon seeds 5 sen

At the main house: Kanezō prepared for treating rice seeds with hot water (doing *yudoshi*) in the early morning, and a little later he and my brother's wife and I finished preparing the *nashiro.* Kanezō pushed the horse-drawn plow, and my brother's wife guided the horse (did *sahitori*). They finished that early, so she removed leaves and other junk from inside the *nashiro* (did *konoha-sarai*). I used a tool called an *iburi* to smooth out the surface of the *nashiro* after the first plowing (see Fig. 0.20). Then I went over it again with a piece of wood 2 *ken* (3.6 m) long, 2 *sun* (6.6 cm) wide, and about 1 *sun* and 5 *bu* (4.5 cm) thick. Next, I stuck miscanthus (*kaya*) rods about 2 *shaku* (60.6 cm) long into the center of each *nashiro*—one or two, or maybe three—and stepped out very carefully to avoid leaving footprints in the mud. Some say the reason for sticking *kaya* rods into the center of each *nashiro* is to make it easy to tell, just at a glance, how many *tawara* of seeds will be needed to plant that paddy—one rod for one *tawara* and two for two, for example (see Fig. 0.21). But there is also a legend about this. A long, long time ago a powerful person who wanted to grow rice in this country sent a fox overseas to steal rice seeds and bring them back. The fox succeeded, but some people over there chased the fox. It ran as fast as it could, but the people were closing in on it, so the fox hid in a miscanthus thicket. The people couldn't find the fox, so they gave up and left. I've heard that because of this legend miscanthus is considered very precious and useful. I've heard that there are other tales about this, too.

Today I heard a story about Mie Yoshida, the twenty-year-old daughter of a local landless family. She fled without her parents' permission to visit her lover in Hokkaidō, but her parents called the cops on her and they caught her somewhere in Aomori. It seems that her lover's family had formally asked her parents to allow her to marry their son, but her parents refused their request, and this was the result.

Returning to the day's work at the main house, my brother's wife went out to the fields in the afternoon to plant burdock and Chinese yams. Kanezō treated rice seeds with hot water (did *yudoshi*). You remove the *tawara* full of seeds from the water that they've been soaking in for a while and let them dry in the sun for a day or two. Then you heat water in a bathtub—just enough for a person to bathe in—and lay two poles across the top of the tub and place a *tawara* on top of them. Then you ladle hot water from the tub over the *tawara* again and again until you're sure that the seeds inside are nice and hot, and then you take that *tawara* off and put another one on the poles and do it again (see Figs. 0.14–0.16). Once you've finished, you place all the warm *tawara* on top of a thick layer of straw in an earthen-floored room, and cover everything with *mogu*, which I believe is a kind of waterweed from Lake Hachirō. The purpose of doing all this is to get the seeds to sprout. Once they've started to sprout, you plant them in the *nashiro* (see Fig. 0.17). This is called *moyashi-maki*. Generally, it takes two or three days for the seeds to sprout.

April 15, Monday

Rainy
Weak south wind
8°C in the morning

Breakfast:	same as most days
Lunch:	*soba nereke*,[42] salted herring, side dishes: same as always
Supper:	same as most days

In the morning I went to the lumber company owned by my younger sister's family[43] to get some boards that I needed for growing watermelons. In the afternoon I worked again on the handbasket I wrote about, and finally finished it. It's rather good, I must say.

Hotbox care: It was rainy, so all I did was lift the mats from the surfaces.

My wife mainly sewed today.

Today's expenditures:

wire	5 sen
candies and sweet potatoes for the kids	5 sen

At the main house: My mother went to another village to join relatives for festival day. My brother's wife sewed and cooked, etc. Kanezō cut hay for the horse early and then later in the morning he went to plow the paddies (to do *ta-uchi*) at Bishamon.[44] He hand-plowed with a three-pronged hoe (*sanbon guwa*). In the afternoon, the master[45] said that Kanezō could take the rest of the day off, but he went out to hand-plow, anyway.

In the old days, a woman could cover a fifty-stroke paddy (about 7 *se* (approx. 694 m^2)) and a man could do a seventy-stroke paddy (about 1 *tan* (approx. 991m^2)), and this would be a full day's work—the rest of the day would be for resting.[46] They say that someone could generally finish this by around one o'clock in the afternoon, or maybe even by twelve-thirty if they were quick. A person would normally take a break around ten or eleven o'clock and eat a rice ball bigger than their own hand and gobble [the equivalent of] a whole pickled daikon with that. People said that such a meal taken in the fields was even tastier than any delicacy from the ocean or the mountains.

Hand-plowing with a three-pronged hoe works in the following way. First, you go out in the morning and get a *wappaga*[47] from the boss [the paddy's owner] and you spread fertilizer all round and start plowing. Each person has his own style of plowing: *furite-uchi, ippon-uchi, battara-uchi,* etc. *Futade-uchi*[48] is a style where you make two rows at the same time. You start on the right foot and grip the hoe with the left hand ahead of the right hand. This is for swinging to the right side of your body. Then you step forward with the left foot and swing and you pull the soil in the paddy on your right over to the right, thus forming a row on that side. Then you switch and place your right hand ahead of the left on the hoe handle and step forward with the right foot as you swing to the left and pull the soil in the paddy to the left over leftward, forming a row on that side. In this way, you can make two nice *futade-uchi* rows. For *ippon-uchi* it's the same, but you just bring the soil up on both sides to form one row. *Battara-uchi* is when you simply leave the ground flat as you plow, and don't make any rows. According to one old man, if you do *futade-uchi* around the edge of any rice paddy, moving along the raised walking paths [ridges] between the paddies, you'll have plowed about one-half of that paddy upon finishing.

A person can cover an area about 1 *ken* (1.8 m) wide when doing *futade-uchi,* and a very fit man can do about 8 *shaku* (2.4 m). In the old days, they had to hand-plow just about every day, so they had lots of blisters on their hands. People say that those really hurt. To get rid of them, they used to burn them with heated metal chopsticks for handling coals on the hearth.[49] The blisters always formed between the muscle and the skin, and they were full of sweat-like liquid and round like soybeans. They also say that in the old days the landless laborers (*temadori*),[50] at the command of the landlords (*oyage*), would go out to plow the paddies and work really hard, competing to finish early, and that some would drop from hunger and exhaustion while others would just totally overdo it and get nosebleeds.

April 16, Tuesday

Sunny
East wind, later out of the north
3°C in the morning, 14°C in the afternoon

Breakfast:	miso soup with cabbage and burdock root in it, pickled daikon, *takuan, misozuke,* etc.
Lunch:	green onions cooked with salted krill, salted herring, side dishes: same as morning (but no *misozuke*)
Supper:	miso soup with dried taro stems and potato in it, leftover salted herring from lunch, side dishes: same as lunch. I ate supper at the main house. We ate *kamabugu*[51] soup with fresh krill caught at the Wakimoto beach. There also was roasted fish (I don't know the name) caught at Minami Iso Village, pickled daikon, and *takuan,* etc.

It's troublesome to gauge daily miso use, so I put it in a container to measure it more easily. It takes us eight days to consume 500 *momme* (2.1 kg) of miso. From now, I'll just go with this.

The summer daikon and green peas (*endō mame*) I planted in my fields are starting to sprout here and there.

I put fertilizer on my wheat crop for the first time this year today. I used one *ka* of night soil (diluted 1:1 with water) and sprinkled it over the wheat. I also warmed up the temporary bed (new hotbox) for the tomatoes and prepared some topsoil (*tokotsuchi*). In the evening I went to the main house to help plant seeds. My wife prepared the green onion and Chinese yam fields, and did some other things, too.

Today's expenditures:

kids' haircuts[52]	10 sen
candies for the kids	2 sen

At the main house: Everyone in the household helped clean up both inside and outside all day long from the morning. Starting at four o'clock, the master, Kanezō, and I planted seeds in the main house's nursery paddies (*nashiro*). In the old days this was generally done later in the year, but now people do it earlier. I heard from an old man that in the past they'd have to go out and start before sunrise and plant seeds all day until after sunset, but nobody around here does that now.

The main house has eight nursery paddies, making for a surface area of about 143 *tsubo* (473 m^2). We planted about 8 *to* (144 L) of *uruchi-momi* [regular short-grain table rice] and about 1 *to* (18 L) of *mochimomi* (*motsumomi*) [glutinous rice for making *mochi*].

April 17, Wednesday

Sunny
2°C in the morning, 12°C in the afternoon

Someone from another hamlet (*buraku*) came to see my hotboxes.

Meals: same as always

I went in the direction of Akita City to buy taro and summer cabbage seedlings. My wife prepared the summer cabbage beds. She also took care of the hotboxes for me today.

Today's expenditures:

taro seeds	3 yen 30 sen
a rubber hose for the sprayer (2 *shaku* (60.6 cm) long)	1 yen 60 sen
train fare	1 yen
gifts for the kids	20 sen
summer cabbage seedlings	2 yen

At the main house: My mother cooked, and my brother's wife watched the store. The master went to plant pine trees in the communal property around the village. Kanezō set up windbreaks around the nursery paddies in the morning. He pounded poles into the ground and attached *sudare* to them. This took until three-thirty. Then he went to his natal village to enjoy the festival (*matsuri*).

April 18, Thursday

Cloudy, light rain, clearing later
Southwest wind
8°C in the morning, 10°C in the afternoon

Meals: about the same as always

I sprinkled a bit of water on the hotboxes today, and only planted some tomato seeds in one of them. I planted most of them at about 3 *sun* (9.1 cm) intervals (on all sides), but I left 4 *sun* (12.1 cm) between some just to see how it goes.

After planting the seeds, I sprinkled some water in that hotbox, and put a straw mat down and closed the wooden *shōji* cover. I planted 432 tomato seeds.

Next, I planted Chinese yams. I dug holes 4 *sun* (12.1 cm) wide and 3 *sun* (9.1 cm) deep, spaced 2 *shaku* 5 *sun* (75.8 cm) apart in the flat field. I filled in the holes with fertilizer mixed with superphosphate of lime and crushed soybean powder (*mamekasu*) and covered all the holes with a thin layer of soil. Over the surface I poured some night soil. Then I spread wood ash over everything and planted the seed yams and covered them with soil and then I was done with that.

distance between rows	2.5 *shaku* (75.8 cm)
distance between each seed yam	7 or 8 *sun* (about 22.7 cm)
amount of fertilizer used	20 *kan* (75 kg)
bean bits	500 *momme* (1.9 kg)
lime	500 *momme* (1.9 kg)
night soil	2 *ka* (diluted 1:1 with water)
wood ash	6 *shō* (10.1 L)
land area planted	about 80 *tsubo* (264 m^2)
weight of seed yams	8 *kan* (30 kg)
number of seed yams	300

Next, I prepared to plant green onions—flat rows with only a bit of night soil as fertilizer sprinkled across them. Area: 15 *tsubo* (50 m^2); row width: 3 *shaku* (90.9 cm); amount of night soil: 2 *ka* (diluted 1:1 with water).

Today my wife and I worked together on everything.

At the main house: Kanezō went out to cut hay for the horse early and then spent the rest of the day hand-plowing the paddies with a three-pronged hoe—the Bishamon paddies. My brother's wife and my mother helped prepare for the village festival, which is tomorrow. They cooked a lot of food.

April 19, Friday

Sunny, later cloudy
West wind
–2°C in the morning, 12°C in the afternoon

A young fellow from another hamlet came over to see my hotboxes.

Breakfast:	miso soup with dried taro stems in it, *misozuke*, pickled daikon, *takuan*
Lunch:	side dishes: same as this morning, and salted krill
Supper:	We ate at the main house and enjoyed the festival. I'll elaborate on the food later.[53]

Hotbox work: I did nothing out of the ordinary for the eggplants, but for the tomatoes I let sunlight in until eight o'clock and later covered the hotboxes with *sudare*. Then I put the lighter straw mats (*komo*)[54] down [on the surfaces] at eleven o'clock, and took them off again at two-thirty, and then removed the *sudare* at three-thirty, let some more sunlight in, and put all the mats back in place at four-thirty.

I planted green onion (*negi*) seeds today. I dissolved 1 *shō* 2 *gō* (2.2 L) of ammonium sulfate in 2 *ka* of water, and later sprinkled it all over the bed I'd prepared yesterday. After that, I planted the seeds (1 *gō* 6 *shaku*—288.6 ml). I spread them around evenly, so there were no areas too thickly or thinly planted. I poked holes in a small tin can and used that—it works really well.

After scattering the seeds, I spread about 3 *to* (54.1 L) of wood ash and then I covered the bed with straw. I used 25 bundles of straw, and I laid poles down across that to keep the straw from blowing away. The straw prevents the seedbed from drying out, and protects the surface from frost.

I then set up *sudare* screens on all sides of the green onion seedbed, to protect it from cold wind.

At the main house: My brother's wife and my mother were kept busy preparing food for the festival. Kanezō plowed the rice paddies at Bishamon. Everyone took the afternoon off.

April 20, Saturday

Not a cloud in the sky (*nipponbare*)
Light southwest wind
0°C in the morning, 14°C in the afternoon

Breakfast:	*wakame* miso soup, side dishes: same as always
Lunch:	salted herring, sides: same as morning
Supper:	I went to the main house to eat leftover food from yesterday's festival. My wife went to her natal home.

A few people who came for the village festival visited, to see my farm.

In the early morning, I started preparing for doing *fumikomi* [stomping] to the temporary bed (new hotbox) for eggplants, and later in the morning I watered the first hotbox a bit. Then I planted eggplant seeds where the tomatoes had been, met the visitors and explained things to them, and did *fumikomi* to the new hotbox. All this took me until three o'clock. Then I rested, together with my wife.

Today's expenditures:

nails	5 sen
kids' allowances	7 sen

At the main house: My mother hosted guests. Kanezō cut hay for the horse in the early morning, and he went out to plow the paddies at Bishamon again in the afternoon. My brother's wife went out to plant burdock and Chinese yams—12 *tsubo* (40 m^2) of burdock and 5 *tsubo* (17 m^2) of yams. She used 3 *ka* of night soil. She prepared tall rows, sprinkled night soil, planted the seeds right on the surface, and then spread dirt over everything—a very easy way. Everyone rested in the afternoon.

April 21, Sunday

Cloudy, clearing later
Strong southwest wind, dying later
9°C in the morning, 14°C in the evening

Breakfast:	miso soup with sea lettuce we got from relatives in the village of Minami Iso and potatoes in it, pickled daikon, *takuan, misozuke,* etc.
Lunch:	squid (caught at Minami Iso, given by my wife's relatives) cooked with green onions and salted krill, pickled daikon, *takuan,* etc.
Supper:	miso soup with various small fish and dried taro stems in it, side dishes same as lunch, roasted salted *yudoyo,*[55] etc. All the fish we ate for supper came from the main house, and all were caught in Lake Hachirō.

Today I worked on making a new hotbox. I set up a frame around the spot where I stamped down straw (did *fumikomi*) yesterday, started warming it up, sifted topsoil and put it in, set glass panes in the wooden *shōji* frame, and made some windbreaks. It kept both my wife and me quite busy. The hotbox I prepared today measures 24 *shaku* (7.3 m) long. With this one, I've finally finished making hotboxes for this year. Compared to last year, I'm 24 *shaku* short on hotbox length, because I couldn't get so much straw. But I still have about 48 *shaku* (14.5 m).

Today's expenditures:

candies for the kids	5 sen

At the main house: My mother cooked. Kanezō went out to cut hay for the horse early in the morning and spent the rest of the morning, and the entire afternoon, plowing paddies. My brother's wife plowed in the morning, and watched the store in the afternoon. They're still working on the paddies at Bishamon.

April 22, Monday

Sunny
Light southwest wind
6°C in the morning, 18°C in the afternoon

Meals: same as always

Snacks: We roasted dried *mochi* and ate them.

I didn't have enough glass for the hotbox *shōji* panels yesterday, so I finished filling in the spaces in the frames with scraps of paper. I covered the paper with wax from melted candles. I got the candles from the local shrine—pieces that were left over after burning them—and I also scraped[56] up melted wax that had dripped from burning candles. I brought all that stuff back home and [melted it and] mixed it with cooking oil.[57] After finishing that I tanned three more rabbit pelts.

My wife prepared rows in the summer cabbage fields (*hatake o ibaru*). I sprinkled water on the eggplant section of one of the hotboxes. I let sunlight into the tomato section until eleven o'clock, and then covered it again with *sudare* screens, then I pulled them back again at two o'clock to let more sunlight in. I covered both the eggplant part and the tomato part with straw mats (*komo*) at five.

At the main house: My brother's wife plowed paddies in the morning, and she headed out to pick bamboo shoots (*sasadake*) in the afternoon. Kanezō cut hay for the horse early and spent the morning plowing paddies. In the afternoon, he loaded the horse with compost (*kyūhi*) and took it out to Bishamon. The plowing of the Bishamon paddies is now finished.[58]

Today, three young ladies from our village headed off to the Tokyo area to do factory work. Two are fourteen and one is seventeen.

April 23, Tuesday

Rain, clearing later
Strong southwest wind
10°C in the morning, 10°C in the afternoon

Breakfast:	miso soup with bamboo shoots, side dishes: no change
Lunch:	green onions cooked with salted krill, pickled daikon, *takuan*, etc.
Supper:	daikon miso soup (stock made from roasted carp), side dishes: no change

Hotbox work: No change in eggplant care, but I removed the screens from [the soil surface of] the tomato sections because they've completely taken root.

I made another handbasket (*tekago*). It's small: finished in one day. While doing that I cleaned the rabbit pelts. My wife prepped the cucumber fields and then made planting rows (did *ibaru*), and later she got ready for fertilizing the cabbage fields.

Today's expenditures:

candies for the kids	3 sen

At the main house: The master was out on some business, so his wife watched the shop and cooked. My mother went to a different village for their festival. Kanezō went out to the paddies early and started filling in a drainage ditch he'd dug last autumn.

April 24, Wednesday

Sunny, later cloudy
Northwest wind, later strong from the southwest
8°C in the morning, 15°C in the afternoon

Meals: basically, the same as always

Someone came from a different hamlet to see my fields.

The summer cabbages have started to sprout.

I temporarily planted eggplants today (1,014 seedlings).[59] I planted them 3 *sun* (9.1 cm) apart horizontally, and 2 *sun* 5 *bu* (6.2 cm) apart going the other way. After taking care of the seedlings, I planted more eggplants.[60]

We finished the cucumber fields today, and also the cabbage fields. My wife and I worked all day on this, from early morning till late.

I watered the eggplant areas thoroughly for two hours before planting. It took from two-thirty to four-thirty. After planting the eggplants, I watered thoroughly again. The rows in the cucumber field are 3 *shaku* (90.9 cm) wide and the surface area of the fields is 25 *tsubo* (83 m^2). I used 40 *kan* (150 kg) of fertilizer. I also mixed 2 *shō* 5 *gō* (3.8 L) of dried anchovy powder with 2 *shō* (3.6 L) of

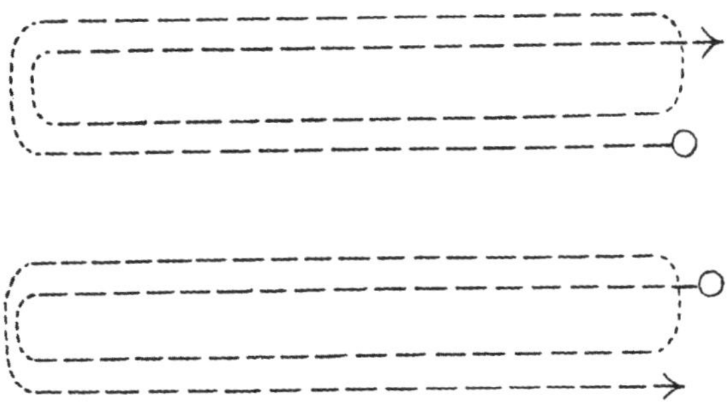

Figure 2.7. Paddy-plowing patterns. One would follow the top pattern first, and then the bottom pattern, while pushing the soil to the center.

calcium superphosphate and spread that around and then covered everything with soil. I didn't plant the seeds yet.

The rows in the summer cabbage fields are 2 *shaku* 5 *sun* (76.5 cm) wide, and there's a total of 70 *tsubo* (231 m^2) of field area. I used 120 *kan* (450 kg) of fertilizer.

I used 5 *shō* (9 L) of dried anchovy powder mixed with 3 *shō* (4.5 L) of calcium superphosphate. I didn't plant the seedlings yet.

At the main house: My brother's wife cooked and spread fertilizer on the rice paddies. Kanezō went out early to spread fertilizer, and today he finally started plowing the paddies with the horse. In local speech, people call it *bakō*.[61] From now on, I'll just use that expression for plowing with a horse. Today he did *bakō* to the paddies at Sanbonmatsu, which is also known as Noda. The Sanbonmatsu paddies are called that because there used to be three big, thick, pine trees standing there. Two of them died, so now there's only one, but villagers still call those paddies the Sanbonmatsu paddies.[62]

To plow, you make passes forming rows along the length of the paddy, about 4 or 5 *shaku* (approx. 1.4 m) wide. But this is for a skilled person. Most regular folk will cut rows maybe 6–7 *shaku* (approx. 2 m) (wide), or sometimes just not form rows and do *battara-uchi*. But I've always been told that it works better for growing rice if you cut narrow rows, 4 or 5 *shaku* wide. The fields are plowed by cutting a path about 1 *shaku* 5 *sun* (45.5 cm) in the soil—parting it on both sides. Later, you push the soil you moved with the plow back over again to the center to form rows. I'll show what I mean with a diagram (see Fig. 2.7). If you do the work as the diagram shows, you'll eventually get rows of soil. The actual width varies from person to person.

April 25, Thursday

Sunny, later cloudy
Strong southwest wind
10°C in the morning, 16°C in the afternoon

Breakfast:	same as everyday
Lunch:	salted herring, side dishes: same as morning
Supper:	We made soy sauce–based soup from crucian carp stock and poured it over *udon* noodles and ate that. Side dishes: same as morning.

We finished eating the 2 *to* (36.1 L) of rice (I'd measured before). It took 17 days.

Today I had to go down to the main part of the village on an errand. I also helped plow the main house's paddies. And I went to my sister's house near the station to get night soil. I also planted autumn cabbage seeds. I used about 1 *shaku* (18 ml) of seeds per *tsubo* (3.3 m^2) of land, and the type I planted was Bandagō. I applied one barrel—about 8 *shō* (14.4 L)—of night soil, diluted 1:1 with water. I also used 4 *shō* (7.2 L) of wood ash, etc. The rows are *handaka-mune* (mid-height) rows.

Hotbox work: I didn't open the *shōji* (panels) on the part I planted yesterday. I did open the tomato side, let sunlight into the temporary eggplant part until eight-thirty in the morning, and then covered it again with the *sudare* screens. I removed the screens at three o'clock and let sunlight in again. At five I covered everything with the straw mats (*komo*). My wife spent the day preparing for hand-plowing (doing *ibaru* to) the fields.

Today's income:

rabbit pelt tanning	50 sen

Today's expenditures:

money given to a poor friend for buying food	5 yen
postage for mailing photographic film	23 sen
postcards	6 sen
udon (six bunches)	16 sen
soy sauce (5 *gō* (902 ml))	13 sen
treats for the kids	7 sen

At the main house: My brother's wife spent the morning spreading fertilizer in the paddies. In the afternoon she went to pick vegetables in the hills—bamboo shoots, thistle (*azami*), dogtooth violets (*katakuri*), etc. My mother cooked. Kanezō also spread fertilizer in the morning, and in the afternoon, he did horse-plowing at Sanbonmatsu (see Figs. 0.27 and 2.8). The master [Saburō's brother] scattered sand in the nursery paddies—the seedlings aren't taking root well for some reason. In other words, the roots aren't going down into the mud so much as going upward into the water and waving about. So, to force them into the soil and bury them, he threw in sand. According to a certain agriculturalist, rice seedling roots waving about in the water is caused by a disease—something like brown rot (*fuhaibyō*), apparently.

Figure 2.8. Plowing the main household's paddies at Sanbonmatsu (April 25).

April 26, Friday

Cloudy, sometimes sunny
Light southwest wind
11°C in the morning, 18°C in the afternoon

Meals: nothing particularly different from the usual

Planting cucumber seeds: I dug holes in the fields I prepared yesterday, put night soil in the holes, then added wood ash, and then put the same soil I use for the hotboxes in the holes—one handful in each. Then I planted the seeds. Finally, I put about 3 *bu* (9.1 mm) of soil over the seeds.

Night soil: I mixed it 1:1 with water to make 1 *ka* and used that. I used about ten seeds per hole. This makes for approximately 230 individual plants (*kabu*). The holes are spaced about 1 *shaku* (30.3 cm) apart.

Hotbox work: I didn't open the *shōji* panels covering the part where I planted eggplants for the second time, but I did open the tomato parts. It's really warm, so the seedlings are getting too tall too fast. As for the other temporary eggplants, I did nothing particularly different today.

At three-thirty I went to my aunt's house to get some firewood—two bundles. My wife plowed fields.

Today's expenditures:
75 salted herrings from Hokkaido 1 yen 50 sen (5 for 10 sen)

At the main house: Kanezō went out early to cut hay for the horse. He plowed fields with the horse today as well. He finished plowing at Sanbonmatsu in the morning, so in the afternoon he went to Sakura. My mother cooked. My brother's wife went out to spread compost (*koe*). She worked alone during the morning, but in the afternoon her eldest son, Ichisuke (see Fig. 0.54), who's fourteen and in the first year of the higher course (*kōtōka*) at school, went out and helped her spread the compost.[63]

April 27, Saturday

A cloudless sky (*nipponbare*) at first, but cloudy from the evening
Light southwest wind
10°C in the morning, 25°C in the afternoon

Meals: same as always

Hotbox work: I did nothing different from yesterday for the eggplant parts. But some of the tomatoes are getting too tall too quickly so I moved them out to a temporary place—the second time to do this (for this season).

Also, in the morning I sprinkled powdered tobacco on the daikon field to get rid of harmful bugs. Clover springtails[64] cause trouble by nibbling on the leaves and making holes in them. I also prepared the summer cabbage fields—16 *tsubo* (53 m^2), about the same as the other day. After that, I started making a new basket at my cousin's request. My wife went to another village to enjoy their festival.

The burdock roots are beginning to sprout.

Today's expenditures:
salt (2 *shō* (3.6 L)) 20 sen
kids' allowances 12 sen

At the main house: My brother's wife cooked and went out to spread compost (*koe*) in the fields. Kanezō also did that—in the early morning and again later in the morning—and in the afternoon he plowed with the people of the main house, at Sakura again.

April 28, Sunday

Rain
10°C at 6:00 a.m., 15°C at 2:00 p.m.

Two villagers came over to visit.

Yoshida Kanegorō's wife gave birth to a baby boy.

Breakfast:	miso soup with thistle and *bonji no koromo*,[65] etc., in it, pickled daikon, *takuan, misozuke,* etc.
Lunch:	salted herring, dried daikon stripes sautéed in oil and then eaten with soy sauce for flavor, other daikon dishes from morning (no miso soup)
Supper:	same miso soup as morning, daikon dishes same as lunch. We also ate loaches I received from my crazy neighbor, simmered with soy sauce for flavor.

I worked all day on the basket I started yesterday. My wife sewed and cooked, etc. For the hotboxes, I simply lifted the *komo* mats at eight o'clock and then put them back in place at four.

At the main house: My brother's wife sewed and cooked. Kanezō cut hay for the horse early and he separated the ridges (*aze o hanasu*) in the paddies he'd spent all day plowing yesterday. In other words, he pulled the soil that had gotten pushed up against the ridges away with a three-pronged hoe.

April 29, Monday

Rain
10°C in the morning, 10°C in the afternoon

Breakfast:	miso soup with thistle, side dishes: same as always
Lunch:	salted herring innards (milt) soaked with salted krill, pickled daikon, etc.
Supper:	basically, the same as morning—no miso soup, but roasted salted herring

Hotbox work: About the same as yesterday, but for the tomatoes I opened the *shōji* about 3 *sun* (9.1 cm) to adjust the temperature, despite the rain, because some are growing too fast.

I spent the morning working on the basket and was able to finish it today. The more I do this, the better I get, it seems. In the afternoon I had to go into the village on an errand. My wife sewed, and she also went up into the hills to pick vegetables.

Today's expenditures:

kerosene (4 *gō* (721.6 ml))	12 sen
mothballs	5 sen
candies and treats	5 sen

Today's income:

pay for rabbit pelt tanning	30 sen

At the main house: The master went to help relatives in another village build a house, so his wife watched the shop. My mother cooked. Kanezō went out early to cut hay for the horse and then worked more on separating soil from the ridges in the paddies. He returned soaking wet from the rain. In the afternoon he wove rope from straw for home use.

The village received money from the government for preparing a communal work shed (*kyōdō sagyōjō*) and a granary (*gōgura*), but it wasn't enough, so a village meeting was held. The solution: cutting down four giant pine and cedar trees in the cemetery and selling them.

In addition, a villager nicknamed Konkiyaku cut down a grand old zelkova (*keyaki*) tree that belonged to the village and sold it for profit, which caused a big fuss. So, a meeting was held and the final decision on that was to have Konkiyaku pay the village 4 yen to make up for his transgression.

April 30, Tuesday

Rain, then hail in the evening
A strong northwest wind from the afternoon
8°C in the morning, 4°C in the afternoon

Breakfast:	miso soup with bamboo shoots, royal fern (*zenmai*) [*Osmunda japonica*], and bracken (*warabi*), etc., in it. Also: chopped Chinese yam and bracken boiled with soy sauce for flavor, pickled daikon, *takuan, misozuke,* daikon, etc.

To prepare bracken for eating, just cut some in the hills and bring it back and stick the cut ends into some ash. This is what you do if you're going to keep it for a while. When you want to eat some, take some out and lay it in a bowl, and cover it completely with ash. Then, pour enough piping hot water over it to fill the bowl, and leave it that way for a full day. Then you take the bracken out and put it in plain water for another day. After that, you can simmer it or just cut it up and eat with soy sauce, etc. For royal fern (*zenmai*), people generally lay it out in the sun to dry soon after picking it, but when you want to eat it, you need to boil it first and then let it sit in cold water for two or three days, and you must change the water two or three times.

Lunch:	roasted salted herring, pickled daikon, *takuan*
Supper:	miso soup containing taro stems mixed with thistle, salted herring innards known as *shirako* (milt) in local speech,[66] simmered with miso, other dishes: same as lunch

Today's expenditures:

candies for the kids	5 sen

Hotbox work: For the eggplants, because of the terribly cold rain, I just left the straw (*komo*) mats in place. Only for the tomato parts, I took the mats off for a bit. There was no way to get any work done outside today, so I just read books indoors and ran an errand or two. My wife did some sewing.

At the main house: My brother's wife sewed and cooked. Kanezō wove rope from straw from the early morning and took the afternoon and the rest of the day off.

Chapter 3: May 1935

May 1, Wednesday

Rain, later mixed with hail
Strong northwest wind
3°C in the morning, 3°C in the afternoon

Breakfast:	miso soup with thistle in it, side dishes: same as always
Lunch:	sides: same as morning
Supper:	miso soup with taro stems in it, salted herring, pickled daikon, etc.
Snacks:	*shida mochi*[1] we received from the main house

Hotbox work: I did nothing different from yesterday.

Today I made (wooden) boxes for growing watermelons.

My wife went out on errands and sewed and such.

Today's income:
payment for some taros I sold 2 yen

At the main house: My mother cooked, and my brother's wife sewed. Kanezō went out early to cut hay for the horse and spent the rest of the day making *fugo*.[2] He made two of these.

The following is a true story about a fellow whose plan to build a house on rented land was ruined. I'll introduce the characters first, using pseudonyms.

Santarō:	the landowner—deeply in debt, terribly poor
Kenji:	the potential tenant—wants to build a house
Bunkichi:	Kenji's "master" [employer]
Sanzaemon:	Santarō's next-door neighbor, the wealthiest man in the village—opposes the construction of a house on Santarō's land next door
Kōnoshin:	Sanzaemon's grandson

Sannosuke:	head of a branch household of Sanzaemon's family—Santarō's relative
Gennoji:	member of a Sanzaemon family branch household—Bunkichi's relative
Jūgorō:	owner of the rice paddies that Santarō farms, Sanzaemon's second son, head of his own branch household—one of the wealthiest men in the village

These are the people who appear in the story.

Santarō has had a lot of bad luck over the years, and he has fallen more and more deeply in debt. It finally reached a point where he couldn't even get by, so he put about one *danbu* (992 m^2) of his residential land up as collateral on a loan. But he continued renting out parts of it to others, and has been getting just enough to live from that. He now rents out 50 *tsubo* (165 m^2) to one person and 40 *tsubo* (132 m^2) to another. The tenants pay him in rice—5 *to* (90.2 L) and 4 *to* (72.2 L) per year, respectively. He still has about 30 *tsubo* (99 m^2) left over, so he has been hoping to find another tenant. Luckily for him, Bunkichi's employee, Kenji, got married and started preparing to establish his own household in the village. But there's not so much empty residential land, so Kenji asked Santarō if he could rent the extra piece of land. Of course, Santarō was more than happy to oblige.

However, news of this soon reached the ear of Sanzaemon. There was no way this greedy landowner, who is considered by most villagers to be nothing more than a materialistic dung heap, would stand by and do nothing. According to what I heard, Sanzaemon claimed that if a house were to be built on that spot and happened to catch fire, his own rice storehouse would be in danger of burning. Thus, he completely opposed Kenji's plan.

Some days later, Sannosuke appeared at Santarō's doorstep. Drawing on various kin connections, Sannosuke tried hard to convince Santarō to give up on renting out the plot of land to Kenji. However, even though Sannosuke is Santarō's relative, he doesn't really trust Sannosuke. Furthermore, Santarō doesn't much like Sanzaemon anyway, so he flat out rejected Sannosuke's proposal. Infuriated, Sannosuke snapped and said, "If that's how you're gonna be, then just go ahead and rent out the land and see what happens!" Santarō, who is quite good-natured and normally not easily angered, was pretty upset about being spoken to in this way by the younger Sannosuke. Unable to remain silent, he shot back at him: "A guy like you should just shut up and stay out of this. It's not your business!"

Soon after that, Sannosuke went to Bunkichi's place and said that Sanzaemon would rent some land to Kenji instead, so Kenji should take that offer and not build on Santarō's property. But Bunkichi and Kenji both said no. Sannosuke gave up and left. A few days later Kenji's wife was summoned to Sanzaemon's home on some pretext. She humbly obeyed, and received a harsh lecture from Sanzaemon: "You two are just getting by day by day on what you can earn by working here and there, aren't you? Just think about it—who is employing and paying you so that you can eat and live? Me—Sanzaemon—and my kin, that's who. Right? So just go ahead and build a house on Santarō's land next door and see what happens. You two won't even be able to eat from the next day. And what about firewood? The wood you burn every day you take from mountain land that belongs to me and my kin. Well, you won't be able to do that anymore. Right? Got it? Now, get out of here!" She went back home in a terrible state.

On the same day, Sannosuke returned to Santarō's house. He said, "If you insist on renting that land, then you're going to lose the right to farm those paddies you farm now—so says Jūgorō [the owner]." But Santarō was unfazed by Sannosuke's threat, saying, "If he thinks he can really rob me of my right to work that land for no reason, then tell him to go ahead and try. The most important thing for me now is to get 3 *to* (54.1 L) or 4 *to* (72.2 L) of rice. How can you really come here with that kind of talk? All I have to do is ask the National Farmers' Cooperative [Zenkoku Nōmin Kumiai] for help, and then there's no reason to be frightened by such threats." Santarō stood firm, and Sannosuke gave up and went away.

Some days later, Sannosuke returned, asking Santarō to rent the plot of land to Sanzaemon instead of Kenji. Sanzaemon, he said, would grow vegetables or something there and pay rent to Santarō. But considering Sanzaemon's usual behavior, Santarō felt uneasy about the offer, so he refused. Sannosuke left once again.

Soon after that, Gennoji barged into Bunkichi's house with a message from Sanzaemon: "If you continue to support Kenji in his bid to build a house on Santarō's property, then Sanzaemon is going to take away the land under your nursery paddies and warehouses." Bunkichi was completely dumbfounded by this, and felt he'd had enough. "Even though Kenji is my employee," he decided, "I shouldn't have to be threatened over trying to help him." He told Gennoji that he'd think about it some and answer later.

Unlike Santarō, Bunkichi didn't have the guts to appeal to the National Farmers' Co-op, so he simply gave his answer to Gennoji a few days later. He said he'd stop Kenji from building a house (on Santarō's land) and promised to search for another place for Kenji to do this. Therefore, Santarō was completely defeated and Sanzaemon raised his hands in joy.

Poor Santarō was deeply vexed and frustrated and fell into a miserable funk over the whole thing. One day, he happened to bump into Kōnoshin, who said, "Santarō—sorry about the recent trouble. My grandfather doesn't have much common sense—he must have said some pretty unreasonable things. Please forgive him. Of course, I know full well that my kin have no legal right to stop anyone from building a house on someone else's property. However, it would be very embarrassing for the household of Sanzaemon to have someone build a house right next to *my* rice warehouse. So, from an ethical perspective, you really should think carefully about renting land to people. From now on, you should consider your moral obligations and not allow anyone else to build on your property."

Santarō related this story to me in tears: "If we poor people have to endure this kind of treatment from multiple generations (of the same family), then we'll always just have to go around with our heads bowed." He lamented the blatant abuse of moral codes.

May 2, Thursday

Rain mixed with hail, sunny later, then rain, then just cloudy
Strong northwest wind
5°C in the morning, 10°C in the afternoon

Meals: same as always

In the morning I collected urine from my own home, dug holes for planting cabbages, then mixed night soil 1:1 with water and filled the holes with that. Before planting cabbages, I mixed 3 *ka* of night soil with 3 *ka* of water and covered the surface of the soil with that. It's been so unseasonably cold lately that the eggplants in the hotboxes have gotten frostbite and the leaves are damaged. This will probably result in somewhat lower yields. Tomatoes are pretty tough, so they haven't suffered any frost damage. Today the eggplants and tomatoes finally saw some sunlight—the first time in four days. Unfortunately, though, my burdocks and some other plants were hit by frostbite.

My wife went up into the mountains to pick wild vegetables. In the afternoon I went southward toward Akita City to buy summer cabbage seedlings.

At the main house: Kanezō cut hay for the horse early and after breakfast he went out to plow paddies with my brother's wife. They started plowing the paddies at a place called Shinden[3] today. My brother's wife watched the shop in the afternoon. Kanezō carried fertilizer out to the dry fields (with the horse), and my mother cooked.

May 3, Friday

Weather: nearly the same as yesterday
Northwest wind
3°C in the morning, 9°C in the afternoon

Meals: same as always

Yesterday I headed toward Akita City to buy summer cabbage seedlings, and stayed overnight at a friend's house. Today I finally made it back home with the seedlings. My wife took care of the hotboxes. Today was also fairly cold, so she only removed the straw *komo* mats for a bit and kept the *shōji* panels shut. But she did open the tomato sections some. In the meantime, she prepared the tomato fields, and did some other tasks.

Today's expenditures:

train fare	1 yen
pens and envelopes	10 sen
one hammer	20 sen
tobacco powder (400 *momme* (1.5 kg))	25 sen
bananas (100 *momme* (375 g))	25 sen
toys for the kids	40 sen
sweets for the kids	25 sen

At the main house: Kanezō spread *marogi-goe* in the paddies during the early morning, and then he plowed paddies with my brother's wife. In the afternoon, she watched the shop and Kanezō hauled fertilizer out to the dry fields. My mother cooked, etc.

May 4, Saturday

Sunny, later cloudy
Southwest wind, later out of the south
8°C in the morning, 12 °C in the afternoon

Meals: same as always

Hotbox work: I took all the *shōji* off the tomatoes today and let in direct sunlight and fresh air. I watered the eggplants a bit and left the *shōji* open about 2 *sun* (6.1 cm).

Today I planted the cabbage seedlings I brought back with me yesterday. I left 1 *shaku* 5 *sun* (45.5 cm) between each plant. I took 1 *to* 5 *shō* (27.1 L) of wood ash and put some of that in each hole I'd dug and filled with night soil the other day, and then I put a handful of compost soil [*baiyōdo*] all around each plant. I planted a total of 625 seedlings.

Later, I thinned out[4] the daikon—the first time this season—leaving four or five plants per cluster. My wife worked with me on all of this.

Today's expenditures:

two *nanadogo gasa*[5] (for working in rainy weather)	20 sen
treats for the kids	5 sen

At the main house: My brother's wife sewed while watching the shop. My mother cooked. Kanezō cut hay for the horse early, then went out to plow in the morning and finished plowing at Shinden. He finished the paddy plowing today. In the afternoon he made fertilizer for the cucumber fields. He dug a hole in the compost and poured night soil in, and then put some wood ash into the hole. This is known as *dohi*. He made about 400 *kan* (1,500 kg) of it. The amount of night soil he used was 5 *ka*, and the ash amounted to 15 *kan* (56.3 kg).

May 5, Sunday

Sunny, later cloudy then a light rain
South wind, later out of the north, dying later
9°C in the morning, 16°C in the afternoon

Meals: same as always

The green onions started to sprout. The autumn cabbages have also begun to sprout.

Hotbox work: About the same as yesterday.

The summer cabbages we planted recently were attacked by cutworms (*nekiri mushi*) so I had to get rid of the ruined ones and replant in the empty spaces. I also prepared the pumpkin fields. My wife thinned out the daikon (in the fields).

While my wife was working in the daikon fields, I went to see the horse-plowing competition hosted by the Wakimoto Village Agriculture Association (Wakimoto Mura Nōkai). I also prepared some Bordeaux mixture,[6] which I'd been asked to make by Yoshida Kumakichi, who also lives in Wakimoto, and I helped him administer it, too. Many of the *nashiro* in the village are suffering from root rot, so the villagers are all in emergency mode and are using a lot of Bordeaux mixture. This disease, *fuhaibyō,* has become quite a problem since people started using a lot of chemical fertilizers, according to my memory. It seems to have happened just about every year since then. In the old days, people say, the seedlings would often stagnate (*yodomu*) and some now say that maybe *that* was actually root rot. But the fact remains that it only happened once every number of years back then and not every year like it does now. That's why I think the viewpoint that this has been caused by chemical fertilizer is not necessarily groundless. What happens to the seedlings is that the roots don't penetrate the soil thoroughly but instead wave around in the water. Then the seedlings turn red and wither, rot, and die.

Today's income:
30 sen, in lieu of 1 *shō* (1.8 L) of polished rice (for helping to administer Bordeaux mixture)

Today's expenditures:

tobacco [a pack of cigarettes]	8 sen (given to the crazy neighbor)
postcards	15 sen
candies/treats	5 sen

At the main house: My brother's wife watched the shop while sewing. My mother cooked. Kanezō went to another village early to borrow a saddle because the main house's saddle had broken and was therefore unusable. After that he spent the entire day transporting fertilizer. He used a device called an *umamokko*[7] to carry the fertilizer (see Fig. 3.1).

May 6, Monday

Sunny
Southwest wind from the afternoon
0°C in the morning, 14°C in the afternoon

Figure 3.1. A man using a horse fitted with an *umamokko* to haul fertilizer. From Yoshida's personal collection: not included in 1938 book.

Meals: same as always

Hotbox work: I watered all parts today, too. Basically, about the same care as always—remove the straw *komo* mats at seven o'clock and open the *shōji* panels at nine. However, I only follow this pattern for the eggplants, on sunny days. I basically always remove the *shōji* over the tomatoes, and partly close the *shōji* over the eggplants at two o'clock, close them completely at three, and put the *komo* mats back in place at five. I always put all the *shōji* back in place over the tomatoes at five o'clock.

I walked all the way around the cabbage fields, checking carefully for cutworm damage early in the morning, and then I attached a rope to my *goshin*. A *goshin* is a box that you strap to your back for carrying fertilizer (see Fig. 3.2).

Figure 3.2. A *goshin*.

During the morning I helped prepare and administer Bordeaux mixture at the main house. In the afternoon, I worked in my own pumpkin and tomato fields. My wife also worked on preparing the pumpkin, tomato, and watermelon fields, etc., both in the morning and the afternoon.

At the main house: My brother's wife went to her natal home to help plant eggplants. My mother cooked. Kanezō carried fertilizer out to the fields in the early morning [probably with the horse, as always] and the rest of the morning he spent helping me and the master spread Bordeaux mixture on the *nashiro*. In the afternoon he went back to hauling fertilizer.

May 7, Tuesday

Sunny
Northwest wind
10°C in the morning, 18°C in the afternoon

Breakfast:	miso soup with bamboo shoots (*sasadake*) and potato in it, pickled daikon, *takuan*, etc.
Lunch:	salted herring, salted krill (*ami no shirokara*), side dishes: same as morning
Supper:	miso soup with thistle, salted herring, daikon from morning

I planted watermelon seeds today. I also planted pumpkin seeds.

In the field I prepared yesterday, I dug holes about 1 *shaku* 5 *sun* (45.5 cm) in diameter and 6 *sun* (18.2 cm) deep and put roughly 1 *kan* (3.75 kg) of fertilizer in each. Then I put about 4 *shaku* (72.2 ml) of a 2:1 mixture of powdered anchovy and calcium superphosphate in each hole. Also, I mixed 1 *ka* of night soil and 1 *ka* of water and used that, too. Then I put about 5 *sun* (15.2 cm) of soil in each hole and then dug small holes in the center of each big hole and dropped a two-handed scoop of hotbox compost into these, mixed it all up well (across the whole field) and put five seeds into each of these holes. Important: I stuck each seed into the soil with the pointed end down. Finally, I covered each hole with about 8 *bu* (2.4 cm) of soil. And last, to prevent field mouse damage and make good use of the sun's heat, I set a 2 *sun* by 2 *sun* (3 × 3 cm) glass plate on top of each hole. It's crucial for each glass plate to be firmly pressed against the soil, because if there is any space between the soil and the glass, the soil beneath will dry out quickly and this is not good. I'm the only one in this area who's doing such a thing.[8]

I planted 47 *tsubo* (155 m^2)—64 holes.

Next, I planted the watermelons. I planted them in nearly the same way as the pumpkins, but the fertilizer was different: I poured roughly 6 *shaku* (108.2 ml) of a 3:2:2:1 mixture of powdered anchovy parts, powdered oil cake, rice bran, and calcium superphosphate into each hole, and a bit more hotbox compost in each than I used for the pumpkins. Then I set a small box just big enough to hold a pane of glass on its top over each and then covered those boxes with glass panes. Of course, each box has fasteners that measure 2 *sun* (6.1 cm) in the front and 4 *sun* (12.1 cm) in the back. At night I'll (use these to) attach a woven straw *noma* mat of that size on each box to protect the plants from the cold air. This is so that I can make good use of the sun's heat and then harvest a bit earlier than I'd be able to otherwise. Again, only I am trying this (around here). I planted 52 *tsubo* (172 m^2) with watermelons. There are 50 holes, but I planted only 21 today. I left about 1 *ken* (1.8 m) of space between each hole. For the pumpkins I left 4 *shaku* (1.2 m) on one side and 6 *shaku* (1.8 m) on the other of each planting hole. My wife went out to sift compost in the morning, and in the afternoon she hand-plowed the tomato fields.

Hotbox work: Basically, I did the same as yesterday, but I didn't water them today.

At the main house: My brother's wife went to help at her natal home again and my mother cooked. Kanezō cut hay for the horse early and spent the rest of the day hauling compost with the horse.

Today I heard the little birds that cry *bobō, bobō* for the first time this year. I don't know their scientific name. Around here we call them *bobodori*. People say that their singing is a sign that warmer weather is coming.

May 8, Wednesday

Sunny
South wind, later out of the southwest
10°C in the morning, 20°C in the afternoon

Breakfast:	same as yesterday
Lunch:	*soba nereke,* side dishes: same as yesterday
Supper:	same as yesterday

We prepared for making miso early this morning.

I watered the hotboxes some, but otherwise did the same things as yesterday. For the watermelons, I lifted the straw *noma* mats at seven o'clock and put them back in place at five. During the morning I finished preparing a field for taros. I have rows 2 *shaku* 5 *sun* (75.8 cm) wide—64 *tsubo* (212 m^2) in total. The compost I'll use has a lot of soil mixed in—more than the amount of compost—so I'll have to use more in total: about 130 *kan* (487.5 kg) plus 6 *shō* (10.8 L) of a 5:3 mix of powdered anchovy and calcium superphosphate. I covered this with about 2 *sun* (6.1 cm) of soil, so I can just use it when I get around to planting taros one of these days.

In the afternoon I worked on my eggplant fields: preparing the land, etc. My wife did pretty much the same work I did today.

Today's expenditures:

salt (1 *to* 1 *shō* (19.8 L))	1 yen
soybeans (1 *to* (18 L))	1 yen 80 sen
brown sugar and candies	8 sen

At the main house: Kanezō cut hay for the horse early, and he transported more compost after breakfast, as I'd expected. In the afternoon, he planted cucumbers and pumpkins.[9] My brother's wife also did the same work in the fields, and she cooked some between tasks. My mother took her lunch with her up into the hills to gather wild vegetables.

Most farmers don't measure the width of the rows in their fields—they just estimate, maybe about 3 *shaku* (91 cm). The main house's fields that were planted today seem to have rows about 3 *shaku* wide, too. The spaces between the plants are also roughly the length of the hoe I use—maybe 3 *shaku*. They leave this much space so they can avoid having to do a lot of work for the cucumbers as they grow—just leave them alone and let nature do its job. For compost (fertilizer) they added about 10 *kan* (37.5 kg) of chicken droppings mixed into what we prepared the other day and used that. The way of applying the fertilizer is somewhat different from my way: dig holes about 1 *shaku* (30.3 cm) in diameter roughly 3 *shaku* apart, put compost in each hole, then put some chicken droppings in each one as well, a bit of soil, then plant the seeds and then cover with soil and that's it. Also (at the main house), they don't prepare special fields for the pumpkins—they always just plant them in the cucumber fields, putting one plant at the ends of each cucumber row.

cucumber field area	63 *tsubo* (208 m^2)
no. of cucumber plants	350
no. of pumpkin plants	40

The type of cucumber they planted is called Jikyūri—a native species. Usually, most of the pumpkins grown are also native, but due to a seed shortage I gave them some Amaguri pumpkin seeds.

May 9, Thursday

Rain, later sunny, then cloudy
Southwest wind
14°C in the morning, 14°C in the afternoon

Meals: exactly the same as yesterday

I didn't work in the fields in the morning due to the rain. Instead, I worked indoors, separating the soybeans—picking out ones to use for planting. In the afternoon I worked on the unfinished parts of my watermelon fields, doing exactly what I did the other day. Later, I plowed my oriental melon (*makuwa uri*) field. In the morning, my wife attached thin ropes to the "umbrella hats" (*nanadogo gasa*)—used for going out in the rain—that I bought recently. In the afternoon she hand-plowed some eggplant fields.

> Hotbox work: I lifted the straw *komo* mats at nine o'clock, and since the sun came out at two, I watered them a bit. I didn't open the *shōji*, but I did take all of them off the tomato sections. I put the mats back in place at five o'clock. For the watermelons, I pulled their *noma* mat covers off in the afternoon and put them back in place at five.

I forgot to write this before, but the type of summer cabbages I've planted is Takenokogata. The watermelons are Shin Yamato No. 1. The pumpkins are Amaguri. The green onions are Aikoku Senjū and Akuminegi.

Today's income:
harmonica (sold) 40 sen

At the main house: Kanezō cut hay for the horse early, and later in the morning he went to Bishamon to do *kizamu*. In the afternoon he hand-plowed dry fields. Doing *kizamu* to paddies means to go back over a plowed paddy with a three-pronged hoe and break up the rows and level the surface. You level a hand-plowed paddy by hand, and you level a horse-plowed paddy with a horse. My brother's wife went out in the morning to the cucumber fields she'd planted yesterday to plant (early ripening) soybeans[10] in the spaces between the cucumbers. My mother cooked.

May 10, Friday

(April 8 according to the old calendar)

Sunny
8°C in the morning, 15°C in the afternoon

Breakfast:	miso soup with bamboo shoots and royal fern (*zenmai*) in it, pickled daikon, *takuan*
Lunch:	roasted salted herring, pickled daikon from morning
Supper:	*meshi mochi*[11] we received from the main house

A young fellow from outside Wakimoto came to see my hotboxes.

In the afternoon I finished preparing my oriental melon fields. The rows are 2 *shaku* 5 *sun* (75.8 cm) wide, and I spaced the individual plants 2 *shaku* (60.6 cm) apart. First, I dug the holes and put about 500 *momme* (1.9 kg) of compost (*taihi*) in each, and about 2 *shaku* of the chemical fertilizer mixture I used on the watermelons [on May 7]. Then I spread a 1:1 mix of night soil with water (total: about 1 *ka*) over everything. Next, I sprinkled just a bit of soil on the surface and spread wood ash—about 4 *shō* (7.2 L)—over everything. Finally, I covered the surface with about a 3 *sun* (9.1 cm) deep layer of soil and that was it for that. The area amounts to 7 *tsubo* (23 m^2). I can plant 42 plants.

Today I accompanied my cousin's wife to Akita City because she got sick and had to be hospitalized. My wife plowed the eggplant fields. I took care of the hotboxes—in the same way as yesterday, except that I didn't water them. I lifted the *noma* mat covers on the watermelons in the morning and covered them again in the evening. The cucumbers have started to sprout, so I took the glass covers off, but I covered them with straw because I thought there might be frost in the morning.

At the main house: Kanezō went to another village early on an errand relating to the broken saddle. He plowed fields during the rest of the morning, and took off in the afternoon.

My brother's wife and my mother worked indoors: pounding *mochi* and doing other food-related chores.

Today is April 8 according to the old calendar. Since the old days, people have said that it's okay to take a daily nap from this day.

Today the villagers pounded and ate *mochi* and took off and relaxed in the afternoon.

May 11, Saturday

Sunny
West wind (but from afternoon)
2°C in the morning, 20°C in the afternoon

Meals: same as always
Snack: *amagasu* [*amazake*—explanation below]

Someone from another hamlet (*buraku*) came to see my fields.

Hotbox work: I watered and otherwise just did the usual things.

It's good that I thought about the possibility of frost this morning, because it was cold last night—there was frost—so the straw I put down on the cucumbers really helped. If I hadn't done that, they'd have been ruined. It seemed cold again tonight, so I put straw over them again.

This morning I finished the tomato fields. I have 27 *tsubo* (89 m^2), with rows 2 *shaku* 5 *sun* (75.8 cm) wide.[12] I used about 25 *kan* (93.8 kg) of the horse-stall compost[13] that I received before from a relative. For chemical fertilizer, I prepared a 2:1:1 mix of oil cake, powdered anchovy, and calcium superphosphate and put about 5 *shō* (9 L) of this on the fields, and then covered the surfaces with a layer of soil about 3 *sun* (9.1 cm) deep. Next, I applied 1 *ka* of night soil and about 2 *shō* (3.6 L) of powdered anchovy to the wheat for the second follow-up fertilizer and did the first *tsuchiyose*. Here in the village, *tsuchiyose* means to spread soil.[14] In the afternoon I planted taros. I dug small holes in the field I prepared yesterday and poured 3 *ka* of night soil mixed 1:1 with water in the holes. Then I put about 1 *to* (18 L) of wood ash in each hole. I pressed one seed taro into each hole and then covered each with about 3 *sun* (9.1 cm) of soil. Note: I put the seed taros in so that the sprouts were facing up. I left 1 *shaku* 5 *sun* (45.5 cm) between the individual plants, and I planted 525 of them. My wife worked along with me, and she also plowed the soybean field.

Today's expenditures:

35 salted herrings from Karafuto[15]	50 sen

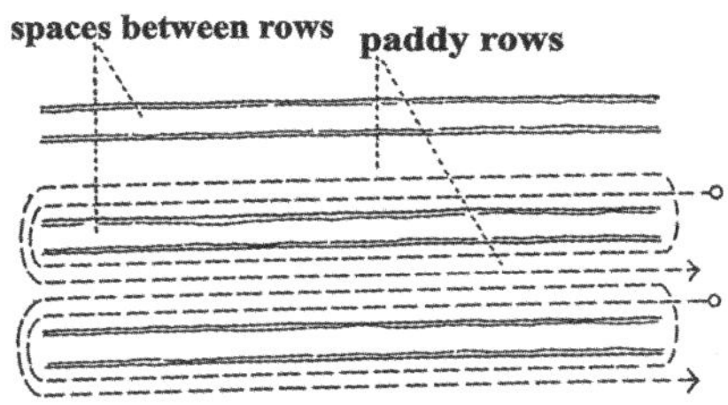

Figure 3.3. Pattern for smoothing paddies using a plow.

loss	5 sen (I had to pay to replace a child's rubber ball; my daughter destroyed it by running around while holding it.)[16]

At the main house: Kanezō cut hay for the horse in the early morning, then he spent the rest of the day smoothing out the rice paddies at Sanbonmatsu with the horse (doing *kizamu*). To do this, you use the same kind of horse-drawn plow that you use when plowing fields. You go back over the paddies again, passing around and around the rows you made before, keeping them at the center, and smashing them as you go. Doing this from both sides, you form new rows, but this time they're basically flat. The illustration shows how to do this (see Fig. 3.3).

It's best to do as shown in the diagram, for good results. But you need to think of one-half of each row becoming a new row after merging with half of another row. It's much easier work than the actual job of plowing.

My brother's wife went out to plant soybeans. My mother cooked.

The *amagasu* we ate as a snack today was *amazake*. You make it by mixing one part of regular rice with one part of malted rice (*kōji*). First, you make *kemishi* (*okayu*—simple rice gruel) with rice. Once it's done you put it in a clay pot or some container, and after it has cooled down you add *kōji* and mix it (*kemashi*) [standard: *kakimawashi*]. Then you cover it with something. You can enjoy it two days later.

May 12, Sunday

Sunny
Strong south wind
8°C in the morning, 22°C in the afternoon

Breakfast:	miso soup with bracken (*warabi*), royal fern (*zenmai*) and such in it, boiled bracken eaten with soy sauce for flavor, pickled daikon
Lunch:	roasted salted herring, *takuan*
Supper:	miso soup with thistle in it, salted herring, pickled daikon, etc.
Snack:	*amagasu*

Hotbox work: Same care as yesterday, and same for the watermelons, too. As for the hotbox tomatoes, from tonight I'll just leave them exposed, with nothing covering them.

Today I planted soybeans together with my wife. The rows are 2 *shaku* 5 *sun* (75.8 cm) wide. I used 15 *kan* (56.3 kg) of fertilizer and 1 *shō* (1.8 L) of calcium superphosphate. I have 14 *tsubo* (46.3 m^2) devoted to soybeans. First, I spread the fertilizer over the field, then I mixed the calcium superphosphate solution and applied it. Next, I covered everything with a thin layer of soil. After that I planted the soybeans—in pairs—in a staggered (*chidori-ashi*) pattern, as shown in the diagram, and covered the beans with a layer of soil about 2 *sun* (6.1 cm) deep (see Fig. 3.4). I spaced the bean clusters about 5 or 6 *sun* (16.7 cm) apart.

I also planted soybeans at the ends of my cucumber rows and all around the perimeters of my watermelons. So, I planted 865 pairs of bean plants, or 1,730 in total. The diagram shows how I planted the soybeans on the ends of the cucumber rows, and around the watermelons (see Fig. 3.5).

In the afternoon, we put our firewood away in the woodshed, and then stacked the new wood we collected this year in the place where the older stuff had been stored. Also, I replanted cucumbers in spots where the seeds had rotted and failed to sprout, and replanted burdocks as well. In the evening, I went to the main house to help stack miscanthus rods.

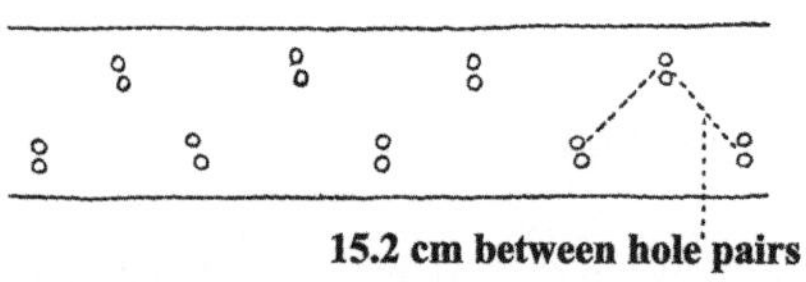

Figure 3.4. Soybean-planting pattern, in rows.

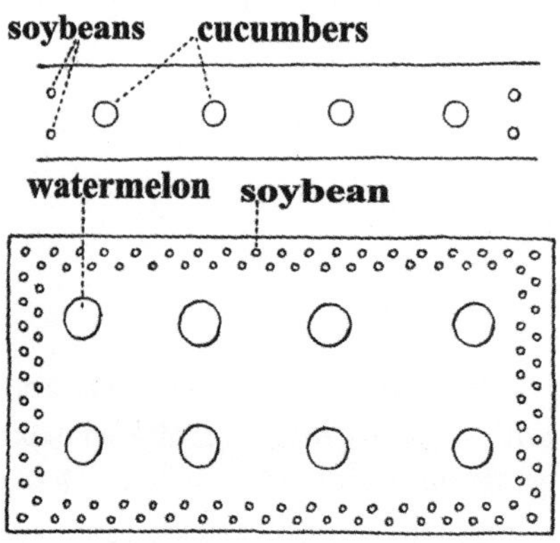

Figure 3.5. Soybean-planting pattern: Ends of cucumber rows and surrounding watermelons.

At the main house: My mother went out to pick vegetables in the hills and my brother's wife cooked while watching the shop. Kanezō cut hay for the horse early, went out to do *kizamu* in the paddies during the morning, and did some hoe chopping and plowing (*ibaru*) in the afternoon.

May 13, Monday

Cloudy, sometimes sunny
Strong south wind
10°C in the morning, 18°C in the afternoon

Meals: same as always

I removed the *shōji* panels covering one section of the hotbox eggplants today. I'll just leave them off. No watering today. For the watermelons, I did about the same things as yesterday—removed the *noma* mats in the morning and put them back in place in the evening.

I went to the hospital again today, because it was the day of my cousin's wife's big operation. My wife covered most of my work today. She took care of the spaces between the pea rows and weeded and then went out to pick wild vegetables in the hills. She got some bracken (*warabi*), royal fern (*zenmai*) and thistle (*azami*), etc. She also did some hoeing[17] in the trenches between the pea plant rows so they'd be uniformly low (see Fig. 3.6).

At the main house: Kanezō cut hay for the horse early. He spent the rest of the day repairing the water reservoir. He used the horse to carry soil and sand. My mother went to visit someone in another village, so my brother's wife cooked and planted soybeans in the fields.

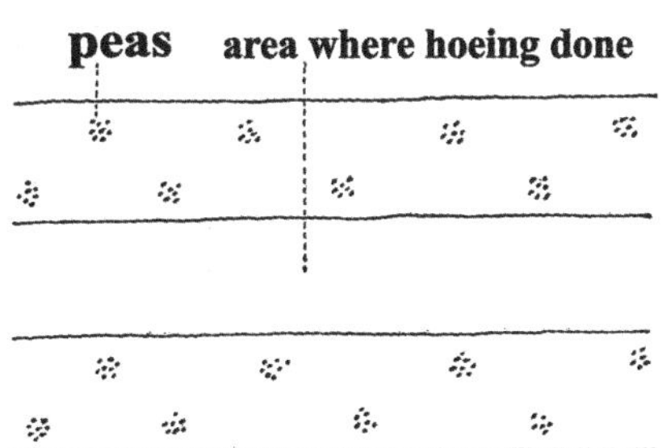

Figure 3.6. Pea plant row-hoeing pattern.

May 14, Tuesday

Rain, later just cloudy
South wind
13°C in the morning, 17°C in the afternoon

Breakfast and lunch:	same as always
Supper:	chicken meat we received from a cousin and tofu simmered together with soy sauce for flavor, side dishes: same as morning

The birds called *karakaraji* in local speech started singing today.

I saw butterflies flying about for the first time this year today.

The Chinese yams began to sprout.

The potatoes started sprouting.

I watered the hotboxes some, but only the parts with the eggplants. There was a little rain, so I left the tomatoes alone. For part of the eggplant section, I opened the *shōji* panels a bit around noon. I left the *noma* mats in place over the watermelons all day.

In the morning my wife and I fertilized the peas again and then covered them with more soil (did *tsuchiyose*). The fertilizer was a 1:1 mix of night soil and water (1 *ka* each)—we just poured it on the surface of the soil around the spots where we'd planted them. We covered them with soil from both sides. In the afternoon my wife thinned out the burdock crop for the first time—to make three-plant clusters. I had to go out on errands.

Today's expenditures:

soy sauce (3 *gō* (541.2 ml))	9 sen
ink	5 sen
tofu (1 block)	6 sen
cotton cloth, sweets for the kids, wheat flour,[18] etc.	10 sen

At the main house: Kanezō prepared *nebawara* in the early morning (by beating it until soft). *Nebawara* is straw that you use for tying rice seedlings together in the nursery paddies (*nashiro*) when you transplant them (to the rice paddies). Later in the morning he started carrying fertilizer out to the fields but then quit because of the rain, and instead went out to do *kizamu* in the rice paddies. In the afternoon he went back to hauling fertilizer. My brother's wife spent the day planting crops—soybeans, red *azuki* beans, *tenko* (a type of *azuki* bean), cowpeas, sesame, etc. All you have to do is spread fertilizer over the surface of a plowed field and plant the seeds and cover everything with soil—quite simple, actually. They have 208 *tsubo* (688 m^2) of soybeans, 120 *tsubo* (397 m^2) of *azuki* beans, 24 *tsubo* (79 m^2) of *tenko,* 7 *tsubo* (23 m^2) of sesame, and 14 *tsubo* (46 m^2) of cowpeas (*sasage*). You can carry about 16 *kan* (60 kg) of fertilizer on a horse at once (1 *da*). Kanezō carried about 50 *da,* or roughly 800 *kan* (3,000 kg). It was actually compost soil (*dohi*) mixed with fertilizer.

May 15, Wednesday

Rain
Southwest wind
13°C in the morning, 11°C in the afternoon

Breakfast:	miso soup with bracken and thistle, boiled bracken eaten with soy sauce, *takuan,* etc.
Lunch:	a chicken and tofu pot dish flavored with miso, salted herring, *takuan,* etc.
Supper:	a chicken and *konnyaku* pot dish flavored with soy sauce, roasted salted herring, *takuan,* etc.

Hotbox work: I lifted the mats at eight-thirty this morning and then put them back in place at five o'clock because of the rain. For the tomatoes and watermelons: nothing different from yesterday. I didn't work outside due to the rain; I prepared a photo album. My wife sewed.

Today's expenditures:

konnyaku (one block)	2 sen
sweets for the kids	2 sen
one red cord (for *geta*)	1 sen

At the main house: Kanezō cut hay for the horse early and my brother's wife did the morning cooking, then they both spent the morning and afternoon doing *kizamu* to the rice paddies at Bishamon. My mother cooked (the remaining meals).

May 16, Thursday

Cloudy, sometimes sunny
Southwest wind
11°C in the morning, 16°C in the afternoon

Breakfast:	miso soup with bracken and potato in it, *takuan, misozuke*
Lunch:	salted herring, *takuan,* etc.

Supper:	same soup from the morning, same side dishes as morning
Snack:	We enjoyed *amagasu* at the main house.

Hotbox work: I removed the *komo* mats at seven o'clock, and then removed the *shōji* panels over the eggplants at nine-thirty. At five o'clock I put the *shōji* and the mats back in place. I didn't do anything at all for the tomatoes today. For the watermelons, I removed the *noma* covers at six this morning, and put them on again in the afternoon.

My wife thinned out the burdock root crop for the first time this morning. I did the first supplemental fertilizer for the summer cabbages by digging small holes on one side of each plant and pouring some ammonium sulfate mixed with water into them. I used 4 *ka* of water, or 6 *to* 5 *shō* (117.3 L), and dissolved 800 *momme* (3 kg) of ammonium sulfate in it. Once I'd poured it all in, I filled in the holes. Next, I planted more watermelons, and I also planted oriental melons in the field that I hadn't yet finished planting. I had already prepared the fields, so I only had to plant the seeds today: dig small holes, drop a double handful of compost in each, place seeds on top, then cover with soil (six seeds per hole for watermelons, ten per hole for oriental melons).

In the afternoon my wife and I started preparing the eggplant fields.

At the main house: My brother's wife took her lunch with her up into the hills to collect wild vegetables. She got bracken (*warabi*), royal fern (*zenmai*), butterbur (*fuki*), *innyo,* thistle (*azami*), *nonoba,*[19] knotweed,[20] *deshika,* and more.[21] My mother cooked. Kanezō dug irrigation ditches for the paddies at Sanbonmatsu starting early, and also later in the morning, and he did *kizamu* to the paddies at Bishamon in the afternoon. He finished that early, so he did a little of the same work at Shinden and then went back to the house.

May 17, Friday

Sunny
Strong west wind
12°C in the morning, 16°C in the afternoon

Breakfast:	miso soup with bracken and royal fern, *takuan, misozuke,* etc.
Lunch:	roasted salted herring, *takuan*
Supper:	same soup as morning, salted krill, *takuan,* etc.

Yoshida Tadaharu's wife gave birth to a baby boy.

Hotbox work: I watered the eggplant parts. From tonight I stopped covering part of the eggplant section nightly with the *komo* mats—glass *shōji* panels only. I also watered the watermelons I'm growing in boxes. And I put some insecticide on the eggplants, tomatoes, and other similar plants to protect them from being eaten by *tabedomushi* (*tentōmushi damashi*).[22] I prepared 4 *shō* (7.2 L) of four-*to*-style Bordeaux mixture[23] and added 6 *momme* (22.5 g) of arsenate (*hisaen*) to that and applied it with a sprayer.

I also transplanted eggplant seedlings (from the hotboxes) together with my wife. I've got 130 *tsubo* (430 m^2) of eggplant fields. We used about 200 *kan* (750 kg) of compost [*taihi:* barnyard manure] and about 5 *kan* (18.8 kg) of wood ash, 4 *kan* (15 kg) of oil cake, and 2 *kan* (7.5 kg) of powdered anchovy. We also used 4 *ka* of night soil diluted with the same amount of water. We began with the fields that we had already prepared rows in and lifted some soil along each row with a hoe, making trenches for fertilizer. Then we filled these with the compost, the oil cake, and the anchovy mix, spread the wood ash on top, poured the night soil mixture over everything, and then covered the trenches with about 4 *sun* (12.1 cm) of soil, and that was it for that. After finishing, we did *tsubo-maki* for about 100 clusters (*kabu*)[24] of eggplants. *Tsubo-maki* means planting (seeds) directly in the fields. We dug small holes about 1 *shaku* 5 *sun* (45.5 cm) apart and poured about 2 *ka* of a 1:1 night soil and water mix, put a handful of compost soil in each one, and then put about 20 seeds in each hole and covered them with soil.

Next, we planted tomatoes. We dug small holes about 1 *shaku* 5 *sun* (45.5 cm) apart, poured in night soil, added wood ash, and then planted a tomato seedling in each hole. When you do this, you must thoroughly water the seedlings in their beds two hours in advance. I brought them from the hotboxes and planted them, and my wife followed me with bamboo poles, sticking one in the ground for each, and then tied the plants to the poles with pieces of straw. Today I also prepared 2 *tsubo* (approx. 7 m^2) of tomato fields as I'd done before, so now I have 264 tomato plants in total.

The cuckoos (*kanko dori* or *kakko*) started singing today.

At the main house: Kanezō cut hay for the horse in the early morning. He carried compost to the paddies after that. In the afternoon he went to do *kizamu* to the Shinden paddies. My brother's wife went out to plant *azuki* beans. People usually don't do *tsubo-maki* for *azuki* and soybeans—instead they normally just scatter them around (do *bara-maki,* otherwise known as *furi-maki*). In general, *tsubo-maki* is planting seeds in spots at set distances from one another [in contrast to *bara-maki*].[25] All of the main house's fields are basically finished now. My mother cooked, etc., today as usual. Tomorrow is the elementary school sports day (*undōkai*) so food had to be prepared for that.

May 18, Saturday

Cloudy
12°C in the morning, 16°C in the afternoon

Breakfast: same as yesterday morning [lunch: explained below]

Yoshida Kiyokichi's wife gave birth to a baby boy.

From today, I removed all the *shōji* panels covering the eggplants in the hotboxes [for the daytime].

I got to work early this morning on my preparations for making miso. I'll also make some for my sister's family, so I had to run and gather the necessary ingredients, clean out the big iron pot, borrow a *koshiki*,[26] make a new *dewa*[27] out of fresh straw to put inside the *koshiki*, fill the pot with water, prepare firewood, and wash and soak soybeans, etc. Most of the work I did with my wife. In the afternoon I went to see the sports day competition (*undōkai*) at my old school. But I came back home early and weeded the eggplant beds.

Today's expenditures:

postage stamps	61 sen
money for the kids to buy food at the sports day event	10 sen

At the main house: Kanezō went out to the paddies early and used a hoe to cut the weeds along the edges of the foot paths between the rice paddies that are for horses to walk on. These workhorse paths are known as *kuro*, and here in the village we call what he did today *nadebada o kejiru*. Later in the morning he carried compost out to the Shinden paddies. He took off in the afternoon.

My brother's wife and my mother cooked food to take out to the sports day event from early morning. They made *mochi* by kneading rice powder with a little water and then boiling this in water. Then they put cooked sweet *azuki*-bean paste inside. These are called *shinko mochi no anbi*, they also made *mugi-magi* with wheat flour and oil, and *abura-age*, or tempura [probably fried tofu]. In addition, they made a dish with bracken (*warabi*), Chinese yam, and cooked herring roe (*kazunoko*) flavored with soy sauce. Another dish they made had dried herring from Hokkaidō and butterbur and *konnyaku* cooked together with soy sauce and sugar—called *nishime*. There were bamboo shoots in that. And they cooked halibut caught at Minami Iso in soy sauce, etc. We ate these dishes for both lunch and supper and then went home.

May 19, Sunday

Cloudy, sometimes sunny
12°C in the morning, 20°C in the afternoon

Breakfast:	miso soup with bracken in it, *takuan*, etc.
Lunch:	roasted salted herring, cooked bracken eaten with soy sauce, *takuan*, etc.
Supper:	*udon* in soup made from dried roasted anchovies, *takuan*, etc. (We ate *udon* for celebration after making miso, but normally *azuki meshi*[28] would be eaten.)

Hotbox work: I watered the eggplants, and left the *shōji* panels off in the daytime. Starting tonight, I'm leaving part of the eggplant sections uncovered, even at night. Some of my box-grown watermelons sprouted, so I covered them with glass. It's an experiment. After all, they really do sprout sooner in boxes than just out on the ground somewhere.

Another rabbit was born today.

My wife weeded the eggplants and cabbages and went up into the woods behind the house to pick wild vegetables and such—especially butterbur.

I woke up at three-thirty and went to the main house to get the big iron pot to make miso. My wife and I started at four o'clock and finished at seven. First, I boiled water in the pot. Once steam started rising from it, I scooped up the soybeans I washed and soaked yesterday and put them in the *koshiki*. There are two ways to put soybeans in a *koshiki*. One way is to put in only a limited amount and wait for the pot to boil again, and then add more soybeans, and wait again, and so on. If you're using 5 *to* (90.2 L) of soybeans, it'll take about five times. The other way is to put all the soybeans in at once. Up until four or five years ago I used the former method, but I reconsidered and decided that the latter way would be better, so since then I've been doing it like that. This time I also did it that way.

After putting the soybeans in, you cover the pot with a wooden lid and put two or three straw mats (*komo*) over the whole thing [see Fig. 0.25]. You must keep the fire strong the whole time, and while monitoring it you have to check the water level and add some when needed. There are two ways to add water to the pot. One is to insert a bamboo tube, with the joints removed, into the pot through the *koshiki* and pour water into the tube. The other way is to insert a piece of bamboo about 1 *shaku* (30.3 cm) long split in half lengthwise into the pot around the edge of the *koshiki*, or between the *dewa* and the *koshiki*, and pour water down that. I've always employed the second method. Some people use "white water" left over after washing rice, and some use water they've boiled separately on the side. In my family we always used "white water," but since I started taking care of it myself four or five years ago, I've been using water I boil separately, beginning around the same time I started adding beans to the *koshiki* in the way I mentioned above.

There is a reason why people sometimes use "white water." They say if you use it, the water in the pot won't boil over. But this isn't necessarily true. I've seen the water boil up over the *dewa* and *koshiki* and cause a lot of trouble. Two or three years ago I heard an expert miso-maker say that it helps if you add just a bit of Shirashimeyu cooking oil—it won't completely prevent the water from boiling over, but it almost completely does. I tried that right away after hearing about it, and have had much less trouble because it rarely boils over.

After steaming the beans thoroughly for an hour, you get what are called "horse beans."[29] These are the tastiest. That is, the soybeans that are only about half-done are better than fully cooked beans. As for calling them "horse beans," I suspect it relates to the old saying that feeding half-cooked soybeans to a horse that's really worn out from hard labor will help rejuvenate it. Probably the name comes from that.

You have to steam the beans until they're the color of a fox. Then, you snuff out the fire. After waiting for an hour, you mash the beans. There are three ways to do this: 1) put them in a big wooden mortar and then five or six adults pound them with big wooden poles, 2) dump them into a big tub and then stamp on them while wearing straw boots, and 3) put them in a mortar and pound them with a mallet. In my family, we always pounded them with wooden poles, but since I started taking care of it, I've been going with the second method. It's more efficient and the result is good [see Fig. 0.26]. If you use a mortar and pestle you can pound about 1 *to* (18 L) at a time. If you stomp on them in a tub you can do about 7 *shō* (12.6 L) at a time. Next, for the malted rice (*kōji*)/salt/soybean mix proportions, for 1 *to* (18 L) of soybeans you use 7 *shō* (12.6 L) each of malted rice and salt. That's called *nana gō ai*. But some people use 1 *to* (18 L) of each, which is known as *jū gō ai*. Most farmers go with *nana gō ai*.

I put the beans in a tub and stomped on them thoroughly and added malted rice and salt in *nana-gō-ai* proportions, plus about 6 *shō* (10.8 L) of (hot) water, and mixed it all up well. Up until two or three years ago I just used the water in the pot [left over from steaming the beans] but the miso specialist said that's no good because it contains a strong toxin. I've been boiling water for miso separately since then. But a lot of people in the village use water from the pot. They say it's great because, since it's red, it turns the miso red. According to the miso master, however, it takes a lot longer for miso to mature enough to be consumed if you use such water. But I can't really say who's right on this issue. The expert said that farmers in the old days didn't use the miso they made in the same year they made it—they'd wait at least three years to use it—so using the red water was fine, but aside from a small number of very wealthy families, most farmers are now so poor that they have to eat their miso in the same year they make it. Therefore, I think the miso specialist's method suits the times well.

After finishing, I put the miso mixture in a different tub (*oke*) for aging. Here, we generally have a little celebration (*irozuke*) to mark the making of a new batch of miso each year—cooking red *azuki*-bean rice (*sekihan*) and offering it to the [spirit of the] (iron) pot and then eating it together.

For the fire under the pot, we use tree roots dug up in the mountains and brought down especially for this.

> At the main house: My brother' wife always cooks breakfast, and my mother does the later cooking. This morning my brother's wife went out to plant daikon in the fields, and in the afternoon, she carried fertilizer out to the Shinden paddies in a *goshin*. Kanezō cut hay for the horse early and then later in the morning went out to do some plowing with the horse. He finished all the paddy plowing today. In the afternoon he carried fertilizer out to Shinden.

May 20, Monday

Sunny, but sometimes cloudy
West wind, dying later
14°C in the morning, 22°C in the afternoon

Breakfast and lunch:	same as always
Supper:	We went to the main house to eat *azuki* rice (*sekihan*). We also ate cooked carp (*funa*) caught in lake Hachirō, salted herring and *takuan*.

Hotbox work: No watering today, but otherwise same as yesterday.

My wife did some hoeing in the summer cabbage fields to loosen the soil a bit. In the afternoon she weeded the onion fields. The pumpkins (*dofura*) started sprouting today.

I went to the main house at four o'clock this afternoon to make their miso for them.

Today's income:

60 tomato seedlings	1 yen 20 sen

Today's expenditures:

salt	10 sen
sweets for the kids	5 sen

At the main house: Kanezō went out to the paddies early to make sluice gates[30] for water control. After breakfast, he carried fertilizer out to the paddies—at Bishamon. He spread some across each paddy. My brother's wife divided the fertilizer for using in the paddies and helped spread night soil. She and I made miso in the evening. My mother cooked.

May 21, Tuesday

Rain, later just cloudy
South wind, later from the southwest
14°C in the morning, 14°C in the afternoon

Breakfast:	miso soup with bracken in it, boiled bracken with soy sauce, *takuan*
Lunch:	miso soup with bracken and egg in it, roasted salted herring
Supper:	miso soup made with stock from cooking carp (*funa*) with bracken in it, roasted salted herring, *takuan*

Sugawara Kinzō's wife gave birth to a baby girl.

Hotbox work: Same as yesterday. From tonight, I'll leave the *shōji* panels off the eggplant parts.

I thinned out the autumn cabbage seedlings. I also set up some windbreaks around part of the eggplant fields, and made bamboo poles for the green peas, Chinese yams, and cucumbers, etc. I also cut grass and such for the rabbits to eat. My wife went out to sell tomato seedlings in the morning and planted cabbages in the afternoon. She also weeded fields and plowed a bit. As usual, the rows are 2 *shaku* 5 *sun* (75.8 cm) wide.

Today's income:

tomato seedlings	55 sen (4 for 10 sen)

At the main house: Kanezō cut hay for the horse early and my brother's wife cooked. My mother made lunch and supper. Today, they finally started plowing the paddies to prepare for planting, starting at Bishamon. Kanezō pushed the plow behind the horse, and in the morning Ichisuke took off from school and helped by leading the horse (doing *sahitori*). My brother's wife separated fertilizer for using in the paddies in the morning, and led the horse instead of Ichisuke in the afternoon.

My brother was complaining this afternoon. Feed for the horse, such as wheat scraps (*fusuma*) and rice bran scraps (*komenuka*), are costing him 10 yen per month—too much to manage, he said.

A schoolteacher is now staying at the home of Yoshida Shigeji, for 18 yen per month, I hear. It seems that another teacher is renting a room in the home of Yoshida Sangobē, but in this case it's the teacher and his wife, and they feed themselves, so they pay only 7 sen per month. Yet another teacher is renting a room in Yoshida Shichisuke's house, for 3 yen 50 sen per month.

May 22, Wednesday

Sunny, sometimes cloudy
Southwest wind
10°C in the morning, 20°C in the afternoon

Breakfast:	miso soup with bracken and *wakame, takuan, misozuke,* etc.
Lunch:	roasted salted herring, *takuan,* etc.
Supper:	miso soup with bracken and potato, roasted salted herring, *takuan,* etc.

Six farmers from a different hamlet (*buraku*) came to see my fields.

Hotbox work: I watered them today.

I set up poles to support my green peas (*endō mame*) in the early morning.

Later in the morning I prepared the fields for planting my hotbox eggplants. I dug holes and diluted 2 *ka* of night soil with the same amount of water and poured it into all the holes. I also dug holes in some unused fields and planted some watermelons, pumpkins, tomatoes, and such—both seeds and seedlings. In the afternoon my wife and I began planting the eggplants. We put one good handful of compost and one seedling into each hole as we went along. It took until well after sunset. We planted 574 eggplant seedlings. My wife went out early this morning to sell tomato seedlings and didn't come back until lunchtime, so I cooked breakfast.

Today's income:

42 tomato seedlings	90 sen

Today's expenditures:

sweets for the kids	5 sen
soy sauce (3 *gō* (541.2 ml))	10 sen

At the main house: Kanezō went out early to start letting water into some of the paddies. After breakfast he went to clean out the water channel for the Enmeiji Temple reservoir. Everyone who relies on it for water for their paddies helped. In the afternoon he went out to shore up the mud ridges between the paddies (to do *kuronuri*). My brother's wife cooked and minded the shop. My brother went out to flood paddies. My mother went up into the hills to collect wild vegetables. Today she mainly got butterbur and bracken—lots of them.

May 23, Thursday

Sunny, later cloudy
Northwest wind
6°C in the morning, 20°C in the afternoon

Breakfast:	miso soup with butterbur, *takuan*, bracken (cooked) with soy sauce
Lunch:	roasted salted herring, *takuan*
Supper:	miso soup with bracken and butterbur, *takuan*, salted krill

I watered the hotboxes today, too. I removed the straw mat covers on my box-grown watermelons in the morning and replaced them in the evening. It was warm in the daytime, so I opened the glass (*shōji*) panels over the hotboxes a bit. I started digging little holes in my summer cabbage fields to put additional fertilizer in the soil early this morning. Later in the morning, I dissolved 800 *momme* (3 kg) of ammonium sulfate in 4 *to* (72.2 L) of water and applied that to the fields. Then I did the first *tsuchiyose* to the summer cabbages—pulling soil up from both sides to build up the rows, with the cabbages in the center. In the afternoon I went to the agricultural warehouse[31] to buy some rolled oats (*tsubushi mugi*) and chemicals. After I got home, I put fertilizer on my daikon fields by pouring 2 *ka* of night soil mixed with the same amount of water. My wife did the second thinning for the daikon and at the same time weeded that field as well. We've arranged the daikon

Figure 3.7. Yoshida Denjirō chopping with a hoe in his paddies at Kogasaki (May 22).

so that they're planted in pairs (*nihondate*), side by side, along the rows. The soybeans have begun to sprout.

Today's expenditures:

rolled oats (1 bag)	2 yen 40 sen
copper sulfate (500 *momme* (1.9 kg))	75 sen
black cotton cloth	10 sen
sweets for the kids	5 sen

At the main house: Kanezō went out early to the paddies to flood them, and later in the morning and again in the afternoon, he went out to Bishamon to shore up the ridges (*kuro*) there. My brother's wife watched the shop in the morning, and went out to collect bamboo shoots (*sasadake*) in the afternoon. My brother went out to flood paddies and my mother cut hay for the horse, stripped the stalks of the butterburs she picked yesterday, and cooked.

May 24, Friday

Sunny
10°C in the morning, 22°C in the afternoon

Breakfast:	Today we started eating a 2:1 blend of polished rice and rolled oats (*tsubushi mugi*). We ate miso soup containing chopped boiled daikon

leaves my wife pulled from the daikon fields when she thinned them out yesterday. Also: *takuan*, etc.

Lunch: roasted salted herring, *takuan*, etc.

Supper: same miso soup from morning and roasted salted herring. We also ate carp (*funa*) we received from the crazy neighbor—roasted and eaten with soy sauce.

Hotboxes and (box) watermelons: same care as yesterday

I did the first thinning of my cucumber crops early in the morning, and later did the first follow-up fertilizing by dissolving 200 *momme* (750 g) of ammonium sulfate in 2 *to* (36.1 L) of water and applying that. Then I piled some soil called *kubitsuchi* ["neck soil"] up at the base of each cucumber plant. I also did the first follow-up fertilizing to the green onions by dissolving about 300 *momme* (1.1 kg) of ammonium sulfate in 4 *to* (72 L) of water and applying that with a watering can. When I did this to the cucumbers, I dug little circular trenches around them before pouring. Next, I prepared a field for medium-ripening (*chūseishu*) summer cabbages. The field measures 45 *tsubo* (149 m^2). The manure had a lot of soil mixed in, so I used 150 *kan* (562.5 kg) of that. I also used 5 *shō* (9 L) of commercial fertilizer.[32] This was a 2:2:1.5 blend of powdered oil cake, powdered anchovy, and calcium superphosphate. My wife and I carried the manure to the fields from the manure pile with a *mokko*.[33] Later, in the evening, we planted more eggplants—266 seedlings (the remainder from two days ago). We planted them like we did the other day, using night soil and compost soil (*baiyōdo*). We diluted 1 *ka* of night soil with the same amount of water before applying it. It took both of us working together all day from early morning to sunset to finish all of this.

Dragonflies started appearing—they're flying about.

Today's income:

10 tomato and 50 eggplant seedlings	50 sen

At the main house: Kanezō was out all night flooding the fields at Sanbonmatsu—he never went to bed. He returned to the house for breakfast, and then he and Ichisuke spent the day hand-plowing the paddies (doing *takaki*). My brother was out flooding paddies all day, so his wife watched the shop. My mother cooked.

Paddy irrigation is getting to be a serious problem. There hasn't been any rain lately, so although it's time for *takaki*, there isn't enough water to do it. Even though tears are pouring from the farmers' eyes over this, rain just doesn't fall from the heavens. I don't know if it's a good thing to say, or a terribly sad thing to say, but when I look out across the vast expanse of paddies at night and see all the little fires of the people struggling to get water into their paddies it strikes me as a beautiful scene. Maybe it's more the latter (sad).

May 25, Saturday

Cloudy, sunny later
12°C in the morning, 24°C in the afternoon

Yoshida Sankurō's wife gave birth to a baby boy.
Yoshida Tokinosuke's wife gave birth to a baby boy.

Breakfast:	miso soup with bracken and butterbur, *takuan,* etc.
Lunch:	soba *nereke* eaten with soy sauce, carp (*funa*) from yesterday cooked with miso, *takuan,* etc.
Supper:	miso hot pot with daikon leaves, narrow bamboo shoots and whelks (*tsubugai*), etc., in it, roasted salted herring, *takuan,* etc.

Watermelons: same as yesterday

I took care of all my *shōji* panels for the hotboxes early. I won't need them anymore, so I removed the glass panes from the frames, cleaned them, and put them away. I also took down the windbreaks I'd set up around the house for wintertime—the reed screens (*sudare*)—and pulled all the poles out of the ground and covered everything with *noma* mats. After that, I cleaned up outside and burned all the junk [leaves, etc.] I'd collected. I'll use the ashes later.

Later, I went to my sister's house near the station to get some night soil. I brought back a barrel full of it. My wife was called over to her natal house to help [for hire]. She guided the horse (did *sahitori*). Therefore, I took care of lunch and supper.

At the main house: Kanezō went out early to Shinden. The people out there got together and divided the water in the irrigation ditches to flood their paddies. Each person took one empty *tawara* with him and filled it with soil; they put them in the ditches to divide the water. Later in the morning, Kanezō went to smooth out the wet paddies (do *kaku*) at Sanbonmatsu. In the afternoon, he worked on shoring up the mud paddy ridges, also at Sanbonmatsu. My brother managed the paddy irrigation all day, and his wife sewed while minding the shop. My mother cooked and cut hay for the horse, etc.

May 26, Sunday

Sunny, cloudy later
6°C in the morning, 24°C in the afternoon

Breakfast:	miso soup with bracken and butterbur (*fuki*), *takuan,* etc.
Lunch:	roasted salted herring, *takuan,* etc.

Supper: miso soup with bracken and butterbur (stock made from roasted herring), roasted salted herring, *takuan*

Box-grown watermelons: same care as yesterday.

My green peas started flowering.

I prepared 1 *to* (18 L) of four-*to*-style Bordeaux mixture and applied it to my green onions, and I set up some support poles for my Chinese yams. They're made of bamboo.

I took down all my hotbox frames and put them away. I covered them with straw *noma* mats for storage.

I applied the third follow-up fertilizer to my wheat crop—1 *ka* of night soil. I also pulled soil up over the roots of the wheat plants with a hoe [did *tsuchiyose*] for the second time. That work took my wife and me all day.

Today's income:
100 eggplant seedlings 57 sen

Today's expenditures:
sweets for the kids 5 sen

At the main house: Kanezō went out early to shore up the Sanbonmatsu mud ridges again. My brother took care of the irrigation, and his wife watched the shop. In the afternoon, she made Ichisuke mind the store so that she could go up into the hills and get some bamboo shoots, bracken, and the like. My mother cut hay for the horse while doing the cooking, and in the afternoon, she went out to check the dry field crops, and she weeded a bit, too.

May 27, Monday

Cloudy, light rain in the evening
South wind, later out of the southwest
17°C in the morning

Breakfast, lunch, supper: nearly the same as yesterday

Box watermelons: Since it was cloudy, I pulled back the straw *noma* mat covers, but I didn't lift the glass panels. They've already started putting out their main leaves—the ones in the fields haven't even started to sprout. After all, my second year of watermelon-raising experimentation may be successful! I thinned out the potatoes, leaving three-plant clusters (*sanbondate*).

I planted my medium-ripening (*chūseishu*) cabbages in the fields today—320 in total. As usual, I dug little holes, and used 2 *ka* of a 1:1 night soil/water mix. I put about 1 *to* (18 L) of wood ash into the holes and then planted the seedlings.

I thinned the pumpkins for the first time, leaving them in a three-seedling cluster (*sanbondate*) pattern.

My wife went to the main house to help level the paddies—she led the horse (did *sahitori*). In the afternoon I went to the main house to pull a *shirogi* through the paddies. You first use a tool (an *iburi*) to basically flatten out a paddy that has already been plowed with a horse and then attach ropes to a big pole—about 5 *sun* (15.2 cm) thick and 2 *ken* (3.6 m) long—and put the rope around your shoulders, and then pull the thing along behind you as you walk through the flooded paddy [see Fig. 0.29].

At the main house: Kanezō went out early today to check on the paddies—to make sure those at Bishamon had been properly irrigated, since the plan was to smooth them out to prepare for planting today. He had to fill any paddies that didn't have enough water, and route water from paddies that had too much into those without enough, etc. In the morning he pulled a *shirogi*, and he pushed the horse plow all afternoon. My brother's wife pushed the horse plow in the morning and then she led the horse (did *sahitori*) in the afternoon. Ichisuke watched the shop, and my brother managed the irrigation again. He complained that it hasn't been going very well due to the lack of water these days. My mother peeled the stalks of the butterburs that my brother's wife picked yesterday, and cooked, etc.

May 28, Tuesday

Sunny
9°C in the morning, 22°C in the afternoon

Two acquaintances of mine came from another hamlet (*buraku*) to see my farm.

The weather was not great today, so I lifted the glass panels covering my box watermelons.

The watermelons and cucumbers in the fields have begun to sprout.

After doing some hoeing in my potato fields, I poured 1 *ka* of night soil over the surfaces of the rows as the first follow-up fertilizing. I also did the first *tsuchiyose* (to the potatoes).

In addition, I did the first follow-up fertilizing to the burdocks by mixing one bucket of night soil with one of water to make 1 *ka* and poured this right over the surfaces of the rows. Other things I did included grinding buckwheat (*soba*) flour with a stone mortar, going out to the paddies to collect whelks (*tsubugai*) (from the irrigation ditches), and cooking both lunch and supper.

Breakfast:	same as yesterday
Lunch:	I made soba *nereke* and we ate that, and we also ate roasted salted herring and *takuan,* etc.
Supper:	same soup as morning, *takuan,* whelks (*tsubugai*) cooked in soy sauce

My wife spent the whole day helping at her natal home—leading the horse (doing *sahitori*).

Today's expenditures:

brown sugar	5 sen

At the main house: Kanezō went out early to irrigate the paddies at Shinden. Later in the morning, and in the afternoon, he hand-plowed (did *takaki* to) the paddies. During the morning my brother's wife led the horse (did *sahitori*), and in the afternoon she did *kadzugu* to the paddies. This means mixing up the soil thoroughly with a hoe in paddies that you can't till with a horse, because they're too deep for a horse to enter. My brother managed the irrigation in the morning and minded the shop in the afternoon. Ichisuke took off from school again today, too, and he watched the shop in the morning and led the horse [did *sahitori*] in the afternoon. My mother cooked, and she also weeded the potato fields a bit.

All the paddy *aragure* plowing has now finished at the main house. *Aragure* means the first rough plowing.

May 29, Wednesday

Sunny, cloudy later
12°C in the morning, 22°C at 1:30 p.m.

Care for box watermelons: same as yesterday.

Both my wife and I went out to help others today—day labor. I went to Kinjirō's[34] place to help pull rice plant seedlings from the nursery paddies (*nashiro*) to prepare for planting them in the paddies (to do *naetori*). My wife went to push the horse-drawn plow at the main house. Villagers are gradually beginning to plant their paddies. Those who did it today are the earliest in the village to plant—the folks at (Yoshida) Suekichi's household. Kinjirō's family will start planting tomorrow. Generally, people pull the seedlings from the nursery paddies (*nashiro*) the day before they plant them in the paddies. Today, four of us pulled 1,350 little seedling bunches from the *nashiro*—three men and one woman. Normally, one person can pull out 400 in a day, but we couldn't each get that many because the *nashiro* weren't in such great condition. We started at six-thirty and took a little mid-morning break (*naka-yasumi*) at about ten o'clock. In the old days, people used to drink *nigorozaké*[35] but today we drank refined saké. We ate cooked *kobumaki*[36] with roasted dried herring from Hokkaidō inside it, a dish with bamboo shoots and butterbur and such cooked with soy sauce, *wakame* from Oga and dried daikon strips mixed and flavored with miso and vinegar [*sumiso-ae*]. We also ate turnips grown at Kinjirō's home with herring roe from Hokkaidō, and herring flavored with soy sauce. And we ate *mochi* pounded in a mortar and covered with soybean powder and brown sugar [*kinako mochi*]. Then we worked again and broke for lunch just before noon. For lunch, we ate regular white rice. Of course, we were served saké again. We ate roasted dried herring with *konnyaku* and such cooked in soy sauce, turnips and herring roe with soy sauce, and pickled cucumbers, eggplants, *takuan,* etc. We took another short break at after four o'clock. We put balls of rice [*damako mochi*] in miso soup made with roasted dried anchovy stock. Also: *konnyaku* and turnips with soy sauce.

We ate breakfast and supper at home—basically the same as yesterday.

At the main house: Kanezō went out to check the paddies early. He did some hand-plowing at Sanbonmatsu and finished that work there today. Ichisuke guided the horse with a bamboo pole (did *sahitori*) and my wife pushed the plow behind the horse. My brother's wife spent the day shoring up mud ridges (between the paddies) at Sakura. My mother cooked and cut hay for the horse and did the first *tsuchiyose* in the potato fields. First, she turned over the surface soil a bit and then pulled fresh soil up onto the rows from both sides with a hoe. Some people just use their hands to move more soil onto the rows instead of using a hoe.

May 30, Thursday

Cloudy, light rain
North wind
14°C at 6:00 a.m., 13°C at 1:00 p.m.

Breakfast and supper: basically, the same as yesterday

Because it was cloudy and drizzly, I only removed the *noma* covers on my box watermelons but left the glass panels in place.

I went to help do *naetori* [pull up seedlings in the nursery paddies] at Kinjirō's place again today. My wife went to help plant paddies at her natal home. For Kinjirō's family, and for my wife's family, it was the start of this year's rice planting. For our mid-morning break at Kinjirō's place we ate red bean rice (*azuki meshi*), *udo* [*Aralia cordata*] picked in the hills and herring mixed with miso and vinegar, cowpeas cooked in brown sugar, *konnyaku,* bamboo shoots, and dried herring and butterbur and such cooked in soy sauce, and more. Of course, we drank saké, too.

Lunch: miso soup with tofu in it, roasted anchovies with salt sprinkled on top, roasted salted herring, dried strips of daikon cooked in soy sauce with whelks (*tsubugai*), pickled cucumbers, *takuan*, etc. For our mid-afternoon break, we ate *damako mochi* (balls of *meshi mochi* in soup or broth). We also ate roasted anchovies with salt again and *konnyaku* and butterbur cooked in soy sauce, etc. Refined saké was offered each time we ate.

There's no one official "way" to remove seedlings from a *nashiro*. Basically, you just lift each bunch out with both hands without damaging the roots in whatever way suits you. The elder folks all say that it's something that can't easily be explained in words. Since it's wet paddy work, you've got to learn through the demonstration and practice. Once you've heard and seen the basics you just have to do it and find the way that works best for you. If the roots of the seedlings in the bunches you've pulled out aren't all the same length, the women who do the planting (the *sodome* [non-standard for *saotome*]) will complain, and if you yank them out roughly, you'll damage the roots or break the seedlings. It really is difficult work. First, you lift out a seedling bunch with both hands and rinse off some of the mud in the water (of the *nashiro*), grabbing the little bunch by the seedlings themselves with both hands and swishing the roots around in the water some to loosen the soil, which then falls off. You take what you can hold in each hand and put these together and count that as one bunch. You cut the straws you'll use for bundling the bunches into pieces about 2 *shaku* (60.6 cm) long and tie them to your waist in advance, then you pull them off as you go, tying those bunches together in little bundles by wrapping the straws around each one twice. It's impossible to explain how to do this in words.

There are roughly 700 little seedlings in each of these tied bundles. You can get this many per bundle when you first start pulling them from the *nashiro*, but by the time you finish pulling the seedlings from all *nashiro*, the number in each tied bundle will certainly be lower. This is because they grow thicker and taller as time passes.

Figure 3.8. Carrying seedlings from the *nashiro* out to the paddies (May 30). The bundle on the horse's back is full of food for the *sodome* to eat at snack time. The man carries more in his hands.

At the main house: Kanezō went out early to check the paddies, and then he smoothed out the paddies all day—pulling a *shirogi* around. Ichisuke took off from school again and led the horse with a bamboo pole (did *sahitori*). They hired a young fellow from another household to push the plow behind the horse for the day. The job of smoothing out the paddies to prepare them for planting (*takaki*) finished today. My brother's wife went to help plant the paddies at the big main house. My mother cut hay for the horse and cooked, and she helped some with the cooking at the big main house.

May 31, Friday

Sunny
Strong northwest wind
9°C in the morning, 17°C in the afternoon

Breakfast: same as yesterday's breakfast

I removed the glass panels covering my box watermelons.

My wife and I both did day labor for other households. I finished my work pretty early, so I came home and did the third crop thinning to the daikon, leaving only one seedling of each cluster.

My wife went back to her natal house to help with planting the rice paddies and I went to Kinjirō's house again to help finish the *naetori* work.

For our mid-morning break, we had saké again, and a dish consisting of strips of dried daikon, *konnyaku* and herring roe mixed with miso and vinegar, one with a kind of fish caught off the shores of the Oga Peninsula that we call *kangoro*[37] cooked in soy sauce together with bamboo shoots and butterbur, strips of dried daikon and small slices of dried herring cooked in soy sauce, *yama mochi,* etc. For lunch we had miso soup with seaweed (*wakame*) in it, cowpeas cooked with brown sugar, pickled cucumbers, pickled eggplants, and *konnyaku* and butterbur cooked with soy sauce, etc. I didn't have an afternoon snack because my stomach was not in good shape; I couldn't eat anything. Instead, I went back for supper. I ate rice covered with soybean powder [*kinako*] mixed with brown sugar, miso soup with *wakame* in it, a dish consisting of butterbur and anchovies cooked in soy sauce, etc.

At the main house: Kanezō went out early to cut hay for the horse. All day he cut *guri* in the hills and carried them back with the horse. *Guri* are patches of topsoil about 1 *shaku* (30.3 cm) wide with grass growing on them that have been cut loose with a hoe. You carry them out to the paddies and use one or two at each water irrigation gate. If you put them under the gates, the soil won't erode and get washed away when the water flows. My brother's wife went to help with paddy planting at the big main house and my mother cooked and cut hay for the horse, etc.

Figure 0.01. Mt. Kanpū and the hamlet of Ōkura.

Figure 0.02. The author and his family. Photo credit: Ogawa Tōru.

Figure 0.03. The author's house and some of the fields in early spring. From Yoshida's personal collection (detail), substituted for photo in original book.

Figure 0.04. The author's house in winter.

Figure 0.05. The author's house in winter, buried in snow.

Figure 0.06. The frame for the cover (*tsuboke*) of the night-soil cistern (March 22).

Figure 0.07. The night-soil cistern cover, after a *noma* mat has been secured to the frame (March 22).

Figure 0.08. Preparing the roof of a house to be rethatched with miscanthus (April 2).

Figure 0.09. Rethatching a roof with miscanthus (April 6).

Figure 0.10. Compost bundles (*marogi-goe*) (April 13).

Figure 0.11. Shoring up the ridges (doing *kuronuri*) in the nursery paddies (*nashiro*): Rough chopping (*arazuke*) (April 7).

Figure 0.12. Shoring up the ridges (doing *kuronuri*) in the nursery paddies (*nashiro*): The final touches (*shiage*) (April 7).

Figure 0.13. Selecting rice seeds using salty water (March 27).

Figure 0.14. Soaking *tawara* filled with rice seeds in a pond, with reeds (*kaya*) and poles on top to press them down (March 27).

Figure 0.15. Rice seed–filled *tawara* drying in the sunshine, after having been soaked in a pond for two weeks (April 13).

Figure 0.16. Doing *yudoshi*: Preparing pre-soaked rice seeds (packed in *tawara*) for planting in the nursery paddies (*nashiro*) by pouring hot water over them, to make them sprout (April 14).

Figure 0.17. Scattering seeds in the nursery paddies (*nashiro*).

Figure 0.18. Nursery paddy (*nashiro*) work: Using a tool called a *gotosarai* to remove large pieces of "trash" (April 14).

Figure 0.19. Nursery paddy (*nashiro*) work: Removing small pieces of "trash" by hand (April 14).

Figure 0.20. Nursery paddy (*nashiro*) work: Using a tool called an *iburi* to smooth the surface (the horse in the background is being used to plow a *nashiro*) (April 14).

Figure 0.21. Nursery paddy (*nashiro*) work: Leveling and smoothing the surface with a long stick (April 14). The rods under the man's belt are for sticking into the *nashiro* to show how many *tawara* of rice seeds will be needed for each—one for one, two for two, etc. These can be seen standing in the neighboring *nashiro*. [The house in the background (center) is the "main house" (Saburō's natal home). The sweets (*wagashi*) shop is in front; the main part of the house stands behind. The house at the far left is the "big main house." Behind the main house, a path led up into the woods and over the mountain, to Saburō's farm and house.]

Figure 0.22. Chopping a paddy: Doing *tazukiri* (April 25).

Chapter 4: June 1935

June 1, Saturday

Sunny
Northwest wind
10°C in the morning, 20°C in the afternoon

Breakfast: *okayu* (rice gruel), *takuan, misozuke* (miso-preserved vegetables)

> Watermelons: Same care as yesterday.

Again, my wife and I both did day labor for the main house. I did *naetori.* My wife helped another family as a representative of the main house. For the main house: *naetori* today, and planting seedlings (*nae*) in the paddies starting tomorrow. Four of us worked on that today—me the only man—and we pulled about 1,600 little bundles from the *nashiro.* We worked more efficiently than we did when I helped at Kinjirō's house.[1] One reason is because the seedling quality was higher. They were also thicker.

For our morning break-time snack, we had saké, small rice balls covered in powdered soybean [*amenogo: kinako*] mixed with brown sugar, butterbur simmered in soy sauce with pieces of dried herring, dried strips of daikon cooked in soy sauce (called *debu*[2] in local speech). We also had a dish of radishes mixed with salt-preserved herring roe [*kazunoko*] and other dishes.

For lunch we ate roasted salted herring, *takuan,* etc.

Our supper consisted of miso soup made with carp stock containing thin bamboo shoots and *wakame, takuan,* roasted salted herring, *debu,* cooked radishes flavored with soy sauce, etc.

This morning I thinned out my daikon crop again and cut grass and straw for the rabbits to eat, among other things. Really busy right now.

Today's expenditures:
cash for the kids to buy something to eat 5 sen

At the main house: Kanezō went out early to inspect the paddies and cut hay for the horse, and since rice planting starts tomorrow, he rolled a *kata* across the paddies.[3] He did the same things in the afternoon. The *kata* measures 7 *sun* (21.2 cm) on each side. There are many different kinds.

My brother's wife helped with *naetori.* My mother cooked and did other things.

In the evening, my brother and his wife bundled all the seedlings we pulled from the *nashiro* today. To bundle seedlings in *sanpa tsunagi* fashion with straw, you take two lengths of straw, each about three times the length of one seedling, and lay them down to form an "X" and lay four seedling bunches down over them and keep stacking them up like that until you've made ten layers, totaling forty bunches. This is called *yonjū marogi.* Some people stack five bunches at a time to make *gojū marogi.* When you stack four bunches at a time you must set the bunches down with the root ends pressed against one another. Recently, due to advances in the scientific study of farming methods, it has been observed that bundling seedlings this way causes some damage to them and leads to poor results in the paddies, so more and more people are simply piling them up in big boxes or baskets and using horses to carry those out to the paddies instead of bundling the seedlings. This is also the case for *naetori.*

People are now starting to bundle seedling bunches when pulling them from the *nashiro* in units of whatever they can grab with two hands together at once, called *hantaba* (half bunches). I've heard that in the old days *naetori* was always done by old folks who supposedly got all their seedling pulling done by drinking homemade hooch. They say that one man would down one or two *shō* (1.8–3.6 L) of saké a day, on average. Apparently, they'd drink for the mid-morning break, at lunch, during the mid-afternoon break, with supper, and even at breakfast. They'd even fill little bottles with the stuff and set them in the streams or channels to cool them and consume it like water, I'm told. For the lunch break they'd quit before the women (*sodome*) and eat quickly and then take a good nap before working again, and then in the evening they'd stop around the time the dew began appearing on the seedlings and drink more.

Today, Kuranuki Kenji (16), a resident of our hamlet (*buraku*) who works for Ishiyama Construction, broke a leg bone while handling heavy stones.[4]

June 2, Sunday

Sunny
North wind
7°C in the morning, 22°C in the afternoon

Breakfast:	miso soup with daikon stems and leaves and bamboo shoots, etc., in it, *takuan, misozuke,* etc.

Box watermelons: Same care as yesterday.

My tomatoes started flowering today.

Today I went to the main house again to help pull seedlings from the nursery paddies, and my wife went to her brother's place to help plant seedlings in the paddies. Before I left in the morning, I thinned out the daikon crop a bit more. That task is over now. During the lunch break I cut grass and such for the rabbits instead of resting.

For today's mid-morning break, we had *azuki meshi* [*sekihan*], saké, dried herring cooked with butterbur, a *nishime* dish of bamboo shoots and *konnyaku* flavored with soy sauce, a kind of fish called *kangoro,* and radishes that we ate with soy sauce. We also ate *debu* made with dried fish from Hokkaidō and strips of dried daikon, etc.

Lunch was almost the same as what we ate for our mid-morning break, but we got some sea bream (*shimadai*) caught off the coast of Oga from relatives, so we ate miso soup with some of that plus some *konnyaku* in it.

Supper: *wakame* miso soup, the (leftover) *nishime* dish from lunch, miso-preserved eggplants and daikon, etc.

At the main house: They started planting the paddies today. This is known as *hatsuyōhi.*[5]

There were 10 *sodome.* They planted 350 *kari*[6] in the paddies at Sanbonmatsu and at Sakura they planted 250 *kari.* They planted about 950 little bunches of seedlings at the former and about 700 at the latter.

Kanezō carried seedlings using the horse. Ichisuke helped by doing *kone uchi:* tossing seedling bundles to the *sodome* from the ridges between the rice paddies when they called out for more (see Fig. 0.32). My brother's wife helped with the planting. My mother cooked. Kieko, my brother's eldest daughter, minded the shop. My brother handled the irrigation. Today it was just me and one other guy doing *naetori.* We picked 850 bunches.

Most of the *sodome* at the main house today were doing *e,*[7] which means helping different households on different days in pre-arranged shifts according to their own schedules.

For planting paddies, the *sodome* usually plant five rows at once as they move along—sticking one little bunch of seedlings in the center and two on each side. Nowadays, people use a *kata* (see Fig. 0.30) for keeping the seedling rows straight, so even beginners can plant seedlings in straight rows, but in the old days when there were no *kata,* they used to run a rope down the middle of the paddy and the most highly skilled woman would start planting well-spaced rows and the others would follow suit. People say it was very hard to do the planting back then.

Although today the rice paddies get dry enough to be called *kanden* (dry paddies), in the old days they were so slushy that folks used to call them *mizuna* (water paddies). There was so much water in them that everyone would have to splash around and get soaked from head to toe just to get the planting done. And now everyone uses a horse to prepare the paddies from the first plowing to the final stage, but I heard that way back they did everything by hand with hoes—turning over the soil again and again until it was soft. It really was backbreaking work.

Another thing—there's a tradition that normally follows the completion of planting in a certain area. All the people invited will come out with handfuls of seedlings known as *tene* ["hand seedlings"] and fill in the spaces in the paddies where seedlings got washed away by the water. It's usually done about a week after the planting has finished. Around here we call it *anajashi.*

Way back, people always planted while walking in reverse through the paddies. But today we use wooden *kata* for making lines in the mud, so everyone plants while moving forward. And before the use of *kata* there was no need to toss bundles of seedlings to the *sodome* from the banks (to do *kone uchi*) because then it was fine to just drop seedling bundles here and there in the paddies in advance. But if you do that today, you'll mess up all the lines you've made with the *kata,* so it's necessary to throw the bundles from the ridges. There was a time when people tried carrying all the seedlings they needed in baskets tied to their waists, but it didn't work out very well so nobody does that these days.

I'm told that in the past the *sodome* always waited for daybreak to begin planting. They'd eat breakfast at the household they were helping that day, and have snacks and lunch as well, and then they'd each take a great big rice ball (*onigiri*) home with them in the evening. And I hear that it used to take 1 *shō* 2 *gō* (2.2 L), 1 *shō* 5 *gō* (2.7 L), or maybe even 1 *shō* 7 *gō* (3.1 L) of unpolished rice for a household to feed the hired *sodome* for a day. Today the *sodome* have breakfast at home and only have to be fed snacks, so the households that hire them usually just pay 2 *shō* (3.6 L) of unpolished rice. Back then, the *sodome* with children would usually bring the children and the family members who were watching the kids as well, and all would have to be fed. And after finishing the planting, the employers had to cook and serve even more food. But that's all over now.

Here in this village, the amount of unpolished rice to be paid to workers for a day's labor is divided into two levels: 2 *shō* (3.6 L) from the paddy-planting time in May through the harvest in autumn, and 1 *shō* 5 *gō* (2.7 L) for the rest of the year. In the old days, it was 1 *shō* 1 *pe* (2.3 L) (1 *pe* = 2 *gō* 5 *shaku* (451 ml)) for the same planting to end-of-harvest period, and 1 *shō* (1.8 L) for the rest of the time, but back then they had to feed the workers more.

These kinds of rules are decided by the heads of the various villages [and/or hamlets]. They're the current officers of the local household heads' [self-governing] association (*koshukai*). These aren't the people who are hired by others—they're the employers.

June 3, Monday

Sunny
South wind
11°C in the morning, 28°C in the afternoon

Breakfast: rice gruel (*okayu*) with *misozuke, takuan,* sesame seed miso [for *okayu*], etc.

I planted autumn cabbages in a temporary spot in the morning and thinned the pumpkin crop for the second time—planted in pairs (*nihondachi*).

Today I went to the main house to help with the seedling transplanting (*naetori*) again. My wife went to help plant paddies again. During the midday break I cut more grass for the rabbits.

It was just me doing *naetori* today so there was no food served for the mid-morning break. Lunch was about the same as yesterday. For supper I had miso soup with tofu, *wakame,* and bamboo shoots in it, and anchovies and radish greens with soy sauce. But the anchovies were roasted. There also was a *nishime* dish with butterbur and bamboo shoots, pickled eggplants, and more.

Today's expenditures:

white sugar	10 sen
soy sauce (2 *gō* (361 ml))	10 sen

At the main house: They finished the paddy planting today. Kanezō carried seedlings from the morning, they got Kin'etsu[8] to come over from the big main house and roll the wooden *kata* through the paddies to mark the planting lines, and of course Ichisuke tossed seedling bundles to the *sodome* from the ridges. My brother managed the irrigation in the afternoon and Kieko watched the shop. My mother cooked. Kanezō went out at night again to deal with the irrigation. Six *sodome* worked today and they planted 250 *kari* in the Bishamon paddies and 150 at Shinden (see Figs. 4.1 and 4.2). They planted 680 bunches of seedlings at the former and 400 at the latter. Of these, 200 bunches were *mochi* rice seedlings.

They fed *yama mochi* covered with soybean powder (*kinako*) and brown sugar to the *sodome* for their snack today.

June 4, Tuesday

Cloudy, later a light rain
Strong south wind
18°C in the morning, 18°C in the afternoon

Breakfast: miso soup with bamboo shoots and *wakame,* pickled daikon and eggplants

Figure 4.1. Six *sodome* dressed for planting rice seedlings. From Yoshida's personal collection: not included in 1938 book.

Figure 4.2. The main house's paddies at Bishamon, just after the planting. From Yoshida's personal collection: not included in 1938 book.

Today was really tough because the wind was so powerful—the glass covers on the watermelons blew off, some eggplant seedlings snapped, and poles for my cowpeas broke.

Today my wife was asked to go help plant another family's paddies by the main house, in exchange for receiving help to plant the main house's paddies (to return *e*). Once again, I went to help with *naetori,* but this time at my wife's big brother's house. It was just me so again there was no snack in the morning or at mid-afternoon.

Lunch today:	miso soup with carp and *wakame* in it, dried daikon strips cooked in soy sauce, *hatahata* (sandfish) that had been preserved in salt, *takuan,* etc.

I ate supper at the main house. Since today was the May festival day (*sekku*) [according to the old calendar] they made *mochi,* so I ate that—with red bean paste, in soup made from dried roasted anchovies and kelp, etc.

At the main house: Kanezō was out all night minding the paddy irrigation—he returned this morning for breakfast. During the day, he was out shoring up the ridges (doing *kuronuri*) for another family's paddies. This was also in exchange for their help with rice planting (returning *e*). My brother's wife went to return *e* by helping to plant another family's paddies. My brother managed the irrigation, Ichisuke watched the shop, and my mother cooked.

There was a labor dispute today between a landlord and a tenant farmer, but the tenant seems to have planted the contested paddies himself. The tenant was Yoshida Kakichi and the landlord was Yoshida Umematsu, and the National Farmers' Cooperative [Nihon Nōmin Kumiai] supported the tenant. The landowner declared that he would take back the paddies and farm them himself, so the tenant said he deserved to either be allowed to continue farming the land according to his tenant rights or receive payment in compensation. The landlord refused to pay: the apparent point of dispute.

June 5, Wednesday

Cloudy, some sun
Northwest wind
15°C in the morning, 20°C in the afternoon

Breakfast:	rice gruel (*okayu*), miso pickled daikon and eggplants, sesame miso
Lunch:	roasted salted herring, *takuan*
Supper:	same as breakfast, *misozuke, takuan,* etc.

Early this morning I had to replace the eggplant and cabbage seedlings that snapped because of yesterday's wind. During the lunchbreak I cut more grass and such for the rabbits.

Once again, I spent the day pulling seedlings out of the nursery paddies (doing *naetori*) owned by a different family. My wife helped her brother's family plant rice.

Today's break-time snacks: red bean rice (*azuki meshi*), a simmered *nishime* dish with sea oak (*arame*) stems and *konnyaku*, a *nishime* dish with carp and butterbur stems, dried daikon strips cooked in soy sauce, etc. And of course, there was saké. The family just started planting their paddies today. I ate lunch and supper at my own home, as usual.

> At the main house: Kanezō cut weeds and such in the early morning, and later in the morning he and Ichisuke took the horse to my brother's wife's natal home and did the final horse plowing (*ta o kaku*). In the afternoon Kanezō took the horse up into the hills to cut hay for it. He returned with two big bundles (*maro*). My brother's wife went to help plant rice in the paddies of another household. My brother managed the irrigation, Ichisuke watched the shop in the afternoon, and my mother cooked.

June 6, Thursday

Cloudy, sometimes sunny
Northwest wind
15°C in the morning, 20°C in the afternoon

Breakfast:	miso soup with bracken in it, *takuan*, miso-preserved daikon
Lunch:	*takuan, misozuke*
Supper:	same as morning

I lifted the glass panels on my box watermelons. It's pretty much the same for these every day, so I won't write about the watermelons daily from tomorrow. I also cut grass, etc., for the rabbits daily. Yet again I spent the day pulling rice seedlings from nursery paddies. My wife complained of sore fingers from planting in the paddies, so she rested today. This condition is called *yubisorade* here in the village. But she stayed at her natal home all day and didn't come back so I had to handle the cooking as well as everything else, which made for a very busy day. Today's break-time snacks were nearly the same as yesterday.

Today's expenditures:

sweets	10 sen

> At the main house: Kanezō cut hay for feeding the horse early, and through the rest of the morning as well. In the afternoon he went out to cut grass and weeds and stuff that people here call *bukkomi-gusa* to be thrown into the horse shed for making compost (see Fig. 4.3). My brother's wife went to help another family plant their paddies as *e* return. Ichisuke watched the shop and my brother worked on building a chicken coop. My mother cooked and such.

June 7, Friday

Sunny, rain in the evening
Northwest wind
15°C in the morning, 26°C in the afternoon

Figure 4.3. A horse loaded with *bukkomi-gusa* in front of Yoshida Kinjirō's house. A whetstone, a scythe, and a water bucket (*tomejo oke*) and such are also attached. [Yoshida Kin'etsu, of the big main house, is leading the horse.]

Breakfast, lunch, and supper: same as always

I went to do *naetori* again today. My wife went out to help plant paddies.

Today's break-time snacks: saké, a *nishime* dish with *konnyaku* and bamboo shoots, a *debu* dish made with dried daikon strips, roasted carp, and boiled radishes with soy sauce, and rice. (The afternoon snacks were basically the same, but there was a *nishime* dish with sardines.)

At the main house: Kanezō didn't start very early today, but he cut hay for the horse in the morning, and *bukkomi-gusa* in the afternoon. My brother's wife went to return *e* again by helping to plant paddies. Ichisuke watched the shop, and my brother worked on the chicken coop. My mother cooked.

June 8, Saturday

Sunny, sometimes cloudy
Northwest wind
16°C in the morning, 18°C in the afternoon

Breakfast, lunch, and supper: same as always

A reporter from the Hōchi newspaper[9] came today.

I spent the morning doing *naetori,* but not the afternoon because I didn't feel so good. I just tended to my own fields instead. My wife was home today, too. She weeded the cucumbers, among other things. In the afternoon I applied the first follow-up fertilizer to the tomatoes. I dug little trenches around the base of each plant and into all of those I poured a 4 *ka* 1:1 mix of night soil and water in which I'd dissolved about 400 *momme* (1.5 kg) of ammonium sulfate. Then I covered everything up. Next, I dissolved 40 *momme* (150 g) of ammonium sulfate in 8 *shō* (14.4 L) of water and sprinkled that over the autumn cabbage seedlings I planted in a temporary spot the other day.

Today's expenditures:

brown sugar	5 sen
candies	5 sen

At the main house: Kanezō cut hay for the horse early and the rest of the day he worked on shoring up the walking ridges between the nursery paddies (*nashiro*) with fresh mud (doing *kuronuri*). Here in this village, once all seedlings have been removed from the nursery paddies, people customarily scoop away the surfaces of the ridges and then rebuild them with fresh mud. They pull out all the weeds and bury them in the new ridges. But this year I've noticed that some people aren't even rebuilding their ridges—they're weeding but not burying the weeds inside.

My brother's wife went again to return *e* by helping another family plant their paddies. Ichisuke watched the shop, and my brother went out to do *anajashi*[10] in the morning. To do *anajashi* to the paddies means to replant rice seedlings where they've vanished or come loose for some reason. In other words, you look around for *kabuhajire*[11] and replant. In the afternoon he completed the chicken coop he's been building. My mother went up into the hills to pick bamboo shoots in the afternoon.

This evening I went to check on the seedlings I'd left sitting in the *nashiro* in the morning and about 40 bundles were missing—stolen. Yoshida Chūzō and seven others were doing *naetori* in the next paddies, so I asked them if they knew anything about that, but they said they didn't. My, my—it is rather odd that people who were working right on the other side of the ridges wouldn't know anything about such a matter, isn't it? The theft must have occurred between two o'clock and five o'clock this afternoon.

June 9, Sunday

Sunny, sometimes cloudy
15°C in the morning, 20°C in the afternoon

Today my wife and I both helped other households again. She planted rice. In the morning I rolled a wooden *kata* (through wet paddies to make planting lines) and did some planting, too. In the afternoon I did more *naetori.* I cut grass for the rabbits in the evening.

Breakfast and supper:	We ate at home.
Lunch:	We ate at other people's homes.

Breakfast:	miso soup with buckwheat leaves in it, *takuan,* miso-preserved eggplant, etc. The buckwheat leaves were from plants that just sprang up in my fields from seeds dropped there last autumn.
Lunch:	red bean rice (*sekihan*), miso soup with butterbur and *wakame* (stock made from dried, roasted anchovies), a *nishime* dish of butterbur, dried herring, and sweet potatoes, etc. Also: strips of dried daikon cooked in soy sauce, red beans cooked in brown sugar, a fish [locally] called *agai* [probably sea bass] caught on the shore of the Oga Peninsula cooked in soy sauce, *takuan,* saké, etc.
Supper:	We cooked carp that we received from the crazy man next door in soy sauce and ate that. Also: cucumbers from the main house that were pickled in malted rice, etc.

I started hanging mosquito nets (*kaya*) in the house tonight because there are a lot of mosquitoes buzzing about now. They're not yet such a problem (down below) in the village.

At the main house: Kanezō went out to cut hay for the horse early, and the rest of the day he worked in another household's paddies—chopping up the soil with a hoe in spots that couldn't be tilled by a horse-drawn plow (doing *kadzugu*). My brother's wife helped with the rice planting in another family's paddies. My mother cooked.

Maybe it's because deer used to live in Oga—I'm not sure—but today two fellows who spoke in a way I couldn't understand very well were walking around the village looking for deer horns to buy. They had a bunch strapped onto their backs. Seeing that made me realize that there really were a lot of deer in Oga in the past.[12]

June 10, Monday

Sunny, cloudy later, then light rain
18°C in the morning, 21°C in the afternoon

Breakfast:	*okayu* (rice gruel), cucumbers pickled in malted rice, *takuan,* miso-preserved eggplant, etc.
Lunch:	cooked rice with hot water poured over it (*yumeshi*), all else: same as morning
Supper:	soup made from leftover carp from last night, all else: same as morning

Today I left the glass panels off the watermelons I'm growing in boxes, but only during the daytime. And even though I caught a cold and didn't feel too good, I pushed myself to work in my fields.

I did the third fertilizing on my summer cabbages—dug a little hole between each plant and poured a mix of night soil, water, and ammonium sulfate into each (see Fig. 4.4). I used a total of 3 *ka* each of night soil and water, in which I dissolved 600 *momme* (2.3 kg) of ammonium sulfate. I also did the second fertilizing to the cucumbers. I dug two shallow holes near the base of each plant—one on either side—and poured in water (2 *to* (36.1 L) in total) with 200 *momme* (750 g) of ammonium sulfate in it. While doing that, I thinned out the cucumbers for the second time, leaving them planted in pairs (*nihondate*). In the afternoon I covered all the holes I'd dug for fertilizing with my hands. Later I pulled off all the little stems that had sprouted at the bases of the branches of the tomato plants. They're just in one row, so I won't get such great results if I don't pluck all the stems. Once I'd done all that, I lugged the night soil from my house's toilet to the cesspit and did other such tasks. Again today, my wife helped another household plant rice.

At the main house: Kanezō went out to cut hay for the horse in the early morning, and the rest of the day he planted paddies owned by another family—to return *e*. My brother's wife went up into the hills to collect butterbur. My brother managed the irrigation, and Ichisuke watched the shop. My mother cooked.

I wrote before about rice seedlings I'd picked getting stolen. Well, when I confronted Yoshida Chūzō and the other six guys who were doing *naetori* just on the other side of the ridge, he crudely said, "Who the hell do you think would steal seedlings from someone like you?"[13] In local talk, the expression he used for "someone like you" (*nga baraida mono*) is pretty insulting; it shows that

Figure 4.4. Some of the author's cabbages. His eldest daughter, Fusae, is in the foreground. Flooded paddies are visible in the background. From Yoshida's personal collection: not included in 1938 book.

the speaker is really looking down on someone. This guy has some property—some wealth—so he always looks down on others and says nasty things. Today, a man of the household whose paddies Chūzō was working in that day showed up with a lot of seedling bunches and said, "Your seedlings were brought to my house the other day, so I'm returning 40 bunches to you now. Please accept them."

As this case shows, seedlings are scarce right now, so people's moral values are getting weak. Even if asked for extra seedlings by needy relatives, they instead turn around and sell them to others. Such cold-hearted behavior really makes me sad.[14]

June 11, Tuesday

Occasional rain
Strong southwest wind, later out of the northwest
17°C in the morning, 11°C in the afternoon[15]

Breakfast: same as always

Today I went to help do *naetori* at my wife's brother's place. My wife went to help plant rice again. My *naetori* work ended today, as did my wife's rice planting.

For lunch today we had miso soup with sea oak (*aosa*) and carp in it, roasted *hatahata* that had been preserved in salt, *nira* [garlic chives]—boiled a bit and eaten with soy sauce—a *nishime* dish made with potatoes, butterbur and carp, *takuan,* etc.

For supper we had some saké, miso soup with *wakame,* roasted flying fish from Oga, more *nira* that was dipped in boiling water and eaten with soy sauce, herring roe that we ate with soy sauce, pond snails (*tanishi*) cooked with dried daikon strips in soy sauce, etc.

Since the *naetori* work finished today, I think I'll try to explain how to tie them into bunches. First, you hold the seedlings you picked in your left hand. Then you pull one strand of straw from your waist, where you're carrying them. Then you press the base of the straw—leaving about 1 *sun* (3 cm) sticking out [with your left thumb pressed against the seedlings] and grab the tip of that straw with your right hand and wrap it around and over your thumb tightly. The second time around, you don't go over your thumb but instead go under it so you can hold it down with your thumb along with the piece you left sticking out and pull it nice and tight. So, you've double-tied one bundle of rice seedlings with only one straw. Then, as you pull your left hand away you insert the end of the straw through the loop you've made, and then with your right hand you pull the short piece you were holding with your left thumb, and you're done.

At the main house: Kanezō went out to cut hay for the horse, and later in the morning he put *katchiki*[16] into the nursery paddies (*nashiro*). In the old days, people would cut a bunch of young branches of trees known as *ganja* and put them in the *nashiro,* but today people normally use straw or *higusa* (dried grass, etc.). But some people still cut *ganja* branches and use them. Today at the main house they put dead grass for feeding the horse that was left over from wintertime into the *nashiro* as *katchiki.*

The scientific name for *ganja* trees is *tani-utsugi.*[17] In the afternoon Kanezō went to cut *bukkomi-gusa,* and my brother's wife spent the day minding the store and cooking. My brother managed the irrigation, and my mother rested because she was sick.

June 12, Wednesday

Cloudy and rainy, strong north wind
12°C in the morning

Breakfast, lunch, supper: same as always

I finished plucking the little stems near the branches of my tomato plants. I also finished planting the autumn cabbages I'd started planting the other day, in a temporary spot. And I replanted some eggplants that had died, and I did the second follow-up fertilizing for my green onions—400 *momme* (1.5 kg) of ammonium sulfate dissolved in 2 *ka* of water, sprinkled on the onions. My wife weeded the green onion patches, picked butterbur in the hills, and such.

Today's expenditures:

haircuts for the kids[18]	10 sen
sweets for the kids	10 sen

At the main house: Kanezō slept in a bit, but later in the morning he went out to cut hay for the horse. In the afternoon he prepared a spot—a *kodzuke*—for piling up compost (*taihi*). My brother's wife planted soybeans along the ridges between the *nashiro* and cooked.

Today a notice made the rounds in the village—from tomorrow, the thirteenth, we'll have a five-day *sanaburi*[19] rest period.

June 13, Thursday

Sunny, later cloudy
Strong north wind, dying later
11°C in the morning, 18°C in the afternoon

Breakfast:	miso soup with bamboo shoots and sea oak (*aosa*) in it, *takuan,* miso-preserved daikon
Lunch:	roasted salted herring, *takuan*

Supper: miso soup with daikon leaves in it, main and side dishes: same as lunch

During the morning, I pulled soil up over the roots of the potato plans with a hoe for the second time (did the second *tuchiyose*). In the afternoon I harvested peas with garden shears—the first time this year. My wife picked green caterpillars off the cabbages and weeded, etc., in the morning. During the afternoon she did various house chores, and in the evening, she went out to sell the peas.

Today's income:

peas (250 *momme* (937.5 g))	25 sen

Today's expenditures:

stamps and postcards	40 sen
fee to mail photo negatives	10 sen
sweets for the kids	5 sen

At the main house: Kanezō went out to cut hay for the horse early, and later in the morning he cut *bukkomi-gusa*. In the afternoon he gathered two bundles of straw (*wara*) and put them in the storehouse. Then he rested. My brother's wife helped another family plant rice, to return *e*. My mother was sick, but she pushed herself to cook and do other house chores.

Today the people of the village were hauling soil and sand and dumping it in the spot where a new common work shed (*kyōdō sagyōjō*) will be built. Every household sent a person, or a horse—or both—for that.

June 14, Friday

Sunny
12°C in the morning, 20°C in the afternoon

Breakfast, lunch, supper: same as yesterday

Someone appeared and asked me to help pick seedlings from their nursery paddies (do *naetori*) so I did that. My wife went toward Funagawa to sell eggplant seedlings and peas.

Today's income:

peas	25 sen
eggplant seedlings	40 sen

At the main house: Kanezō went out to cut hay for the horse early. Then he used the horse to carry soil and sand to the common work shed spot. He rested in the afternoon. My brother's wife minded the shop and cooked, and my brother went out to check the rice paddies.

June 15, Saturday

Cloudy, but also sunny
Northwest wind
17°C in the morning, 20°C in the afternoon

Breakfast:	miso soup with sea oak and taro stems in it, *takuan,* sesame miso, etc.
Lunch:	buckwheat plant stems dipped in miso, *takuan,* etc.
Supper:	same as morning, plus roasted salted herring

The husband of my younger sister, who lives in Tokyo, came over to visit.

I spent the day trying to make a round basket[20] with bamboo strips. I did make a small one, but it didn't go well because it was my first time. My wife went to help another family plant rice.

At the main house: Kanezō went out to cut hay for the horse early and spent the rest of the day cutting *bukkomi-gusa.* My brother's wife pulled soil up over the roots of the potato plants in the field with a hoe for the second time (did the second *tsuchiyose*); my mother cooked. Today my younger sister's husband [and family] was visiting from Tokyo so there was a lot going on over there.

June 16, Sunday

Cloudy, sometimes sunny
Weak southwest wind
18°C in the morning, 22°C in the afternoon

Breakfast:	miso soup with daikon leaves in it, *takuan,* sesame miso, etc.
Lunch:	roasted salted herring, *takuan,* etc.
Supper:	miso soup with daikon leaves in it, *takuan,* etc.

I put follow-up fertilizer on my tomatoes for the second time early this morning. I dug little holes between the plants, diluted 2 *ka* of night soil with the same amount of water, and poured that in. Then I mixed up the soil a bit. Next, I did the third follow-up fertilizing on my cucumber plots and at the same time thinned them out, also for the third time this year—leaving each remaining plant to stand alone (*ippondachi*). As usual, I dug little holes between the plants and poured in one bucket of water in which I'd dissolved 1 *ka* of night soil, and then of course I turned over the soil a little. After that I put follow-up fertilizer on the pumpkins—the first. I dug a circular trough around each plant and poured in a mix of one bucket of water and 1 *ka* of night soil and as I filled in the troughs with soil I spread more soil around each plant in a circular motion with my hand—or *shudo o mawashita*[21] as people say. I also did the third thinning to my pumpkin crop, leaving the plants to stand alone (*ippondachi*).

Then I added arsenate (*hisaen*) and lime casein and such to some Bordeaux mixture, and sprinkled this on my summer cabbages, tomatoes, autumn cabbages, brown potatoes, etc. I used 1 *to* (18 L) of four-*to*-style Bordeaux mixture, 15 *momme* (56.3 g) of arsenate, and 5 *momme* (18.8 g) of lime casein, and I mixed it all up.

At the main house: Kanezō carried barnyard manure (*umayagoe*) to the compost pile and dumped it there early in the morning, and then he took the horse up into the hills to cut and gather hay for the horse to eat—spent the rest of the morning on that. He took off in the afternoon. My brother's wife went out to pick weeds in the main house's rice paddies. My mother cooked and such.

June 17, Monday

Cloudy, rain later
Southwest wind
17°C in the morning, 25°C in the afternoon

Breakfast:	miso soup with daikon leaves, *takuan*, sesame miso, etc.
Lunch:	*udon* (soup made from dried roasted anchovies), a *nishime* dish with *konnyaku* and bamboo shoots (*takenoko*), etc.
Supper:	*mochi* made with regular rice (*meshi mochi*) covered with red bean paste [*o-hagi*], dried daikon strips [*kiriboshi daikon*] cooked (in soy sauce), etc.

This morning, I pulled soil up over the roots of my tomato plants with a hoe (did *tsuchiyose*) and did the same to the cucumbers. I took it easy in the afternoon, and my wife cooked. In the afternoon, since it's *sanaburi* time, the villagers gathered at the shrine and ate and drank. The older women went to the temple, and the younger women and the men had a wild time and made a bunch of noise at the shrine. No old men were there, though.

Today's expenditures:

brown and white sugar	8 sen
soy sauce	3 sen
udon noodles	10 sen
konnyaku	6 sen

At the main house: Kanezō went out to cut hay for the horse early, and then he took the horse up into the hills to cut *bukkomi-gusa,* and he spent the rest of the morning doing that. My brother's wife and my mother cooked. In the afternoon, everyone but my brother went out for *sanaburi*—a kind of party held in celebration of the completion of the rice planting.

June 18, Tuesday

Cloudy but also sunny
Strong southwest wind
18°C in the morning, 18°C in the afternoon

Breakfast:	*okayu* with sesame miso, miso-preserved cucumbers, *takuan,* etc.
Lunch:	salted roasted herring, *takuan,* a dish with flying fish, bamboo shoots (*takenoko*), and *konnyaku* cooked together in soy sauce, etc.
Supper:	miso soup with bamboo shoots and *konnyaku,* salted herring, *takuan,* etc.

Today I replanted eggplants that had withered and died, and replanted some cabbages. I poured about 6 *shō* (10.8 L) of a roughly 1:1 mix of water and night soil on my autumn cabbages, which I'd planted in a temporary spot. And I stuck some poles in the ground for some of my tomatoes and tied them up. Also, I did the second thinning on my burdock crop, leaving them to stand in pairs (*nihondachi*). I also made a round sieve like I did the other day, and I picked some daikon for selling tomorrow, and washed them and bundled them in pairs. This was the first time to harvest daikon this year. In the morning, my wife washed clothing, cooked butterbur, and picked peas, which she went out to sell in the afternoon.

Today's income:

peas	45 sen
daikon	24 sen

At the main house: Kanezō went out early to cut hay for the horse, and the rest of the day he cut *bukkomi-gusa,* using the horse (to haul it). In the morning, he cut one horse load (1 *da*), and cut the same amount in the afternoon. My brother's wife weeded the main house's paddies again. My brother went out to check the paddies, and Ichisuke minded the shop. My mother cooked. My younger sister and her family headed back to Tokyo this morning. My mother looked a bit sad.

June 19, Wednesday

Cloudy
11°C in the morning, 16°C in the afternoon

Breakfast:	miso soup containing daikon and burdock root from thinning the crop yesterday, *takuan,* sesame miso, etc.
Lunch:	roasted flying fish, *takuan,* etc. The fish was from Oga. I bought it yesterday and packed it in salt.
Supper:	same soup as morning, side and main dishes: same as lunch

I took the *noma* mats and the glass covers off my box watermelons last night. I'm just leaving them exposed now. They've gotten pretty big. The eggplants are starting to flower—three days earlier than last year.

Today my wife went out very early to sell vegetables, so I cooked breakfast and took care of the kids. During the morning, I stuck poles in the ground to support the tomato plants, which I tied to the poles with pieces of straw. I also put up some small supports for my cucumbers. After preparing lunch, I went out to cut grass and weeds for making compost. My wife had returned by lunchtime, so she weeded the eggplant beds and did some other things in the afternoon.

Today's income:

peas	15 sen
daikon	55 sen

At the main house: They started weeding the rice paddies today (see Fig. 4.5). Kanezō cut hay for the horse early but spent the rest of the day weeding the paddies, along with my brother's wife. My brother oversaw the irrigation. Ichisuke minded the shop in the afternoon, and my mother cooked and such. They weeded the Bishamon and Shinden paddies today. Four women and one man worked.

June 20, Thursday

Cloudy, sometimes sunny
West wind
12°C in the morning, 19°C in the afternoon

Breakfast:	miso soup with butterbur in it, *takuan*, preserved plums [*ume-zuke*] (received from the main house)
Lunch:	roasted salted herring, *takuan*
Supper:	basically, the same as morning

I started digging little holes next to my eggplants early—one hole per plant—for doing the first follow-up fertilizing. I diluted 3 *ka* of night soil with 3 *ka* of water and poured in 800 *momme* (3 kg) of ammonium sulfate and used that. But I also used one bucketful of that on my box watermelons—the first follow-up fertilizing for them. I dug little trenches around the base of each plant to do that. In the afternoon I cut more grasses and weeds for compost and mixed soil with that and stomped it down, picked more daikon and peas for selling tomorrow, prepared supper, etc. My wife was helping to weed the main house's paddies today.

Figure 4.5. Weeding in the paddies beneath my house.

At the main house: Kanezō cut hay for the horse in the early morning, and then cut *bukkomi-gusa*. In the afternoon he took the horse out to cut more hay for feeding it. My brother's wife weeded the paddies. Ichisuke minded the shop in the afternoon again, and my brother took care of the irrigation. They finished weeding the paddies at Sanbonmatsu, and also finished the Sakura paddies. This marks the end of the first weeding. Eight people worked today. Yesterday there were five. My mother cooked and such.

June 21, Friday

Sunny
14°C in the morning, 25°C in the afternoon

Breakfast:	miso soup with butterbur, *takuan,* sesame miso
Lunch:	raw buckwheat plant stems dipped in miso, roasted salted herring, *takuan,* etc.

Supper: miso soup containing bamboo shoots. In it we put a kind of seaweed called *kuromo* [*Hydrilla verticillata*] from the coast of Minami Iso in Oga. We got this seaweed from my wife's brother, who apparently received it from someone living over there. We also ate leftover herring from lunch and *takuan*. The stock for the soup we had for supper was made from roasted dried fish from Hokkaidō called *konago*.[22] These we also got from my wife's brother's family. And we ate grated daikon with soy sauce, etc.

I weeded my soybean crop in the early morning today, started making breakfast from six o'clock, and then thinned out the soybean crop starting at eight. When I planted them, I did so in pairs, so to thin the crop I pulled out one of the sprouts as I went along leaving the remaining plants to stand alone (*ippondachi*). Right after finishing that I pulled soil up over their roots for the first time (did the first *tsuchiyose*). Ideally, in a perfectly level field, one should pull soil up from both sides. But my soybean beds are on a steep slope, so pulling soil up from below is very hard. Therefore, I pulled soil only from above to cover the roots. After that I prepared lunch. In the afternoon I filled in the fertilizer holes I dug next to my eggplants yesterday, and picked daikon and snow peas and such for selling tomorrow. I washed the daikon and cut off part of the leafy tops and stripped off the little roots that run out of the daikon. Then I bundled them in pairs. We'll sell three of these bundled pairs for 5 sen. I gave the leaves I'd cut off to the rabbits. Up until yesterday we sold 100 *momme* (375 g) of peas for 10 sen, but the price has really dropped—today 1 *shō* (1.8 L) sold for 10 sen. My wife went out very early and came back by lunchtime, and in the afternoon, she weeded the eggplants, etc.

Today's income:

daikon	60 sen
peas	28 sen

Today's expenditures:

snacks for the kids	10 sen

At the main house: Kanezō went out and cut hay for the horse early and then spent the rest of the day cutting *bukkomi-gusa*. In the evening he cleared some manure compost from the horse stall, and my brother's wife went to weed another family's paddies to return *e*. My mother cooked. My brother went out to check the paddies in the afternoon, and Kieko (my brother's daughter) minded the shop.

Yoshida Shigezō, whose family runs a rice store, bought about 1 *chō* 3 *dan* (12,892 km^2) of paddy land from the owner, who lives in Tsuchizaki.[23] It's land that Takakuwa Yasuke has been farming as a tenant. So, Shigezō declared that he was going to farm the land himself. But Yasuke, determined to fight to the end if necessary, asserted his right as a long-term tenant to keep farming the land, and rejected Shigezō's move. The dispute, which dragged on for several months, has finally been settled. I'll explain how.

Yasuke received 1,150 yen [from Shigezō], and then his debts of about 80 yen to the rice shop, and of about 20 *tawara* (1,443.2 L) worth of rice that he owed to the landlord were both canceled, and Yasuke quit farming. Then, the land was granted to Shigezō's second son, Yoshida Yasuke, for him to farm.

Also, Yoshida Shigezō took back some rice paddy land he'd been letting others farm for himself, and then allowed the tenants to use different paddy land. The tenants' names are Yoshida Sangorō and Takase Takezō.

The paddy land Yoshida Shigezō let Takase Takezō farm, which his son Yasuke had been farming, is located in a different village. The land he let Sangorō farm—an area of 4 *tanbu* (4 km^2)—is one that Yoshida Minokichi has been farming as a tenant. Shigezō bought that from someone and granted Sangorō the rights to farm it. For Minokichi's tenant tights, he got to use 1 *tanbu* (992 m^2) of the 5 *tanbu* (5 km^2) parcel of paddy land (to grow rice for his own household) plus about 150 yen in cash, and he quit farming for a living.

June 22, Saturday

Rain
18°C in the morning, 20°C in the afternoon

Since it rained from morning, I didn't work in the fields.

Breakfast:	We put a seaweed called *tsurumo,* that we received when we got *kuromo,* in miso soup that had bamboo shoots and butterbur in it and ate that. Generally, *tsurumo* is chopped up fine and eaten in soup. We also had *takuan* and grated daikon with soy sauce on it.
Lunch:	salt-preserved buckwheat plant stems, salted krill, *takuan,* etc.
Supper:	miso soup made with *konago* [fish] stock with bamboo shoots, butterbur, and peas, etc. We put *kuromo* in that and ate it. We also put salted krill in grated daikon and ate that. Also: *takuan,* preserved buckwheat stems, etc.

These days, the four of us are consuming 9 *gō* (1.6 L) of polished rice and 4 *gō* (721.6 ml) of rolled oats [mixed] per day, for a total of 1 *shō* 3 *gō* (2.3 L).

I went to the main house to grind buckwheat seeds in a mortar (*fugushi ishi*) this morning. If you pour buckwheat seeds in the hole in the stone while grinding, the fine powder will come out all mixed up with bits of chaff. Then you sift that with a sieve called a *konaroshi* and only the powder will be left. I ground 3 *shō* (5.4 L) of seeds and got 1.5 *shō* (2.7 L) of powder. In the afternoon, my wife made rope called *seoinawa* to use for carrying a basket on one's back. To make it, she used a thick string called a *hosobiki* you can buy at a store. We already had that in the house—left over from something before—so she used it. After that she picked vegetables to sell. It takes about 220 pea pods to fill a 1 *shō* (1.8 L) container. Weight: 130 *momme* (487.5 g). To fill a 100 *momme* (375 g) container, you'd need about 180 pea pods.

Figure 4.6. Knocking leaf beetles off the rice plants with a bamboo pole in paddies beneath my house.

The taros in my field have pretty much all begun to sprout—just today. My wife sewed some in the morning and went out to sell vegetables in the afternoon.

Today's income:

peas	13 sen
daikon	10 sen

At the main house: Kanezō went out early to cut hay for the horse. Later in the morning he went out to the paddies at Sanbonmatsu to knock *doro-oi mushi* (*katsugi mushi*) [leaf beetles] off the rice plants. He walked along, holding a long bamboo pole with some leafy bamboo branches on the end of it, waving the pole around to brush the bugs off the plants (see Fig. 4.6). This makes the bugs fall off into the mud. But even if you do this, they'll soon climb back up onto the plants. However, there appears to be a new invention on the horizon—something that will catch the bugs, not only knock them off. In the afternoon Kanezō rested. My brother's wife rested in the morning, but she weeded the dry fields and such in the afternoon. My mother cooked, etc.

June 23, Sunday

Sunny
West wind
16°C in the morning, 18°C in the afternoon

Breakfast:	bamboo shoot miso soup with *tsurumo* in it, *takuan,* preserved buckwheat plant stems, etc.
Lunch:	roasted salted herring, *takuan,* preserved buckwheat stems, etc.
Supper:	miso soup with bamboo shoots and *kuromo* in it, side dishes and such: same as lunch

I prepared breakfast this morning. After eating, I went to the market at Nagane to buy *jikatabi*[24] and secondhand shirts. In the afternoon I prepared a package to send to an acquaintance in Tokyo and picked vegetables for selling tomorrow. My wife went out early this morning, returning by lunchtime. In the afternoon she weeded the Chinese yam beds.

Today's income:

daikon	60 sen
peas	42 sen

Today's expenditures:

jikatabi for my wife	75 sen (size: 9.7 *mon* (23.3 cm), brand: Tsurigane Jirushi)
2 secondhand shirts	40 sen (for me)
snacks for the kids	14 sen

At the main house: Kanezō went out early to cut hay for the horse. Then, both in the morning and in the afternoon, he took the horse up into the hills to cut *bukkomi-gusa.* My brother's wife went to help weed another family's paddies—to return *e.* My mother cooked and such. My brother attended a household heads' association (*koshukai*) meeting. There was a need to discuss problems relating to the communal work shed (*kyōdō sagyōjō*) and the granary (*gōgura*).

June 24, Monday

Sunny, sometimes cloudy
14°C in the morning, 20°C in the afternoon

Breakfast:	bamboo shoot miso soup, grated daikon with soy sauce poured on it, *takuan,* preserved buckwheat plant stems
Lunch:	roasted salted herring, side dishes: same as morning
Supper:	daikon miso soup, salted herring, *takuan,* etc.

I cooked breakfast this morning. After breakfast I did the first follow-up fertilizing on my Sadaya medium-ripening (*chūsei*) cabbages. As usual, I dug a little hole next to each and poured a mixture of night soil, water, and ammonium sulfate, etc., into the holes. The amounts: 1 *ka* each of night soil and water, and 400 *momme* (1.5 kg) of ammonium sulfate, and such. Later I cut grass and weeds for making compost. In the afternoon, I filled in the holes I'd dug next to the cabbages, cleaned out the rabbit hutch, and turned over the compost heap. Then I picked vegetables to sell. After that I cut some unwanted branches off my eggplant plants, etc. My wife got home at about ten o'clock this morning, and in the afternoon, she weeded the dry fields, etc.

Today's income:

daikon	60 sen
peas	36 sen

Today's expenditures:

postage on package to Tokyo	75 sen
fish for my mother since she got sick [a *mimai* gift]	10 sen

At the main house: Kanezō cut hay for the horse early and cut *bukkomi-gusa* the rest of the day. My brother's wife stayed home and minded the shop. My brother attended an Agricultural Support Association (Shinōkai)[25] meeting. He learned how to make fertilizer—particularly compost.

June 25, Tuesday

Cloudy, sometimes sunny
Northwest wind
18°C in the morning, 21°C in the afternoon

Breakfast:	miso soup with bamboo shoots in it, *takuan,* miso-preserved cucumbers, etc.
Lunch:	grated daikon with soy sauce on it, *takuan,* etc.
Supper:	bamboo shoot miso soup (stock made from *konago* fish) with *tsurumo* in it, roasted salted herring, *takuan,* etc.
Snack:	*shitanami* [small turban shells]

I cut more unwanted branches from my eggplant plants early. I left the leaves, but I removed all branches except for three at the top (see Fig. 4.7). In the old days we didn't do this, but due to advances in plant science we've learned that this should be done, so I'm doing it.

I made breakfast at seven o'clock, and we finished eating at eight-thirty. After I woke the kids (and fed them) and folded up the *futon*[26] and such, I headed down to the Wakimoto Station area on an errand, on which I spent the morning. In the afternoon, I went to buy grass and stuff for composting, and picked some vegetables to sell. I also cooked both lunch and supper. My wife went out early today to sell vegetables, and then she went to see her cousin in Minami Iso Village, where she collected some *shitanami* shellfish, *tsurumo* and *kuromo* (seaweed), and such, and then she came home in the evening.

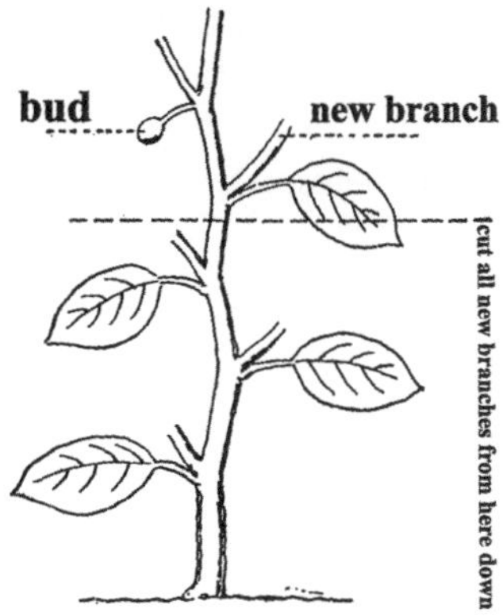

Figure 4.7. How to properly prune eggplant plants.

Today's income:

daikon	60 sen
peas	40 sen

Today's expenditures:

kerosene	10 sen
candies for the kids	10 sen

At the main house: Kanezō cut hay for the horse in the early morning, and later in the morning he cut *bukkomi-gusa*. He spent the afternoon making compost with my brother. To do this, you spread some *bukkomi-gusa* around on the floor of the stable and then after the horse has had some time to stamp it down you take it out to the compost pile and mix grass and soil with it and pile it up. You pour water on it daily for four or five days, and then you pull some more straw that has been well stamped from the stable (*umayagoe*). You then move the pile you've been watering to another spot and pile it up there. Then the next time you pull stamped straw from the stable you move the stuff you've piled up again. This is known as *honzumi,* and it makes real good "true" compost.

This way I just explained, it seems, is the "official" way of making compost, according to the latest scientific research. But it wasn't this way in the old days. I'll explain how we did it when I was young. First, we'd go up into the hills and cut some short grasses and straws and bundle them tightly. This is called *hako marogi* [box-bundling]—probably because you are tying what you cut in box-shaped, squarish bundles. We'd stand all the straws and grasses on end, and bundle them together tightly, about 2 *shaku* or 2 *shaku* 5 *sun* tall and about 2 *shaku* wide (60.6–75.8 cm by 60.6 cm). To tie them up we'd make a thin rope by winding four straws together and place a bunch of these straws and grasses—called *kotsuta-gusa*—held tightly together with both hands on the

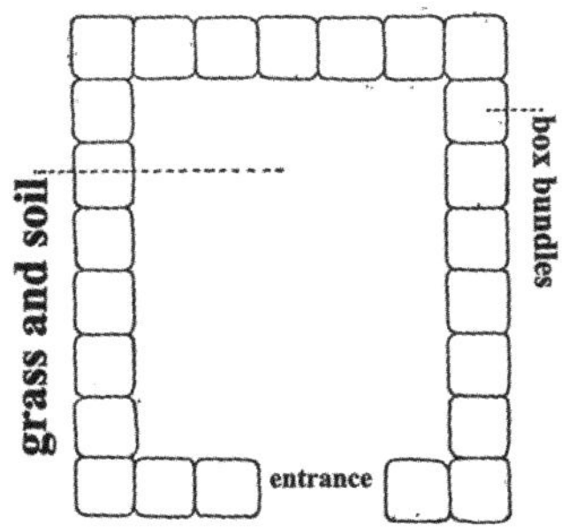

Figure 4.8. A box-bundle enclosure.

rope so that the rope lay under the middle point of the bundle and tie it tightly. Normally, people would use a hoe to lift one of these bundles.

Three or four of these *kotsuka-gusa* would be tied together to make one *hako-marogi* (box bundle), and we'd take them back home that way—three on each side of the animal and one on top. A person would then take these box bundles to the compost pit (*kotsuge*) and place them all around it (see Fig. 4.8). Inside this enclosure, we'd put straw from the horse stable (*umayagoe*), soil, and *bukkomi-gusa*. We'd start with straw, then put about 5 *sun* (15.2 cm) of soil on top, and then lay bundles of straw down all the way across, one after another, and then more soil, and the pile would just get taller and taller. And daily, we'd pour horse pee or water on this pile, which would eventually reach a height of 6 or 7 *shaku* (approx. 2 m), or maybe even 1 *jō* (3 m). This is how we used to make compost.

Today my brother's wife spent the day weeding the rice paddies of another household—to return *e*. My mother cooked and such.

June 26, Wednesday

Sunny
Northwest wind
16°C in the morning, 22°C in the afternoon

Breakfast:	bamboo shoot miso soup with *kuromo* in it, *takuan*, etc.
Lunch:	*maguro* (tuna) I got from someone I know in Funagawa,[27] cooked with bamboo shoots in soy sauce
Supper:	daikon miso soup, leftover fish from lunch, *shikoi* made from *kuromo*, etc. The *shikoi* was made by pouring boiling water on *kuromo* and mixing it with grated daikon and miso, etc.
Snack:	*shitanami* (a kind of shellfish from the sea)

I stuck some big bamboo poles in the ground to support my cucumbers early. Then I made breakfast. Later I cut unneeded stems from the eggplants and plowed their beds a bit (did *chūkō*). In the afternoon I cut compost straw and picked more vegetables for selling. My wife weeded the pumpkin crop starting in the afternoon.

Today's income:

daikon	50 sen
peas	20 sen

Today's expenditures:

soy sauce (3 *gō* (541.2 ml))	9 sen
snacks for the kids	5 sen

At the main house: Kanezō cut hay for the horse early, and cut *bukkomi-gusa* the rest of the morning and in the afternoon. He always waters the compost pile (*kotsuge*) in the early evening. My brother's wife was in the fields all day. She covered the potato plants' roots with soil for the second time (did the second *tsuchiyose*), weeded the Chinese yam and burdock beds, and saw to the needs of the Chinese yam plants. My mother cooked.

June 27, Thursday

Sunny
Southwest wind
18°C in the morning, 30°C in the afternoon

Breakfast:	exactly the same as yesterday morning
Lunch:	mackerel cooked in soy sauce, *takuan,* etc. The mackerel was from Minami Iso Village in Oga.
Supper:	miso soup with daikon and mackerel head cooked in it, *kuromo shikoi, takuan,* salted mackerel, etc.

In the morning I dug a hole near the spot where we usually dump water. This produces *tsuchi hiryō* [soil compost]. If you cut the grass you've dug up at the water dumping spot and shove it in the hole, in a year you'll have some real good compost. After making breakfast I opened the lids of my box watermelons and picked extra growth off my tomato plants while tying them to support poles. In the afternoon I weeded my wheat crop and expanded my compost making spot. I also picked vegetables for selling. I was pretty busy all day. My wife returned before noon. She weeded the pumpkin fields and such, and washed her own laundry, etc., in the afternoon.

Today's income:

daikon	60 sen
peas	5 sen

Today's expenditures:

used kimonos for the kids	15 sen
2 mackerel	12 sen

At the main house: Kanezō cut hay for the horse early, and he took the horse to cut more after breakfast. In the afternoon he cut *bukkomi-gusa*. My brother and his wife made sweets [*wagashi*] in the morning, and in the afternoon my brother went out to see to the paddy irrigation. His wife went out to the dry fields and covered the roots of the soybean plants with more soil for the first time this season (did the first *tsuchiyose*)—she planted a little first and then scooped soil onto the plants' roots with her hands. My mother was a bit sick, but she cooked and such anyway.

June 28, Friday

Cloudy, a light rain later
Strong south wind, changing to a strong southwest wind later
23°C in the morning, 21°C in the afternoon

Breakfast:	We put *tsurumo* into bamboo shoot miso soup, and ate *takuan*, etc.
Lunch:	roasted mackerel that had been packed in salt, *takuan*, grated daikon with soy sauce
Supper:	We put *tsurumo* into miso soup made with *konago* fish stock, which had bamboo shoots in it. Also: roasted salted mackerel, *takuan*

Between ten and twenty members of the Ishikawa Association came from Toyokawa Village[28] today.

I was up and ready to work early, waiting for the sun to rise, because it really looked like rain; it wouldn't have been good if I hadn't been able to get everything done. First, I dug little holes in my eggplant fields—on the opposite side of each plant as before—and poured in a mixture of 2 *ka* of night soil diluted 1:1 with water, with 600 *momme* (2.3 kg) of ammonium sulfate dissolved in it. I finished this, and made breakfast. Then I covered the eggplants' roots with soil (did *tsuchiyose*). I worked as hard as I could without a break on that, even as rain fell, and finished covering the roots by 1:00 p.m. The wind was strong even from the morning, so many eggplants were snapping, creating extra work for me. I just ate lunch standing up and then went back to work in wet clothes—putting up poles to support some of the eggplants and tying them up. All this kept me very busy. In the evening, I tried to fix the wooden washtub. The band that goes around it, holding it together, had come off, so I split a piece of bamboo and worked with that, but it just didn't go well at all. My wife returned by noon, and weeded the onion beds and such in the afternoon.

Today's income:

daikon	60 sen
peas	20 sen

At the main house: Kanezō did exactly what he did yesterday. My brother's wife went to help weed the paddies of the big main house to return *e*. My brother turned over the compost pile in the afternoon, and my mother cooked and such.

June 29, Saturday

Rain
South wind
20°C in the morning, 20°C in the afternoon

Breakfast:	exactly the same as yesterday
Lunch:	roasted salted mackerel, *takuan*
Supper:	We put *tsurumo* in miso soup made with *konago* fish stock that also had bamboo shoots in it. Also: roasted salted mackerel, *takuan*, etc.

Rain was falling this morning, so I didn't work in the fields early. And I got wet working in the rain yesterday, so I felt a little sick and couldn't do so much anyway. After breakfast I went to the main house to help transport some polished rice, and I finished fixing the washtub by fitting it with a new band. In the afternoon I spread some straw on the soil around the watermelons that I'm growing in boxes, and weeded the watermelon crop in my field. These tasks I did during breaks in the rain.

Today my wife went to help weed the paddies of the main house, so I made lunch and supper.

The cucumbers in my fields started flowering.

At the main house: Kanezō didn't work in the early morning, but after breakfast he went with my brother's wife to weed paddies. Kanezō weeded, and knocked *katsugi* bugs off the rice plants, and cut weeds that get in the way along the ridges between the paddies, etc. My brother went to make sure the paddies had the right amount of water in them in the afternoon, and Ichisuke minded the shop. Today they weeded the paddies at Bishamon, and the Shinden paddies.

June 30, Sunday

Light rain
Southwest wind
19°C in the morning, 20°C in the afternoon

Breakfast:	daikon miso soup, *takuan*
Lunch:	roasted salted mackerel, *takuan,* etc.
Supper:	daikon miso soup (stock made from *konago* fish), salted mackerel and salted krill mixed together and simmered, *takuan*

I built a new hutch for the rabbits today, despite the rain. I'd been thinking of doing this for about a month, and finally got to it. My wife helped weed the paddies of the main house all day, so I made lunch and supper.

Today's expenditures:

snacks	2 sen
nails	5 sen

At the main house: Kanezō cut hay for the horse early, and he spent the rest of the day making compost. My brother's wife weeded the paddies, and my mother cooked. Ichisuke minded the shop. My brother made compost [with Kanezō] in the morning, and arbitrated in a property boundary dispute between Yoshida Kiyomatsu and Yoshida Toyoji.[29] Paddies weeded today: those at Sanbonmatsu, and the Sakura paddies.

Chapter 5: July 1935

July 1, Monday

(June 1 according to the old calendar)

Rain, later just cloudy
Northwest wind
17°C in the morning, 21°C in the afternoon

Breakfast:	bamboo shoot miso soup, *takuan,* etc.
Lunch:	roasted salted herring, *takuan*
Supper:	*udon* noodles in soup made from *konago* fish and flavored with soy sauce, *takuan,* roasted salted herring

I worked from morning to noon today on the rabbit hutch, and while doing that I made breakfast and such. In the afternoon I rested. Today was June first according to the old lunar calendar, so we ate dried *mochi* we'd made by hanging leftover New Year *mochi* and relaxed. In the evening, as usual, I picked some vegetables for selling. During the afternoon I went to my old school to hear Colonel Tominaga, one of the leaders of the Kōdō-ha Teishintai, speak.[1] My wife went out in the morning to sell vegetables and came back before noon. She rested in the afternoon, and did some sewing. One thing I did this morning was thin out my oriental melon crop, leaving each plant to stand alone (*ippondachi*), and at the same time I plucked off the tops [to improve production].

Today's income:

daikon	60 sen

Today's expenditures:

udon (8 bunches)	20 sen
candies	2 sen

At the main house: Kanezō went to cut hay for the horse early. After breakfast he took the horse to cut more. He rested in the afternoon. My brother's wife weeded the rice paddies in the morning—undone parts from yesterday. That completed the second weeding. My mother rested, and my brother's wife took care of the cooking. She has recently begun to cook breakfast every morning instead of my mother.

July 2, Tuesday

Light rain, later cloudy and then sunny
20°C in the morning, 25°C in the afternoon

Breakfast:	leftover *udon* noodles from yesterday evening, leftover rice, *takuan*, etc.
Lunch:	roasted salted herring, *takuan*, etc.
Supper:	miso soup with daikon in it, *takuan*, salted krill

Despite the morning rain, I hauled compost soil out to the fields for planting autumn cabbages. The spaces between the ridges around my wheat beds is perfect for spreading fertilizer, so I put that soil there. The field area comes to 35 *tsubo* (116 m^2). I brought about 100 *kan* (375 kg) of compost soil, and I mixed in about 5 *shō* (9 L) of an equal mixture of calcium superphosphate and powdered dried anchovy. First, I spread the soil. Then, I put the calcium superphosphate mix on top of that, and then a layer of soil about 2 *sun* 5 *bu* (7.6 cm) deep and planted the cabbage seedlings. I left 1 *shaku* 7 *sun* (51.5 cm) between each plant, and I planted 194 seedlings. This took me from early morning until nine-thirty, so I gave up on cooking breakfast. I just fed dried *mochi* to the kids and then cooked up a combined breakfast/lunch—a brunch. In the afternoon I had to go into the village on an errand. Later I picked vegetables for selling tomorrow. My wife came back before noon, and she cleaned up inside the house in the afternoon. We started selling summer cabbages today—11 days later than last year.

Today's income:

daikon	35 sen
summer cabbages (1 *kan* 300 *momme* (4.9 kg))	45 sen
peas	3 sen

Today's expenditures:

sweets	5 sen

At the main house: Kanezō cut hay for the horse early and spent the rest of the day cutting *bukkomi-gusa* (see Fig. 5.1). My brother's wife went to weed another family's rice paddies to return *e*. My mother cooked and such.

July 3, Wednesday

Rain, later sunny, then cloudy
South wind, later out of the southwest
20°C in the morning, 25°C in the afternoon

Breakfast:	miso soup with peas, *takuan*, etc.
Lunch:	roasted salted herring, *takuan*, etc.

Figure 5.1. Yoshida Kin'etsu using a horse to haul *bukkomi-gusa*. From Yoshida's personal collection: not included in 1938 book.

Supper:	daikon miso soup (stock made from *konago* fish), salted herring with grated daikon in it
Snack:	dried *mochi*

I started working on the rabbit hutch early this morning again. By seven o'clock I'd pretty much finished the body. After that I made breakfast, and at nine-thirty I started weaving *noma* mats from straw. I put them on top of the rabbit hutch to form the roof, and the hutch was finished. Then I transferred the rabbits from their boxes into the new hutch, separating them in the process. Eight rabbits that were born forty days ago were in a box together, but I separated them into two groups of four. Four born three months ago I divided into two pairs. Then I spread grass and straw on the floor of the hutch for them. By the time I was done it was twelve-thirty—time for lunch.

The rabbit hutch is 6 *shaku* (1.8 m) across the front and 4 *shaku* (1.2 m) deep, divided into eight compartments. Each is 1 *shaku* 5 *sun* (45.5 cm) wide and 2 *shaku* (60.6 cm) deep. On the front of each compartment, I stuck alternating strips of bamboo and cedar I'd split—about 8 *bu* (2.4 cm) apart—into the ground and then nailed them at their tops to the hutch's wooden body. I did all sides of the hutch this way, but I used strong pieces of bamboo and cedar. Each compartment has a door. These are on the long sides, or the two fronts, of the hutch. I made the roof so that the rain will collect in the center and run off. I folded a thin tin plate and laid it on the roof, and placed *noma* mats on top of that. The tin plate prevents water from running down inside the hutch from the center (see Fig. 5.2).

For feeding the rabbits, I made the hutch so that I can drop food in easily without opening the doors—to save time (see Fig. 5.3). The long sides of the hutch are 4 *shaku* (1.2 m) tall, and the height at the center is 3 *shaku* (90.9 cm). I made it this tall because I thought it might be nice to let the droppings pile up inside to a depth of 1 *shaku* (30.3 cm) or so and get well stomped and composted before cleaning out the hutch each time.

I'd been thinking that I wanted to build a hutch like this from about a month ago.

In the afternoon I did the second follow-up fertilizing and mid-row hoeing for my Sadaya cabbages. As usual, I dug little holes on the opposite sides of the plants from where I'd dug holes for the

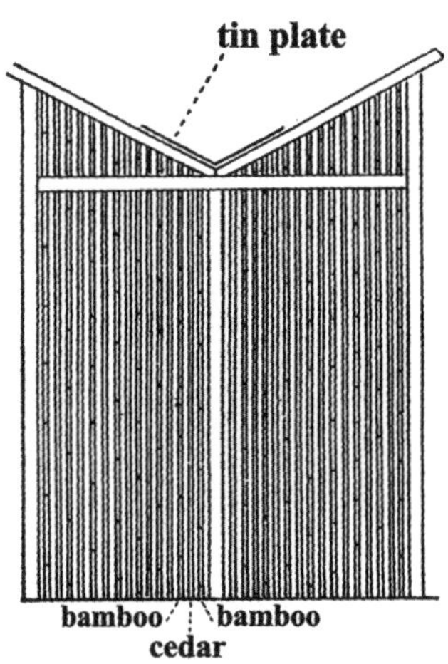

Figure 5.2. End view of the rabbit hutch.

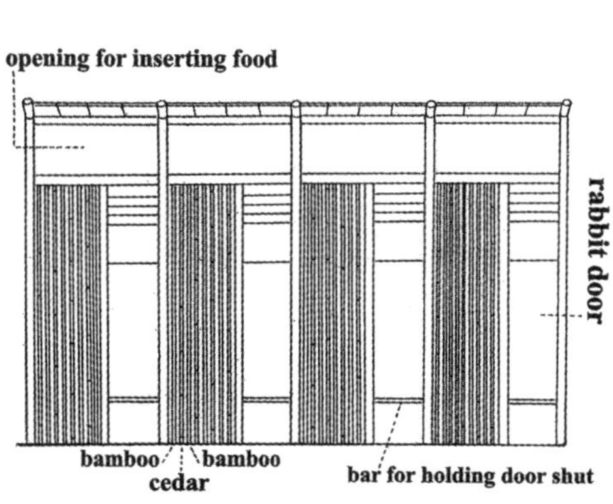

Figure 5.3. One long side of the rabbit hutch.

first follow-up fertilizing. Then I did the first follow-up fertilizing, and some mid-row hoeing, for my oriental melons—digging circular troughs around the plants. After that I did the first follow-up fertilizing on my watermelons—digging circular troughs and sprinkling soil around the plants' bases by hand. Later I picked vegetables for selling and did some other things. Basically, I really was quite busy. My wife came back by lunchtime. She weeded the taro field and picked caterpillars off the summer cabbages in the afternoon.

This evening, the cicadas called *kaka-zemi* started singing.

Today's income:

daikon	25 sen
peas	10 sen
cabbages (2 *kan* (7.5 kg))	55 sen

Today's expenditures:

postage stamps	8 sen

At the main house: Kanezō did just about the same things as yesterday. So did my brother's wife. My mother cooked and such, as usual.

July 4, Thursday

Cloudy, sometimes sunny
Southwest wind, later out of the south
20°C in the morning, 26°C in the afternoon

Breakfast:	miso soup with peas in it, *takuan*
Lunch:	roasted salted herring, *takuan*
Supper:	daikon miso soup with shaved seaweed from Hokkaidō I'd received from an acquaintance, boiled cabbage with soy sauce for flavor
Snacks:	dried *mochi*

About yesterday's fertilizer, I took 1.5 *ka* of night soil and mixed the same amount of water with that to make for 3 *ka,* dissolved one *shō* (1.8 L) of ammonium sulfate in it and used that for the cabbages, oriental melons, and the watermelons. Today I pulled soil up over the roots of the cabbages and oriental melons I'd fertilized yesterday (did *tsuchiyose*), and from about seven o'clock I started making breakfast. From nine I went looking for a rabbit that had escaped. The hutch wasn't exactly perfect; that one was missing when I went to peek this morning. I searched for about an hour but couldn't find it, so I plowed half the watermelon field. I searched again around noon, and then my eldest daughter, Fusae, and two or three other kids appeared coming from a direction I hadn't expected with the missing rabbit. They'd caught it and brought it back.

In the afternoon, I weeded the watermelon fields I plowed in the morning. My wife returned by lunchtime, and she collected the weeds I'd picked and carried them to the compost pit after lunch. I also thinned out some eggplant seedlings that sprouted from seeds I'd planted in a small field, and picked vegetables for selling.

Today's income:

daikon	20 sen
cabbages (4 *kan* (15 kg))	1 yen 5 sen

Today's expenditures:

envelopes	5 sen
sweets and candies	5 sen

At the main house: Their activities were just the same as yesterday, except that Kanezō cut one horse load (*da*) of *hakomarogi* [box bundles] in the morning.

July 5, Friday

Cloudy, later a light rain, then sunny
Southwest wind
20°C in the morning, 26°C in the afternoon

Breakfast:	bamboo shoot miso soup, *takuan*
Lunch:	same as yesterday's lunch
Supper:	bamboo shoot miso soup with *wakame* from the coast of Oga that we received from the main house, *takuan*

In the early morning, while picking extra stems off the tomato plants, I tied the tall tops to poles with straw. I started making breakfast at six o'clock. At eight-thirty I fed soybeans I'd soaked in water for a while and chopped up fine to a juvenile dove I caught yesterday. After that, I cut some lower leaves off the tomato plants with shears to improve the air flow. I threw the leaves on the compost pile and put weeds from the pumpkin patch there, too. These days I don't have to cut grass especially for the rabbits because I have plenty of daikon and cabbage leaves. In the afternoon I went over to the irrigation pond in Momokawa hamlet (*buraku*)[2] in Wakimoto because I heard that some creature nobody has ever heard before has been making noise there lately, but I didn't hear anything. Too bad. My wife came back after lunchtime, so I made lunch. In the afternoon my wife picked caterpillars off the cabbages and picked vegetables to sell tomorrow.

Today's income:

daikon	20 sen
cabbages (4 *kan* 300 *momme* (16.1 kg))	1 yen 5 sen

Figure 5.4. Hauling firewood at the hamlet (*buraku*) firewood area. Pine and cedar branches are attached to a horse in this way. [The man appears to be Saburō's friend Sonoji.]

Today's expenditures:
snacks and candies for the kids 5 sen

At the main house: Kanezō cut hay for the horse in the early morning, and cut *bukkomi-gusa* after breakfast. In the afternoon he made compost with my brother. My brother's wife went out again to weed another family's paddies to return *e*. She got home pretty early in the evening, so she piled one load (*da*) of compost on the horse and headed out to the fields to plant taros. She planted 60—dug holes in a plowed field, spread compost, planted the taros, and then covered everything with soil. My mother went to another village to visit someone in the morning, returned in the afternoon, and cooked supper.

July 6, Saturday

Cloudy, sometimes sunny
Southwest wind
20°C in the morning, 26°C in the afternoon

Breakfast:	bamboo shoot miso soup, *takuan*
Lunch:	same as yesterday's lunch
Supper:	daikon miso soup (stock made from *konago* fish) salted krill, *takuan*

I did the first follow-up fertilizing to the eggplants early. I'd planted the seeds directly in the ground—hadn't started them in a hotbox. As usual, I dug a small hole on one side of each plant. Then I replanted some taros that had rotted in the field—replaced them with good ones—so ended up with four rows for which I had no seed taros. I finished all of that at about seven o'clock, and then started making breakfast. At nine I applied the first follow-up fertilizer to my Sadaya cabbages—since I'd planted them a bit late—digging a hole on each side of the plants. Later, I weeded them. After that, I filled in the holes I'd dug earlier for fertilizing the eggplants. On both the eggplants and the cabbages I used one bucket of night soil and one of water mixed, and I dissolved 200 *momme* (750 g) of ammonium sulfate in that. I also cut some grass for the rabbits, and made lunch, and in the afternoon, I cut weeds around the outer boundaries of my farm to use for making compost and picked vegetables for selling. My wife picked more caterpillars off the cabbages in the afternoon, and then went down to the station at around four o'clock to sell vegetables, etc.

Today's income:

daikon	20 sen
peas	10 sen
cabbages (6 *kan* 500 *momme* (24.4 kg))	1 yen 87 sen

At the main house: Kanezō cut hay for the horse early, and after breakfast he hauled unpolished brown rice down to the rice polishing place with my brother to have it processed for eating. Later he went to the wood shop and got some sawdust. In the afternoon, he cut *bukkomi-gusa*. My brother's wife spent the day out in the dry fields, weeding the soybean crop and pulling soil up over the plants' roots for the first time this season (doing the first *tsuchiyose*). My mother cooked and such.

July 7, Sunday

Rain, later sunny
Northwest wind
19°C in the morning, 24°C in the afternoon

Breakfast:	bamboo shoot miso soup with kelp shavings (*tororo konbu*) in it, *takuan*
Lunch:	same as yesterday's lunch
Supper:	*udon* in soup made from dried roasted anchovies, boiled cabbage with soy sauce, *takuan*, etc.

Today I heard *jijime* singing for the first time this year. *Jijime* are cicadas that sing *ji-ji-*.

The rain was too heavy this morning to work outside, so I did some reading. I started making breakfast from six o'clock and cut straw for making compost from eight. I made lunch and then rested some in the afternoon. My wife came back by lunchtime, and she also rested in the afternoon.[3] But we did pick vegetables for selling in the evening.

Today's income:

daikon	20 sen
peas	5 sen
cabbages (4 *kan* 600 *momme* (17.25 kg))	1 yen 25 sen

At the main house: Kanezō cut hay for the horse early, and also after breakfast—this time using the horse to carry it. In the afternoon he took off and went to see the events at the festival at Tennō.

My brother's wife was out there [in Tennō] from the morning. My mother cooked and such.

July 8, Monday

Cloudy, sometimes sunny
Southwest wind
19°C in the morning, 23°C in the afternoon

Breakfast:	pea miso soup, *takuan*, leftover *udon*, and cabbage and such from last night
Lunch:	exactly the same as yesterday's lunch
Supper:	daikon and bamboo shoot miso soup (stock made from *konago* fish) with shaved kelp (*tororo konbu*) in it, leftover salted herring from lunch, *takuan*, etc.

Satō Sonoji and five women of the village came to visit today.

I cut compost grass and weeds today from early in the morning—spent the whole day on that. But I also made breakfast, picked vegetables for selling, and took down the windbreaks protecting the green onions. And I gave them their fourth follow-up fertilizer, too. I mixed 1 *ka* of night soil with the same amount of water and just poured it over the surface. At about nine o'clock I manually pollinated my pumpkin plants. Their flowers started blooming today. I picked off some of the male flowers, being careful not to knock the pollen off, and pressed them onto the pistils of the female flowers. It's enough if the pollen just gets onto the pistils, but you must do as thorough a job as possible when pollinating by hand. My wife returned by lunchtime, and she went to help another family weed their rice paddies in the afternoon.

Today's income:

daikon	10 sen
cabbages (5 *kan* 800 *momme* (21.75 kg))	1 yen 32 sen

At the main house: Kanezō cut hay for the horse from early this morning, and after breakfast he cut *marogi-gusa*. In the afternoon he cut *bukkomi-gusa*. My brother's wife took her lunch out with her and worked in the dry fields. She weeded the soybean crop and scraped soil up over their roots for the first time this year (did the first *tsuchiyose*), etc. My brother went out to check the paddies in the afternoon, and Ichisuke minded the shop. My mother cooked.

July 9, Tuesday

Cloudy
Northwest wind, later out of the southwest
20°C in the morning, 21°C at 2:00 p.m.

Breakfast:	pea and seaweed *wakame* miso soup, *takuan*
Lunch:	roasted salted herring, grated daikon cooked with salted krill, *takuan*, etc.
Supper:	potato and *wakame* miso soup, *takuan*, salted herring, etc.

Satō Saburō's wife gave birth to a baby boy today.

The oriental melon plants have started blooming.

Again today, I cut grass and weeds for composting (*hiryō-gusa*) from early morning. I started making breakfast at seven o'clock, and then hand-pollinated the pumpkins again from nine-thirty. Then I cut more *hiryō-gusa*, and in the afternoon I picked caterpillars off the cabbages and picked ladybugs off my eggplants. Later, I went out to buy some farming chemicals, picked vegetables for selling, and made supper, etc.

My wife went out again this afternoon to help with the paddy-weeding of another household.

Today I dug up and ate some of my brown potatoes for the first time this year.

Today's income:

daikon	10 sen
cabbages (6 *kan* 600 *momme* (24.75 kg))	1 yen 50 sen

Today's expenditures:

stamps	3 sen
agricultural chemicals	75 sen

At the main house: Kanezō cut hay for the horse early. After breakfast he cut box-bundle grass and straw (*hakomarogi-gusa*). In the afternoon, he made compost with my brother. My brother's wife scraped soil up over the soybean plants' roots again (did *tsuchiyose*) and weeded and such. She caught a cold, so she rested in the afternoon. My mother cooked.

July 10, Wednesday

Sunny
North wind
20°C at 6:00 a.m., 28°C in the afternoon

Breakfast:	pea and seaweed (*wakame*) miso soup, *takuan*, etc.
Lunch:	same as yesterday's lunch
Supper:	same as yesterday's supper

The principal of Tarusawa Elementary School[4] and his wife came to visit today, and so did a number of villagers.

Again, I cut grass and weeds for composting from the early morning. I started making breakfast from six o'clock. From eight I cleaned the toilet, and then I got ready for applying the chemicals [I bought yesterday]; I had to wash out the big tub, check the equipment and fix it up a bit, and measure the chemicals. Then I made lunch.

In the afternoon, I applied the chemicals, with my wife. We sprayed the cabbages, eggplants, tomatoes, cucumbers, etc. I used 6 *to* (108.2 L) of chemicals. Two *to* of this was casein lime, into which I mixed some arsenate (*hisaen*). For 1 *to* of water I used 8 *momme* (30 g) of casein lime, and 20 *momme* (70 g) of arsenate, and I sprayed it on the cabbages. For the rest, I mixed up four-*to*-style Bordeaux mixture—5 *momme* (18.8 g) of casein lime, and 15 *momme* (56.3 g) of arsenate per *to* (18 L)—and sprayed this on my eggplants, tomatoes, cucumbers, etc., to protect them from disease and bug damage. In the evening, I picked more vegetables for selling.

Today's income:

daikon	12 sen
cabbages (6 *kan* 700 *momme* (25.1 kg))	1 yen 20 sen
money from selling the dove chick I'd been keeping	30 sen

Today's expenditures:

stamps	10 sen

At the main house: Kanezō cut hay for the horse in the early morning and then spent the rest of the day cutting *bukkomi-gusa*. My brother surveyed the rice paddies and his wife rested. My mother cooked and such.

July 11, Thursday

Very sunny and pleasant [*kaisei*]
Weak north wind
16°C in the morning, 28°C in the afternoon

Breakfast:	cowpea (*sasage*)[5] miso soup with kelp shavings in it, *takuan*
Lunch:	miso soup made by boiling a chicken carcass that we got from my younger sister's household, salted herring, *takuan*, etc.
Supper:	miso soup made from chicken carcass stock with daikon and peas in it, salted krill, *takuan*

I plowed the watermelon field some and weeded it early, and started making breakfast at around seven o'clock. At nine I went back to weeding the watermelon field, hauled the stuff I pulled up over to the compost pit, and then spread straw around the bases of the watermelon plants. After that, I weeded around the soybean plants I'd put in around the edges of the watermelon field and scraped soil up over their roots for the second time this season (did the second *tsuchiyose*). In the afternoon, I cut compost grass and weeds, and picked vegetables for selling, etc. My wife had to go to the school for something in the afternoon, but she came home pretty early, and weeded among the daikon, etc. The economic situation isn't very good—vegetables aren't selling so well. My wife wound up having to bring back home what she'd taken out with her this morning.

Today I harvested eggplants and cucumbers for the first time this year.

Today's income:

daikon	10 sen
cabbages (6 *kan* 400 *momme* (24 kg))	1 yen 10 sen
eggplants	10 sen

Today's expenditures:

a *koshimaki*[6] for my wife	50 sen
deworming medicine[7]	5 sen
candies for the kids	5 sen

At the main house: Kanezō cut hay for the horse early, and after breakfast he cut *bukkomi-gusa*. In the afternoon he cut hay for the horse again. My brother's wife rested again today.

July 12, Friday

Cloudy
16°C in the morning, 26°C in the afternoon

Breakfast:	cabbage and daikon miso soup, *takuan*
Lunch:	roasted salted herring, *takuan*
Supper:	miso soup with potato and daikon in it, *takuan,* tuna roe cooked in soy sauce. We got the dish from the main house, and the tuna roe originally came from relatives in Minami Iso Village in Oga.

Early this morning, I removed more wasteful growth from the tomato plants and attached more of them to poles. I started making breakfast at seven o'clock. At nine I started manually pollinating the pumpkins again, and worked on cutting the tallest parts off the oriental melon plants to prevent them from growing too high, and cut some mugwort (*yomogi*) to burn for keeping mosquitoes away. In the afternoon, I cut compost grass and weeds—as I always do—and picked vegetables for selling. My wife returned by lunchtime, and she weeded among the daikon plants and such until about four o'clock. Then she went down to the village again to try to sell more vegetables.

Today's income:

daikon	10 sen
eggplants	40 sen
cucumbers	40 sen
cabbages (8 *kan* (30 kg))	1 yen 65 sen

Today's expenditures:

snacks for the kids	5 sen

At the main house: Kanezō cut hay for the horse early and spent the rest of the day cutting box-bundle grass and straw (*hakomarogi-gusa*). He hauled some firewood and such in the evening. My brother's wife took her lunch with her out to the dry fields to work. She weeded the red bean (*azuki*) crop, and scraped soil up over the plants' roots, etc. She also did some plowing with a hoe and spread soil around the plants by hand.

July 13, Saturday

Sunny
South wind
18°C in the morning, 26°C in the afternoon

Breakfast:	pea and daikon miso soup, *takuan*

Lunch:	shark meat cooked with cabbage in soy sauce, *takuan*. We got the shark from the main house yesterday.
Supper:	daikon and pea miso soup, roasted salted herring, boiled cabbage eaten with soy sauce for flavor, *takuan*

I caught a little cold, so I slept in this morning—got up at six o'clock and made breakfast. At eight I started working outside: cutting weeds and grass, hand-pollinating the pumpkin and watermelon plants, and preparing poles for supporting the eggplants. In the afternoon a cabbage merchant came, so I picked some cabbages and sold them to him, and then I walked out to some distant rice paddies to take photos [see Fig. 0.33].

My wife washed laundry in the afternoon, and went out again to try to sell vegetables, among other things.

Today's income:

daikon	10 sen
cucumbers	10 sen
cabbages (18 *kan* (67.5 kg))	3 yen 20 sen
scrap iron (sold)	5 sen

Today's expenditures:

soy sauce (2 *gō* (360 ml))	6 sen
sweets for the kids	5 sen

At the main house: Kanezō cut hay for the horse again early and cut box-bundle grass and straw the rest of the day. My brother's wife weeded the red bean fields and scraped soil up over the plant's roots during breaks in her cooking activities in the kitchen. My mother went to a relative's house for a ceremony to celebrate the birth of a grandchild (*mago-ue no iwai*).

For the past several weeks people in the hamlet of Momokawa here in Wakimoto Village have been hearing cowlike cries in the nighttime around a reservoir there. Residents of that hamlet have been gathering in great numbers of course, and people—even hundreds in a day—have even been coming from other towns and villages in the hope of seeing the creature during the daytime, and hearing it at night. I heard an old man say that if this had happened in the old days, folks would have just assumed that it was the spirit of the pond, some earthly deity, or maybe some strange monster, and it might have become the source of some new legend, and that nobody would have dared to try capturing it. And it seems that this is true—not a single village elder tried to catch this thing. But young fellows these days don't care at all about that—they went right out to try to catch it. Another sign of the times, I thought. So today, after several weeks of this, a thirtyish fellow in his prime named Sōma Hisakichi, who lives near the train station, went out to the pond with a tool called a *yajiri* (*yashiri*) and captured this mystery creature by spearing it. Now the villagers are all excited and going to see it. After its capture, it was revealed to be a rather small "monster"—don't know where it came from, but it seems to be a bullfrog . . . dark green and about the size of an adult's hand. And so, the problem has been solved.

July 14, Sunday

Rain
South wind
28°C in the morning, 23°C in the afternoon

Breakfast:	snow pea and daikon miso soup, *takuan*
Lunch:	roasted salted herring, *takuan,* etc.
Supper:	miso soup with potato, peas, and daikon in it plus salted herring left over from lunch, *takuan,* etc.

I set up poles for supporting the eggplants early this morning, and I tied the plants to them. Then I started making breakfast, at seven o'clock. From nine I hand-pollinated the pumpkins again, but it probably didn't work well because of the rain. On such a day, you should really cover the flowers with leaves to prevent them from getting wet before they open but I forgot to do this because I was so extremely busy. After doing the pumpkins, I went to the main house to pound some straw for tying up plants and brought it back with me and returned to setting up support poles and tying the eggplants to them. It took me until one o'clock, but I finished. Then: lunch preparations. After that, I went into the village to read the newspaper since it was just so rainy, and then I harvested vegetables for selling from five o'clock. My wife returned by lunchtime. She took it easy in the afternoon—did some sewing indoors.

Today's income:

daikon	10 sen
eggplants	30 sen
cucumbers	30 sen
cabbages (3 *kan* 800 *momme* (14.25 kg))	72 sen

Today's expenditures:

various sweets (some made from glutinous rice like *mochi,* etc.)	10 sen

At the main house: Kanezō cut hay for the horse early, and made compost with my brother later in the morning: turning it over, pulling it from the horse stall and mixing it with soil, and piling it up again. In the afternoon Kanezō cut more hay for the horse. My brother's wife rested in the morning, and went to the beach to collect seaweed in the afternoon.

July 15, Monday

Rain
Strong southwest wind
24°C in the morning, 22°C in the afternoon

Breakfast	cowpea (*sasage*), daikon and potato miso soup, *takuan,* etc.
Lunch:	roasted salted herring, *takuan,* etc.
Supper:	daikon miso soup, boiled cabbage with soy sauce for flavor, salted krill, *takuan,* etc.

I rested in the morning because of the rain. My wife cooked breakfast, and after eating she went out to sell vegetables. I hand-pollinated the pumpkins again but it doesn't seem to have gone very well today, either. Later, I went down to Wakimoto's post office to order some seeds. In the afternoon, after preparing lunch and eating, I wove rope for carrying things on my back (*seoinawa*), harvested vegetables for selling, and planted new seedlings in place of some that had withered and died, etc. My wife rested some in the afternoon, and went over to her natal home to visit.

Today's income:

24 eggplants	35 sen
24 cucumbers	30 sen
cabbages (3 *kan* 800 *momme* (14.25 kg))	60 sen

Today's expenditures:

money for seeds	1 yen 25 sen
stamps and postcards	25 sen
snacks	3 sen

At the main house: They had planned to do the third rice paddy weeding today but they gave up and went home because of the rain. Kanezō cut hay for the horse early, and after breakfast he made a *kura mushiro* for the horse. This is a [woven] *mushiro* [mat] to lay over a horse's saddle. In the afternoon he went out to cut *bukkomi-gusa.* My brother's wife rested in the morning, and went out to do some weeding and such in the dry fields in the afternoon. My brother walked around the rice paddies to check them.

July 16, Tuesday

Rain, later sunny and then cloudy
Northwest wind, dying later
18°C in the morning, 22°C in the afternoon

Breakfast:	cowpea miso soup, *takuan,* etc.
Lunch:	exactly the same as yesterday
Supper:	pea and mushroom miso soup, salted krill with grated daikon in it, *takuan.* We received the mushrooms from a relative's child; it was the first time to eat these mushrooms, called *dzusabo-dake* in local speech, this year.

Early this morning I did the second follow-up fertilizing to my Sadaya cabbages, which I'd already moved from a temporary, to a permanent, spot. Next, I did the first follow-up fertilizing and such to my autumn cabbages—digging a little hole on one side of each plant, as usual, and pouring in a mixture of 1 *ka* of night soil, 1 bucket of water, and 300 *momme* (1.1 kg) of ammonium sulfate. Later, I plowed some in my Sadaya cabbage beds. I started making breakfast at seven o'clock, and pollinated the pumpkin plants again while doing that. From nine o'clock I scraped soil up over the roots of the Sadaya cabbages (did *tsuchiyose*), filled in the holes from fertilizing the autumn cabbages, and then prepared lunch. In the afternoon I went down to the village with cabbages, cucumbers, and daikon and such, that some people had asked for. Then I weeded around those eggplants I planted directly in the ground, with my wife, and then dug a little hole on one side of each and did the second follow-up fertilizing. We also plowed a little and scraped more soil up over the plants' roots (did *tsuchiyose*).

Today's income:

daikon	13 sen
cucumbers	50 sen
eggplants	50 sen
cabbages (5 *kan* 600 *momme* (21 kg))	1 yen

Today's expenditures:

sweets made from candy[8] for the kids	5 sen

At the main house: Kanezō cut hay for the horse early. They started the third paddy weeding today, finishing the Sakura paddies and the ones at Sanbonmatsu. Kanezō helped with that, and he cut and cleared weeds and grass along the ridges and pathways among the paddies. My brother's wife weeded paddies, and my brother surveyed the paddies in the afternoon. My mother cooked. A total of nine people did the weeding today.

July 17, Wednesday

Rain, later just cloudy
19°C in the morning, 23°C in the afternoon

Breakfast:	daikon and pea miso soup, boiled cabbage with soy sauce for flavor, *takuan*
Lunch:	roasted salted herring, *takuan*, etc.
Supper:	We ate at the main house.

I rested this morning due to the rain. My wife prepared breakfast. She didn't go out to sell vegetables today. Instead, she sewed in the morning and went to the main house to weed the rice paddies in the afternoon. I hand-pollinated the pumpkins again in the morning, and fixed up the cabbage fields. In other words, I got rid of a bunch of leaves that had been left on the ground after harvesting. I did more of the same work in the afternoon. I gathered up all those leaves and carried them to the compost dump and piled them up, and I harvested more vegetables for selling. I also went over to the main house to help pound *mochi*.

At the main house: Kanezō cut hay for the horse early, and after breakfast he cut weeds on the ridges between the rice paddies. Cutting away weeds that interfere with the growth of the rice seedlings is called *kuroharai* here in this village. My brother's wife went out to the dry fields in the morning to plant rapeseed (*nanohana*). In the afternoon, Kanezō and my brother's wife went out to weed paddies together—to Bishamon. If they had been able to work from morning, they could have finished the paddy weeding. But they couldn't, so they weren't able to finish. But since other people's schedules are involved, they went ahead and pounded *mochi* and had a little celebration (*torige-iwai*) to mark the completion of the paddy weeding. The four of us ate there, too.

We ate *yama mochi* with mashed red bean paste (*tsubushi azuki*). As for the side dishes, we poured soy sauce over cabbage and ate that, along with *momi* cucumbers, and a *nishime* dish with bamboo shoots, cowpeas, and mackerel (*saba*) cooked together in soy sauce, and some other dishes. *Momi* cucumbers are cucumbers that have been sliced thinly and then flavored with miso and vinegar—all mixed up together by hand (*momi*).

July 18, Thursday

Rain
16°C in the morning, 18°C in the afternoon

Breakfast:	cowpea and cabbage miso soup, *takuan*. (From today we quit mixing oats with our rice; we'll just eat pure rice now: 1 *shō* 3 *gō* (2.3 L) of rice daily.)
Lunch:	same as yesterday's lunch
Supper:	cowpea and cabbage miso soup (stock made from *konago* fish), roasted salted herring, *takuan*, etc.
Snack:	cucumbers for the kids (but we give them candies and sweets every day)

Two young fellows from the village came over today.

A quiet rain fell steadily today from the morning. There has indeed been a lot of rain this year, and everyone has been worrying about the effect of the low temperature on the crops.

In the morning, I picked some pumpkin leaves and put them over the flowers so they wouldn't get hit by the rain. At six o'clock I started making breakfast, and after eating I changed the grass and straw on the floor of the rabbit hutch, putting down some clean straw for them. Later, I hand-pollinated the pumpkins more, and went down to the village with cabbages and cucumbers someone had ordered. After I got back, I prepared for getting cucumber seeds (called *tane-o-tsukeru*). In other words, I marked seed cucumbers by hanging pieces of straw from their vines. Then I picked cucumbers and eggplants for which I'd received some orders, and made lunch. After lunch, I took the vegetables that had been ordered with me and went down into the village. After getting back, I picked more vegetables for selling tomorrow, and that was the end of my work today. My wife said she felt like she'd caught a cold, so she rested all day.

Today's income:

daikon	10 sen
cucumbers	36 sen
eggplants	60 sen
cabbages (6 *kan* 200 *momme* (23.25 kg))	1 yen 25 sen

Today's expenditures:

paper for mounting photos	10 sen
sweets made from candy (*amegashi*)	5 sen

At the main house: Kanezō cut hay for the horse early. He cut *bukkomi-gusa* after breakfast and in the afternoon. My brother's wife went to help weed the rice paddies of another household, to return *e*. My brother went out to inspect the paddies in the afternoon. My mother cooked.

July 19, Friday

Sunny
15°C in the morning, 22°C at 5:00 p.m.

Breakfast:	cabbage miso soup, *takuan*, etc.
Lunch:	roasted salted herring, *takuan*
Supper:	miso soup (stock made from *konago* fish) with daikon and peas in it, salted herring, *takuan*

Yoshida Kiyomatsu's wife gave birth to a baby boy. Yoshida Shichisuke's wife gave birth to a baby boy, too.

I weeded the soybean field early this morning, and then made breakfast. I hand-pollinated the pumpkin plants from eight o'clock, and then pulled soil up over the roots of the soybean plants for the second time in the season (did the second *tsuchiyose*). Later I weeded the daikon field. In the afternoon, my wife and I weeded together. I picked vegetables for selling in the evening.

Today's income:

12 daikon	10 sen
32 eggplants	50 sen
cabbages (3 *kan* 500 *momme* (13.1 kg))	57 sen

At the main house: Kanezō cut hay for the horse early, and later in the morning he cut *bukkomi-gusa*. In the afternoon, he took the horse up into the hills to cut straw and grass for feeding it. No horse really likes to eat straw and grass from up in the hills (*yama no kusa*), but they all really love the grass and straw that grows in and around the rice paddies. Up until last year, there was an open space (*yazu*) in this village where most people cut straw, grass, and such to feed their horses. But this year it was converted to paddy land, so the *yazu* is gone. Villagers have been complaining some about not being able to cut straw and grass to feed their horses there anymore. My brother's wife went to help another family weed rice paddies again—returning *e*. My mother cooked and such.

July 20, Saturday

Cloudy
18°C at 6:00 a.m., 24°C at 2:00 p.m.

Breakfast:	cabbage miso soup, *takuan*, etc.
Lunch:	*takuan*, roasted salted trout
Supper:	miso soup with potatoes and peas in it, *takuan*, salted cabbage, etc.

In the morning, I carried weeds and such that I'd pulled out of the daikon fields (to the compost heap) and piled them up to use later as fertilizer. Then I prepared breakfast. At about eight o'clock I started hand-pollinating the pumpkin plants again, and then I gently smoothed out the soil between the rows in the daikon field, and picked some green peas. These are pea pods that I left on the plants until today so I can use them as seeds (next year). My wife returned by lunchtime, and she cooked lunch. In the afternoon we weeded the pea field, cleaned up and put away the poles that supported the pea plants, and carried the plants themselves to the compost pile. We also carried fertilizer to the daikon field, which I will soon convert into a space for growing napa cabbages (*hakusai*). We also picked vegetables for selling.

Today's income:

daikon	10 sen
32 eggplants	50 sen
cabbages (5 *kan* (18.75 kg))	78 sen

Today's expenditures:

sweets made from candy (*amegashi*)	10 sen
1 salted trout	30 sen

At the main house: Kanezō cut hay and grass for the horse early. Later in the morning, he took the horse up to a cliff (*mama*) to get a bunch of sand, and used the horse to carry a load back to the compost heap. This is for mixing with barn floor straw manure and fresh straw to make compost, which is what he did in the afternoon. First, he pulled the old, trampled straw and manure out of the barn and piled it up, and then he piled up the straw he needed and the sand he'd brought. Then he overturned the older compost that he'd piled up before and readied it for using in the fields. This kept him very busy. My brother also participated in this activity during the morning and in the afternoon. His wife helped another family with their paddy weeding again, to return *e*. My mother cooked and such.

July 21, Sunday

Light rain
19°C at 6:00 a.m., 23°C at 2:00 p.m.

Breakfast:	cabbage and pea miso soup, *takuan*
Lunch:	roasted salted herring, salted cabbage and daikon, *takuan*
Supper:	miso soup with potato and eggplant—we put *igeshi* seaweed in before eating it—pickled daikon with salted krill (on it), *takuan*, etc. The *igeshi* came from the shore of Minami Iso (in Oga). My wife brought it back from there with her, after receiving it from an acquaintance in Funagawa.

I jumped out of my *futon* at three-twenty this morning. My wife went out soon after that to sell vegetables, and I got to work in the [new] napa cabbage field. I put the compost from yesterday into troughs I'd dug in the field, finishing that by seven o'clock. I carried it on a large bamboo colander/sifter (*koezaru*),[9] and transported the loads one by one to the field and spread it throughout the troughs evenly. After that, I made breakfast. At about eight-thirty I started hand-pollinating the pumpkins again, and then later I put some commercial fertilizer in the napa cabbage fields. I used 1 *to* 5 *shō* (27.1 L) each of powdered anchovy and powdered oil cake, and 8 *shō* (14.4 L) of calcium superphosphate, etc., and mixed it thoroughly, and spread 1 *to* 3 *shō* (23.5 L) of that on top of the

Figure 5.5. How I filled the troughs with compost and fertilizer.

compost I put in the napa cabbage fields this morning. But I only spread this over one-half of the compost. Over the other half, I spread 3 *to* (54.1 L) of wood ash (see Fig. 5.5). Once all of this was done, I used a hoe to pull soil up over everything from both sides, so that all the fertilizer was covered with 3–4 *sun* (approx. 11 cm) of soil in the end. All in all, I used about 130 *kan* (487.5 kg) of fertilizer.

In this way, by one o'clock, what had been a daikon field was one big step closer to becoming a napa cabbage field.

We finished eating lunch at one-thirty. My wife had made it back by noon and cooked. In the afternoon, my wife and I weeded the green pea field, and we carried the weeds to the compost heap using a two-person *mokko.* Then we picked vegetables for selling—wiped eggplants with a cloth, washed the cucumbers, picked extra leaves from the cabbages, and washed them, and put everything into a big carrying basket (*shoikago*) that goes on a person's back. After that, I cut off some tomato branches that were diseased (had gotten *fuhaibyō*) and disposed of them.

It really was a heck of a lot of work—all of this.

Today's income:

32 eggplants	57 sen
25 cucumbers	25 sen
cabbages (5 *kan* (18.75 kg))	90 sen

At the main house: Kanezō cut hay for the horse early, and later in the morning he cut *bukkomi-gusa.* In the afternoon he used the horse to carry a load of sawdust back with him from the lumber mill for mixing with compost. In other words, this is one ingredient for making fertilizer. My brother's wife went to help another family weed their rice paddies (to return *e*) again, and my mother cooked and such.

July 22, Monday

Rainy and cloudy
Southwest wind
21°C at 6:00 a.m., 26°C at 1:30 p.m.

Breakfast:	green pea miso soup with *igeshi* in it, pickled daikon, *takuan*
Lunch:	roasted salted herring, pickled daikon, *takuan*
Supper:	cowpea and daikon miso soup with *igeshi* in it, pickled daikon with salted krill mixed in with it, *takuan*
Snack:	cucumbers

I couldn't work in the fields due to the rain, so I organized stuff in the storage shed. I also tidied up in the house. My wife made breakfast. After eating, she went out to sell vegetables. I picked caterpillars off the cabbages, hand-pollinated the pumpkins, pruned the oriental melon plants so the fruits can grow better, harvested some wheat, and prepared the soil of the pea field for other use. It was the second time to prune the oriental melon plants. The first time, I cut them from their bases, leaving only three joints. The second time, there were three branches on each plant, so I cut them to leave two joints on each (see Fig. 5.6).

I made lunch [and we ate], and in the afternoon I worked on making rows in the field for planting green onions because the pea field is changing into an onion field. I made the rows 2 *shaku* 7 *sun* (81.81 cm) wide. Once the rows were done, I dug troughs [running parallel to them] for planting onions—3 or 4 *sun* (approx. 11 cm) deep—right up next to one side of each row by scooping up a little soil with a hoe and piling it up next to the trough. I will plant the green onions in the troughs. I'll draw a cross-section diagram to show what I mean (see Fig. 5.7).

After that was done, I picked vegetables for selling. Then I went to the main house to get some straw and hauled it back. My wife returned at three o'clock, and carried compost to the green onion field using a *goshin* [see Fig. 3.2] strapped on her back.

Today's income:

50 eggplants	90 sen
25 cucumbers	25 sen
cabbages (2 *kan* 500 *momme* (9.4 kg))	40 sen

Today's expenditures:

white cloth, 2 *shaku* 2 *sun* (36.4 cm) long	8 sen (to make an Etchū *fundoshi*[10])
candies	2 sen

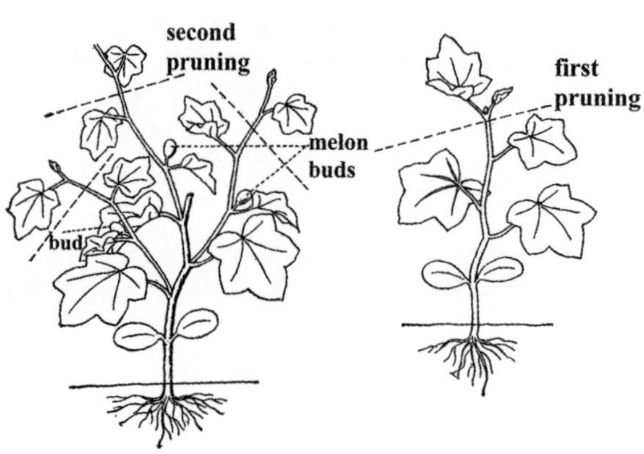

Figure 5.6. How to prune oriental melon plants.

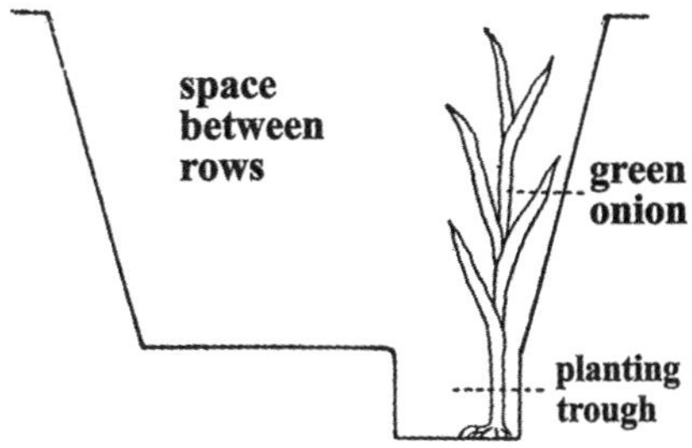

Figure 5.7. How green onions will be planted in the troughs.

At the main house: Kanezō did the same morning work as always. Then he spent the rest of the day cutting *bukkomi-gusa*. My brother's wife went out to weed the rice paddies. The Shinden paddies weren't done, so she weeded them herself. My mother cooked.

July 23, Tuesday

Light rain, later just cloudy
Southwest wind
23°C in the morning, 24°C at 1:00 p.m.

Breakfast:	cowpea and eggplant miso soup, *takuan*, pickled daikon, etc.
Lunch:	roasted salted herring, salted daikon and cabbage, *takuan*, etc.
Supper:	miso soup made with *konago* fish stock and green onions in it, *takuan*. We had salted daikon and cabbage, so we put salted krill on it and ate that. The green onions we ate tonight were ones that hadn't made the cut for being planted in the field.

I woke up at three-thirty this morning and got started right away, picking onion seedlings in the onion nursery bed, and sorting them into three groups: large, medium, and small. Those of the first two groups I kept for planting, and those of the last group I set aside for eating. I started making breakfast at six o'clock. From eight I hand-pollinated the pumpkins and after that I planted the onions—standing them up one by one in the troughs I dug yesterday and then gently pouring some soil into the troughs. There is roughly 1 *sun* (3 cm) of space between each onion. After putting in the dirt, I mixed water with night soil and poured that over it. Then I mixed up a bunch of commercial fertilizer (*kinpi*), just as I recently did for the napa cabbage fields, and spread that over the layer of night soil. On top of that, I spread a layer of compost about 2 *sun* (6.1 cm) deep, and then sprinkled wood ash over everything, and covered the surface with straw (see Fig. 5.8).

I was so busy with all this that I just kept working all day—ate lunch while standing without a break. My wife came back by lunchtime, and she sorted more onion seedlings in the afternoon. I picked vegetables for selling after the sun went down—in complete darkness.

Today's income:

60 eggplants	1 yen 10 sen

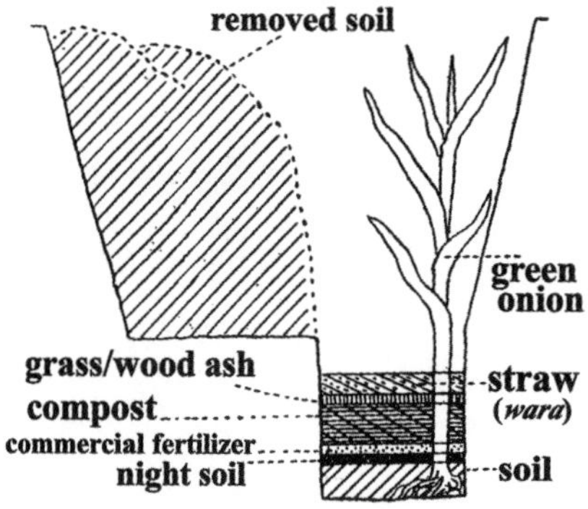

Figure 5.8. How green onions were planted in the troughs.

30 cucumbers	22 sen
cabbages (4 *kan* 200 *momme* (15.75 kg))	65 sen

Today's expenditures:

mosquito repellent (*katori senkō*)	6 sen
candies and sweets (snacks)	4 sen

At the main house: Kanezō cut hay for the horse early, and he took the horse up into the hills later in the morning to cut more hay for it. In the afternoon, he cut box-bundle grass and straw (*hakomarogi-gusa*). My brother's wife weeded the main house's rice paddies again, finishing the Shinden paddies, so all the paddy weeding is done now. In the afternoon, she went out to the dry fields and picked about one *kamasu*[11] of cucumbers. My mother cooked.

July 24, Wednesday

Cloudy
West wind
18°C at 7:00 a.m., 21°C at 1:30 p.m.

Breakfast:	green pea and eggplant miso soup with *igeshi* in it, *takuan,* etc.
Lunch:	mackerel fry cooked with green onions in soy sauce, *takuan*
Supper:	miso soup with mackerel fry, potatoes, cowpeas and such cooked in it, roasted salted trout, *takuan,* etc.
Snack:	raw cucumbers

Again today, I did onion-planting work from the very early morning. First, I weeded the fields and made more rows. At seven o'clock I prepared breakfast, and from nine I hand-pollinated the pumpkins. Later I sorted green onion seedlings again. Just like yesterday, I ate lunch while standing, and after that I cut grass for feeding the rabbits and planted more green onions in the fields. At last, I finished planting the onions in the evening, and then I picked vegetables for selling tomorrow. My wife came back by noon, and she made lunch, and in the afternoon, she weeded the fields, made rows for planting, and hauled compost—just like me. We have now planted onions in a field that was for green peas, in the onion nursery bed itself, and in one part of a summer cabbage field. All in all, I used about 150 *kan* (562.5 kg) of compost, and 1 *to* 5 *shō* (27.1 L) of commercial fertilizer (*kinpi*). As for wood ash, I used 3 *to* (54.1 L). Night soil used amounted to 3 *ka,* mixed with the same amount of water. I also used 30 *pa*[12] of straw and such.

Today's income:

60 eggplants	1 yen
25 cucumbers	20 sen
cabbages (3 *kan* 200 *momme* (12 kg))	50 sen

Today's expenditures:

soy sauce	9 sen
candies	3 sen

At the main house: Kanezō went up into the hills before sunrise to cut *bukkomi-gusa.* It took all day. He cut and brought back three horse loads. One big grassy hill ("mountain") that had been closed off in accordance with village rules was opened for use today. In local speech people say, *Tsuyo kara bugukusa-yama age da* ["The *bukkomi-gusa* mountain is open from today"]. My brother's wife went to help at another household—to return *e.* She said she pulled soil up over the roots of eggplants with a hoe (did *tsuchiyose*). My brother overturned compost, etc. My mother cooked.

July 25, Thursday

Sunny
18°C at 6:00 a.m., 26°C at 2:00 p.m.

Breakfast:	cowpea and eggplant miso soup, *takuan,* etc.
Lunch:	*momi* cucumbers, roasted salted trout, *takuan,* etc.
Supper:	cowpea and cabbage miso soup (stock made from *konago* fish), roasted salted herring, *takuan*

I prepared for planting napa cabbages early this morning—digging small holes 2 *shaku* (60.1 cm) apart in the field—enough to plant about 390 plants. But I only planted about 200 today. I poured about 1 *to* (18 L) of night soil mixed with the same amount of water into the holes and then made breakfast. From nine o'clock I started putting one scoop of a soil/compost mix into each little hole that I'd poured night soil into earlier and then I planted the napa cabbage seeds—14 or 15 in each hole. Then I filled each hole with about 2 *bu* (6.1 mm) of soil and that was the end of that. I planted two kinds: Yamato Wase Hōtōren from Hokkaidō, and Chinese Tsūfū—about a hundred of each type. I ate while standing up again today, and right after that I started getting ready to spread pesticide. First, I washed out the wooden tub and got some water, then I measured the chemicals. My wife helped by boiling water and such.

Then I made about 2 *to* (36.1 L) of four-*to*-style Bordeaux mixture and spread it on the tomatoes. Next, I prepared about 1 *to* 5 *shō* (9 L) of lime casein and arsenate and applied it to the cabbages and soybeans. My wife's job was to pump the can while I sprayed. We finished that pretty early, so I moved on to something else: about four rows that had been planted with taros didn't do well—the plants didn't take root—so I weeded them, re-plowed them, and rebuilt the rows, applied compost and commercial fertilizer, covered everything with a thin layer of soil, spread wood ash over the surface, and put down a layer of soil about 2 *sun* (6.1 cm) deep, and when I was done, I had four rows ready for planting napa cabbages. I used 15 *kan* (56.3 kg) of compost, 4 *shō* (7.2 kg) of commercial fertilizer, and 6 *shō* (10.8 kg) of wood ash, etc. I hadn't planned to grow napa cabbages there, but that's what I'm going to do. Later, I picked vegetables for selling. My wife did the same things I did. In the evening, I weeded a bit around some of the tomato plants.

Today's income:

eggplants	77 sen
cucumbers	30 sen
cabbages (3 *kan* 800 *momme* (14.25 kg))	75 sen

Today's expenditures:

stamps	6 sen
black cotton cloth	10 sen
candies	3 sen

At the main house: Kanezō cut hay for the horse early, and later in the morning, and in the afternoon, he cut *bukkomi-gusa*. My brother's wife washed laundry in the morning, and in the afternoon, she went out to the fields to plant green onions—seedlings I'd raised. The way she planted them isn't like the way I do it; she simply made rows and put them in the ground and spread soil over the roots. She says she's planning to put some fertilizer on them later. I hear farmers typically planted green onions like this in the old days.[13] My mother cooked today.

July 26, Friday

Cloudy, sometimes sunny
20°C at 6:00 a.m., 22°C at 2:00 p.m.

Breakfast:	nearly the same as yesterday morning
Lunch:	mackerel cooked in soy sauce with green onions, *takuan*, etc.
Supper:	potato and daikon miso soup (stock made from *konago* fish), roasted mackerel eaten with soy sauce for flavor, *takuan*, etc.

Early this morning I dug little holes in the four rows I prepared yesterday for growing napa cabbages, put a handful of compost soil into each, and then dropped about 20 seeds into every hole. The type: Chinese Tsūfū Hakusai. After finishing that I made breakfast. From nine o'clock I picked some leaves off the tomato plants. Having too many leaves prevents the tomatoes from getting enough light and ripening well, so removing some of them helps. Also, I judged that the plants had all put out enough tomatoes, so I cut them back. In the afternoon I weeded the tomato fields with my wife's help. We picked vegetables for selling in the evening. Today I decided to suspend my hand-pollination of the pumpkin plants, and instead hand-pollinated the watermelons. The compost soil I used today I dug up from the dishwashing area of the main house—such soil tends to have at least a small amount of extra nutrients in it.

Today's income:

45 eggplants	65 sen
30 cucumbers	25 sen
cabbages (3 *kan* 400 *momme* (12.75 kg))	50 sen

Today's expenditures:

kerosene (4 *gō* (720 ml))	12 sen
roasted dried anchovies	5 sen
a new lamp wick	1 sen
sugar (white)	5 sen
candies	5 sen

At the main house: Kanezō cut hay for the horse early, and later in the morning he cut box-bundle grass and straw (*hakomarogi-gusa*). In the afternoon he went up into the hills to cut more hay for the horse. When he cuts such grass and straw in the morning he usually goes out with a container on his back and cuts around the rice paddies or the dry fields. In the afternoons, he normally takes the horse up into the hills to cut it. My brother's wife was at home in the morning—she and my brother baked bread—and in the afternoon she went out to the fields and harvested some potatoes. My mother went to another village to visit a sick acquaintance, so my brother's wife cooked instead of her.

July 27, Saturday

Sunny
South wind
22°C at 7:00 a.m., 28°C at 1:00 p.m.

Breakfast:	miso soup with *dzusabo-dake* mushrooms that we received from the main house and eggplant in it, *takuan*
Lunch:	cabbage with soy sauce for flavor, roasted salted trout, *takuan*
Supper:	potato and green bean miso soup (stock made from *konago* fish)

Yoshida Kinzaburō's wife gave birth to a baby girl.[14]

I started early, working on my summer cabbage fields; I'll convert them to napa cabbage patches. First, I hauled all the leftover cabbage leaves in the fields back to the house for feeding to the rabbits, and then lightly plowed the spaces between the rows in the fields. I started making breakfast at around seven o'clock, and weeded from nine. I finished at around twelve-thirty. My wife got back by noon, and she made lunch. I spent the afternoon working on the new napa cabbage fields, without a break. After that, I overturned the soil in the fields and pulled out all the leftover cabbage stems and roots that were still in the ground. Next, I plowed the fields well and smoothed them out, and then I dug fertilizer troughs in the fields—roughly 3 *sun* (9.1 cm) deep and about 2 *shaku* 5 *sun* (75.75 cm) apart. Once this was done, I filled them with compost—about 120 *kan* (450 kg) in total. On top of this, I spread about 1 *to* (18 L) of commercial fertilizer, like a couple of days ago. I sprinkled some soil on top of that, and then 2 *to* (36.1 L) of wood ash. Then I spread 2 or 3 *sun* (6.1–9.1 cm) of soil over everything. It was quite a lot of work. My wife and I did this together until late in the evening. Then I harvested vegetables for selling tomorrow.

Today's income:

60 eggplants	50 sen
33 cucumbers	25 sen
tomatoes (300 *momme* (1.1 kg))	15 sen (the first tomato sales of the year)
cabbages (2 *kan* 500 *momme* (9.4 kg))	40 sen

Today's expenditures:

salt	10 sen

At the main house: Kanezō went out early to cut hay for the horse. He also dug up a bunch of potatoes and hauled them, along with stems and such from the plants, back to the house, etc. They said at the main house that three rows worth of potatoes—about 2 *to* (36.1 L) or so—were stolen from the field last night. My brother's wife went out to the fields to harvest potatoes today, and she also cooked, among other tasks.

July 28, Sunday

Rain from morning, later just cloudy
Strong southwest wind
22°C in the morning, 26°C in the afternoon

Breakfast:	cowpea and white onion miso soup, *takuan*, etc.
Lunch:	roasted salted trout, *takuan*, etc.
Supper:	miso soup with *dzusabo-dake* mushrooms that we got from my wife's brother's family and daikon in it, salted trout, *takuan*, etc.
Snack:	cooked lily bulbs (*yamayuri*)[15]

Early this morning, I took down the windbreaks I'd set up to protect the eggplants and packed everything in the shed for storage. Then I started making breakfast at six o'clock. After breakfast, I went up into the hills and walked around to dig up lily (*yamayuri*) bulbs—known locally as *yoro*. I got about 1 *shō* (1.8 L) of lily bulbs. We washed them and ate them. I didn't work in the fields today because of the rain. After I finished making lunch, my wife returned [from selling vegetables]. In the afternoon she shelled peas. I had to go into the village on an errand, and I went out and tied up miscanthus that has been growing; the stuff really gets in the way if you let it spread out too much. I also harvested vegetables for selling and marked some eggplants for getting seeds later. When I do this, I look for a healthy plant that produces a lot of fruit, and individual fruits that are nicely shaped . . . although a certain agronomist says that the second fruits are best for this.

Today's income:

50 eggplants	70 sen
33 cucumbers	25 sen
37 tomatoes (1 *kan* (3.75 kg))	65 sen
cabbages (3 *kan* 200 *momme* (12 kg))	60 sen

At the main house: Kanezō went out early to cut hay for the horse, and he spent the rest of the day cutting *bukkomi-gusa.* My brother's wife sewed at home during the morning, and in the afternoon, she went out to the fields and picked cucumbers. She also planted more green onions that she got from me and selected herself for planting. And she cooked, etc.

July 29, Monday

Cloudy, sometimes sunny
Southwest wind, later strong
24°C at 6:00 a.m., 26°C at 2:00 p.m.

Breakfast:	miso soup with cowpeas and eggplant in it, *takuan,* etc.
Lunch:	tuna we got from an acquaintance in Funagawa cooked with cabbage in soy sauce, *takuan*
Supper:	potato and cowpea miso soup, *takuan,* pickled daikon, roasted salted herring
Snack:	tomatoes, cucumbers

I cut grass and straw for making compost early this morning. Then I made breakfast. After eating, I harvested potatoes. As I dug them, I set aside especially large and well-shaped ones that were disease-free for using as seed potatoes later. I harvested a total of about 40 *kan* (150 kg) of potatoes. After finishing that, I plowed the ground between the rows a bit and weeded, and then laid all the weeds I'd pulled up and the potato plant stems and such down on the surface of the soil I'd plowed. I pulled soil over this from both sides to make rows for planting daikon. My wife came back by noon, and she cooked lunch. Again today, I ate lunch mostly while standing, and then I went straight back to making rows for planting daikon. In other words, my potato plots are going to become daikon fields. Once I'd finished forming the rows [for daikon], I started preparing for planting napa cabbages by stretching a rope across that field and digging little holes about 2 *shaku* (60.1 cm) apart from one another. After that, I took a *goshin* with me to the main house, dug up more soil from under the dishwashing area there, filled the *goshin* with it, and carried it back home strapped to my back. I did this three times. Later, I harvested vegetables for selling, and I cut some more grass and such for the rabbits. My wife weeded the burdock field some and she also weeded around the autumn cabbages, among other things.

Today's income:

eggplants (75 large, 36 small)	1 yen 30 sen
28 cucumbers	20 sen
40 tomatoes (1 *kan* (3.75 kg))	50 sen
cabbages (2 *kan* 800 *momme* (10.5 kg))	35 sen

Today's expenditures:

postage for mailing a manuscript	15 sen
candies for the kids	5 sen

At the main house: Kanezō cut hay for the horse early. Later in the morning, he cut *bukkomigusa*. In the afternoon he hauled soil, and he cleared out the horse stall and put fresh straw and grass in it, too. In the evening he went out to cut hay for the horse again. My brother dealt with the compost all day—overturning it and hauling it, etc. My brother's wife was out in the fields all day—weeding the soybeans and pulling soil up over the roots of the plants for the second time of the season (doing the second *tsuchiyose*), etc. She returned with a load of cucumbers on her back in the evening. My mother cooked, etc.

July 30, Tuesday

(July 1 according to the old lunar calendar)
Cloudy, sometimes sunny
Strong southwest wind
24°C at 6:00 a.m.

Breakfast:	cowpea miso soup, *takuan*. We got a tray of red bean rice (*sekihan/azuki meshi*) from the main house.
Lunch:	roasted salted herring, pickled daikon and *takuan*
Supper:	I'll explain below.

I planted napa cabbages early this morning—in the field I'd left partly unplanted the other day and in the remaining parts of the former cabbage fields I prepared recently. What I planted in part of the cabbage fields were special Ibaraki Kekkyū Hakusai.[16] And in the remainder of the field I'd already planted some with napa cabbages the other day, I planted Ibaraki Hakusai and Takai Hakusai. I'm going to raise 42 of the special cabbages, and I planted about 110 of the Ibaraki type and about 80 of the Takai type. First, I took 1½ *ka* of a 1:1 mix of water and night soil and poured that in all the holes I'd dug the other day. Then I put a good scoop of the compost soil I carried back with me yesterday into each hole as well. I sprinkled the seeds on top of that and covered everything with dirt. This took until about seven o'clock. After that, I made breakfast. After eating I did the first crop-thinning to the napa cabbages I'd planted previously—the very first napa cabbages—leaving five or six sprouts at each spot.

Next, I went into the village to buy some things. In the afternoon I went to my old school; an emergency summons came yesterday from the commanding officer of the local division of the Akita Regiment [Akita Rentai-ku Shireikan] about a roll call for potential reserve soldiers to be held tomorrow, so I had to be there to practice for that. I got back fairly early in the evening, and then I picked vegetables for selling. I have to wipe each eggplant, cucumber, tomato, etc., with a cloth, so it's a lot of trouble. My wife came back by lunchtime, and she cooked in the afternoon because we invited my mother and younger sister, and my wife's mother. All the dishes were made with vegetables from my farm. First, there were *udon* noodles in soup that had been made by boiling dried roasted anchovies. There was also a bowl of cucumber slices that had been mixed with sugar and salt. Another dish was eggplants cooked in brown sugar and oil with miso, etc. And there was another dish: small sea bream bought in Funagawa, roasted, and served with boiled cabbage flavored with soy sauce. We also had pickled daikon. For soup, there was a soy sauce–flavored broth with sea bream, *konnyaku*, and dried pieces of wheat gluten (*fu*) in it.

Today's income:

65 eggplants	84 sen
27 cucumbers	20 sen
37 tomatoes (1 *kan* (3.75 kg))	50 sen
cabbages (2 *kan* 600 *momme* (9.75 kg))	40 sen

Today's expenditures:

udon (800 *momme* (3 kg))	40 sen
soy sauce (5 *gō* (900 ml))	14 sen
sugar (white and brown)	10 sen
konnyaku and *fu*	5 sen
fish	15 sen
oil	10 sen
stamps	9 sen
soap	5 sen

At the main house: Kanezō cut hay for the horse early, and later in the morning he went up into the hills to cut more. He took the afternoon off, since today is July 1 according to the old lunar calendar. My brother's wife went out to sell cucumbers in the early morning, and after breakfast she went out to the fields to prepare a plot for growing daikon—that is, to form rows. She was at home sewing in the afternoon. My brother went to our old school this morning for the inaugural ceremony of the election cleanup alliance.[17] Ichisuke minded the shop.

July 31, Wednesday

Sunny
Strong southwest wind
24°C at 6:00 a.m., 28°C at 2:00 p.m.

Breakfast:	yesterday's leftover rice in hot water [*yuzuke*], cabbage with soy sauce, *takuan,* pickled daikon
Lunch:	tuna we received from my wife's natal household cooked in soy sauce, *takuan,* pickled daikon
Supper:	cowpea and potato miso soup, roasted salted herring, *takuan,* pickled cucumbers

I started spreading tobacco powder on my napa cabbage fields early this morning. This is for killing garden springtails (*jinomi mushi,* a.k.a. *kiboshi maruto mushi*). After that, I did the first thinning on the napa cabbages that I'm growing as an experiment, among other chores. I also cut grass for the rabbits and made breakfast. I finished eating at six o'clock, left the kids behind in the fields, and then I went to the school at Futto Village,[18] for the roll call for reserve troops. There really is a strong sense of national crisis now; we've already had four official roll calls and now they issue another draft order. The feeling of war is in the air. I got home by two o'clock, and then I planted daikon. The rows had already been made. I planted almost the same way as when I planted my spring daikon. I poured one bucket of water into 1 *ka* of night soil and used that. I planted enough seeds to grow about 300 daikon. Then I harvested vegetables for selling. Later I went to Tayazawa to pay a visit to my [maternal] grandfather (*mago-oyaji*), who has been sick. My wife weeded around the autumn cabbages in the afternoon.

Today's income:

80 eggplants	1 yen
60 tomatoes (1 *kan* 800 *momme* (6.75 kg))	90 sen
cabbages (3 *kan* 300 *momme* (12.4 kg))	60 sen

At the main house: Kanezō cut hay for the horse early, and the rest of the morning he cut *bukkomi-gusa.* In the afternoon, he carried night soil out to the dry fields again. He put it into four 1 *to* 5 *shō* (27.1 L) barrels (*taru*), strapped them to the horse, and took them to the fields like that. My brother's wife worked on preparing a daikon field in the morning, and in the afternoon, she planted daikon. As I'd done, she used a potato field for this, but she planted daikon differently from my way. In other words, my way is different from the usual way of planting daikon around here. People of my village plant daikon in the same old way folks have been doing it for many years, and the way they do it at the main house is also the old-fashioned way. They just dig holes in a drunken fashion along the length of the rows, pour in some night soil, then some dirt, plant seeds on top of the dirt, and then cover everything with more dirt using a hoe.[19]

Chapter 6: August 1935

August 1, Thursday

Cloudy
Strong southwest wind
24°C at 6:00 a.m., 26°C at 2:00 p.m.

Breakfast:	miso soup with *dsusabo-dake* mushrooms we received from the main house and cowpeas (*sasage*) and such in it, *takuan,* etc.
Lunch:	roasted salted trout, *takuan,* salted cucumbers and daikon, etc.
Supper:	cowpea and potato miso soup, pickled cucumbers, *takuan,* etc.

I prepared for planting more napa cabbages early this morning. I dug holes in the rows I made before and poured a mix of 1 *ka* of night soil and a bucket of water into them. I waited to make sure the liquid had fully vanished into the soil and then put a scoopful of compost soil into each hole and smoothed everything out. Once that was done, I made breakfast, and then I started planting napa cabbage seeds from about nine o'clock: the same type I planted before—Chinese Chiifu Hakusai, Takai Hakusai, and Ibaraki Hakusai, and such. At the end, I sprinkled soil carefully over everything by breaking it up with my hands and letting it fall through my fingers. I've planted enough seeds to grow 250 cabbages. After finishing, I went into the village on an errand. My wife got home by noon, made lunch, and weeded the taro field in the afternoon. I did the second follow-up fertilizing to the autumn cabbages—dug a little hole beside each plant and poured in a mixture of 1 *ka* of night soil, one bucket of water, and 200 *momme* (750 g) of dissolved ammonium sulfate. Then, I plowed a bit around the cabbages and pulled soil up over their roots for the first time in the season (did the first *tsuchi-yose*). After that, I picked vegetables for selling.

Today's income:

61 eggplants	70 sen
34 cucumbers	13 sen
46 tomatoes (1,700 *momme* (6.4 kg))	75 sen
cabbages (1,600 *momme* (6 kg))	25 sen

Today's expenditures:

snacks for the kids	10 sen

At the main house: Kanezō did what he usually does early every morning. Later in the morning he hauled night soil—just like yesterday. In the afternoon he cut *bukkomi-gusa*. My brother's wife planted daikon in the morning, like yesterday. She also planted some napa cabbage seeds, in the same way she planted the daikon seeds. In total, three night soil–hauling trips were made, meaning that twelve 1 *to* 5 *shō* (27.1 L) barrels were used. But a bit of this amount was applied to the green onions. In the afternoon, my brother's wife weeded around the soybean plants, and she also pulled soil up over their roots with a hoe for the second time (did the second *tsuchiyose*). My mother cooked and such. My brother hiked up Mt. Kanpū. A resident of another village infringed upon our village's rights to the mountain's land; he went to confront this man.

August 2, Friday

Cloudy and sunny
20°C in the morning, 28°C at 2:00 p.m.

Breakfast:	cowpea eggplant miso soup, salted cucumbers, *takuan*
Lunch:	potatoes cooked with salt, roasted salted trout, *takuan,* pickled daikon, etc.
Supper:	cabbage and potato miso soup (stock made with dried roasted anchovies), *takuan*

I went out and walked around and checked my napa cabbage and green onion fields early this morning, and I washed the big tub carefully to prepare for spreading chemicals (*kusuri*). Then I prepared breakfast. After eating, I cut compost grass and weeds, cut grass and straw to feed the rabbits, hauled weeds I'd pulled out of the fields to the compost pile, and pruned the oriental melon plants some more. I just cut off all the branches that had sprouted, leaving only one joint on each. Then I fixed up the sprayer, and measured the chemicals for later. Then I made lunch. In the afternoon I made 4 *to* (72.2 L) of four-*to*-style Bordeaux mixture with my wife—mixing in 20 *momme* (75 g) of lime casein and 70 *momme* (262.5 g) of arsenate (*hisaen*), and we sprayed this on the eggplants and soybeans. Once this was done, I harvested vegetables for selling. My wife weeded the taro field.

Today's income:

61 eggplants	73 sen
60 tomatoes (2 *kan* (7.5 kg))	70 sen
cabbages (1 *kan* 600 *momme* (6 kg))	30 sen

At the main house: Kanezō did the same early morning work as always, and he spent the rest of the day cutting *bukkomi-gusa*. My brother's wife weeded the soybean fields and cooked.

My brother attended an officers' meeting of the household heads' association (*koshukai*) because he needed to confer with the others about the problem I mentioned yesterday.

August 3, Saturday

Sunny
Weak west wind
22°C at 6:00 a.m., 28°C at 2:00 p.m.

Breakfast:	exactly the same as yesterday morning
Lunch:	cooked potatoes—same as yesterday
Supper:	cowpea and cucumber miso soup (stock made from dried roasted anchovies), *momi* cucumbers, *takuan*, roasted salted trout, etc.

Yoshida Shinichirō's wife gave birth to a baby boy.
Yoshida Shinichirō's wife died in childbirth.

Early this morning, I weeded in the taro field, and spread tobacco powder over some of the napa cabbages. I also harvested eggplants and sold them [to someone]. Then I made breakfast. After eating, I went back to weeding around the taros—just going through and pulling every weed out with the left hand, and with the right hand . . . a lot of trouble, really. After the weeding, I dug a little hole next to each plant, for doing the first follow-up fertilizing. Then I poured a 1:1 mix of night soil and water into all the holes—3 *ka* in total. Next, I cut grass for the rabbits. My wife made lunch. In the afternoon, she weeded the Chinese yam field. I used a hoe to pull soil up over the roots of the taro plants for the first time in the season (did the first *tsuchi-yose*) and lightly hoe-plowed between the plants. After that, I did the first thinning on the late-planted napa cabbage crop. Finally, I harvested vegetables for selling.

Today's income:

113 eggplants	1 yen 15 sen
85 tomatoes (3 *kan* 300 *momme* (12.4 kg))	1 yen

Today's expenditures:

stamps	6 sen
candies	4 sen

At the main house: For Kanezō, it was the same early morning work as always. Later, he cut *bukkomi-gusa,* and in the afternoon he used the horse to haul sawdust from the lumber mill. My brother's wife went out to the dry fields and weeded the soybean crop and pulled soil up over the plant's roots with a hoe (did *tsuchi-yose*). My brother went out in the afternoon to see to the paddy irrigation. My mother cooked and such.

August 4, Sunday

Sunny
Weak west wind
20°C in the morning, 29°C in the afternoon

Breakfast:	cucumber and cowpea miso soup, pickled cucumbers, *takuan,* etc.
Lunch:	miso-preserved (*misokagi*) cucumbers,[1] pickled cucumbers, *takuan,* roasted salted trout, etc.
Supper:	cabbage and potato miso soup (stock made from dried roasted anchovies), pickled cucumbers, *takuan*

Five young fellows came over from a different hamlet.

Grasshoppers and other insects I can't identify have started chirping and singing.

Early this morning, I continued pulling soil up over the roots of the taro plants for the first time this season, and I spread more powdered tobacco over the napa cabbages. Then, around seven o'clock, I started making breakfast. After eating, I cut more grass for the rabbits, and later I weeded some in the summer cabbage fields and overturned the soil a bit, to prepare to plant daikon. My wife made lunch today. In the afternoon she went out to buy rice, and she also took some cucumbers over to a relative's house. To prepare for planting daikon, I formed rows in the field I'd prepared in the morning. Then I dug holes and poured a 1:1 mix of night soil and water—two buckets in total—into all the holes. The types of daikon I planted today included Aokubi and Hōryō. After all this, I harvested vegetables for selling. I'm going to grow 200 daikon.

Today's income:

80 eggplants	95 sen
70 tomatoes (3 *kan* (11.3 kg))	1 yen
cabbages (1 *kan* 500 *momme* (5.6 kg))	25 sen

Today's expenditures:

rice (2 *to* 5 *shō* (4.5 L))	6 yen 50 sen
soap	5 sen
candies	3 sen

At the main house: Same schedule and activities as always for Kanezō. My brother and his wife went out to debug the rice paddies. They poured oil into flooded paddies, waited for the oil to spread all the way across the surface of the water, and then my brother's wife just went through the paddies, lightly brushing the rice plants with a stick and knocking the bugs off into the oil, which kills them. For this you use a big piece of bamboo with tiny holes punched in the bottom. In the afternoon my brother went to another village on an errand. His wife went to her natal household to help in the fields there. This was to return *e.* My mother cooked and such.

August 5, Monday

Sunny
22°C in the morning, 28°C in the afternoon

Breakfast:	cabbage and cowpea miso soup, *takuan*. Red bean rice (*sekihan*): ate some at the main house.
Lunch:	*momi* cucumbers, roasted salted trout, salted cucumbers, *takuan,* etc.
Supper:	cowpea and eggplant miso soup (stock made from dried roasted anchovies), side dishes: same as lunch

I applied follow-up fertilizer to my Chinese yams early this morning: poured two buckets of a 1:1 mix of night soil and water over the surface of the field. Later I planted more napa cabbage seeds in those fields—to replace plants that dried up or got eaten by bugs. I also cut grass for the rabbits and washed out my tub (*oke*) for making more insecticide. After all this, I made breakfast. Once I'd eaten, I watered my napa cabbages. Then I made some *rokueki*[2] and poured a little around the base of each napa cabbage plant. This was a mixture of 15 *momme* (56.3 g) of flea powder (*nomitoriko*) and 5 *shaku* (90 ml) of kerosene that I'd covered tightly and allowed to sit for over 24 hours and then added to 1 *to* (18 L) of water in which I'd already dissolved 20 *momme* (75 g) of soap. After applying the insecticide, I took some pumpkin leaves and used them to cover the soil over the seeds in the napa cabbage fields that hadn't sprouted—the fields I'd watered earlier in the morning—to block the rays of the midday sun. In the afternoon I took a break from farmwork. My wife returned by noon, and she also rested a bit. But I picked vegetables for selling in the evening.

Today's income:

74 eggplants	70 sen
80 tomatoes (2 *kan* 800 *momme* (10.5 kg))	80 sen
cabbages (1 *kan* 600 *momme* (6 kg))	25 sen

At the main house: Same as always in the early morning for Kanezō. Later, he went up into the hills to cut hay for the horse, as usual. He took the afternoon off. My brother and his wife went to the paddies to pour oil into them again [to kill bugs], and they finished that work during the morning. They also took off in the afternoon, but she did go out to pick some cucumbers in the afternoon. My mother cooked and such.

August 6, Tuesday

Rain
20°C in the morning, 20°C in the afternoon

Breakfast:	cowpea and eggplant miso soup, salted cucumbers, *takuan,* etc.
Lunch:	potatoes cooked in salty water, roasted salted trout, salted cucumbers, *takuan,* etc.
Supper:	same as morning

I cut grass for the rabbits early this morning, and then made breakfast. Then, I did the second thinning to the first napa cabbages I'd planted, leaving three or four seedlings in each spot. At last, nearly all the third-group cabbages I planted, and also the daikon, have sprouted. I made lunch, and took a break in the afternoon because of the rain. But I picked vegetables for selling in the evening. My wife made it back by lunch, and she weeded in the eggplant fields in the afternoon, despite the rain.

Today's income:

80 eggplants	75 sen
52 tomatoes (1,500 *momme* (5.6 kg))	62 sen
cabbages (2,400 *momme* (9 kg))	40 sen

Today's expenditures:

tax	76 sen
sweets made from candy (*amegashi*)	5 sen

This evening, the chief priest of the temple at Futto Village came to the Enmeiji Temple here in our village to give a talk, and in a mere two hours he covered what would have taken a university professor three years to explain in lectures. Whatever he said, it was way over the heads of the local farmers, according to what I heard from someone there.

At the main house: Same as always today for Kanezō in the early morning. He went out to cut *bukkomi-gusa* later, but he couldn't go up into the hills in the afternoon due to the rain, so he went out among the paddies and dry fields to cut hay, which he hauled home on his back. My brother's wife went out early in the morning to sell cucumbers, returned around eight o'clock, and weeded the rice seedling beds (*nashiro*) after breakfast. In the afternoon, she went out to the fields to pick more cucumbers, and she did some weeding before going home. My brother went to check on the paddies in the afternoon. Ichisuke minded the shop, and my mother cooked and such. Something I forgot to write: Hay for the horse has to be cut up into small pieces after bringing it home. Kanezō usually takes care of this, but my mother sometimes helps.

Apparently, the main house had nine *kamasu* of potatoes after harvesting. One *kamasu* generally holds about 3 *to* (54.1 L) of potatoes. The local branch of the Agricultural Support Association (Shinōkai) did a survey of the potato harvest, and the main house was number one in our hamlet, with a reported amount of about 1 *kan* 750 *momme* (6.6 kg) of potatoes dug from one three-meter-long row. Somewhere (in a different hamlet) in the same village, a household apparently got 2 *kan* 200 *momme* (8.3 kg) from one row of the same length. That must be the record for this year.

August 7, Wednesday

Sunny and cloudy
18°C at 7:00 a.m., 22°C at 2:00 p.m.

Breakfast:	exactly the same as yesterday
Lunch:	roasted salted trout, salted cucumbers, *takuan,* etc.
Supper:	miso soup with cabbage and potatoes in it, side dishes: same as lunch
Snack:	tomatoes, cucumbers, and such—same as every day

I weeded the eggplant fields from early morning. I used a hoe to loosen the soil a bit between the rows for walking. Then I cut grass for the rabbits and made breakfast. After eating, I replanted daikon where the seedlings had failed, walked around and surveyed the napa cabbage and watermelon fields, and then weeded the eggplant fields again. My wife got back by lunchtime, and she made lunch, and then she weeded among the eggplants, too. But first she helped me spread arsenate on the soybeans and the autumn cabbage crop. After that, I also weeded. In the evening, I picked vegetables for selling.

The *men-men* cicadas[3] started singing today.

Today's income:

140 eggplants	1 yen 18 sen
56 tomatoes (1 *kan* 700 *momme* (6.4 kg))	55 sen
30 cucumbers	15 sen
3 pumpkins (600 *momme* (2.3 kg))	15 sen

Today we sold pumpkins for first time this year.

Today's expenditures:

candies	2 sen

At the main house: Kanezō cut hay for the horse early. Later, he took the horse out to cut more of the stuff. He cut *bukkomi-gusa* in the afternoon. He also carried unpolished brown rice down to the plant to have the chaff removed, and hauled the polished rice back. My brother also went to the rice-processing plant, etc. His wife went to her natal home to help in the dry fields. Of course, this was to return *e*. My mother got sick, but she still cooked today.

August 8, Thursday

Sunny and cloudy
Northwest wind
19°C at 7:00 a.m., 24°C at 2:00 p.m.

Breakfast:	same as yesterday morning
Lunch:	roasted salted herring, side dishes: same as morning
Supper:	same soup as this morning, side dishes: same as lunch

A cousin from Tayazawa came over.

I surveyed my napa cabbage fields early this morning, and planted seeds again in spots where nothing had sprouted, or where the sprouts had been eaten by bugs, and then I prepared to make an insecticide called *rokueki:* fetching water, washing the tub (*oke*), measuring soap. Then I cut some grass for the rabbits, and finally started making breakfast at six-thirty. After eating, I made *rokueki,* and I used it on all my napa cabbages except for the ones I planted first. After that, I weeded in my eggplant fields. My wife got home by noon, and she made lunch. In the afternoon, we weeded the eggplant fields together, and carried the weeds [to the compost pile], etc. I picked vegetables for selling in the evening.

Today's income:

90 eggplants (26 very big)	70 sen
75 tomatoes (2 *kan* 800 *momme* (10.5 kg))	75 sen
cabbages (3 *kan* (11.5 kg))	48 sen

Today's expenditures:

mochi made from wheat flour	5 sen

At the main house: Kanezō cut hay for the horse early, and cut *bukkomi-gusa* after breakfast. In the afternoon, he hauled two barrels (*taru*) of night soil to the fields, planted green onions with my brother's wife, and poured the night soil on the newly planted onions, etc. In the evening, he cut more hay, and carried cucumbers home with him on his way back. My brother's wife had to go to a different village on an errand relating to the family *wagashi* shop in the morning, and in the afternoon she did as I wrote above. She went back home a bit early and made supper; she did all the cooking for the day. My mother had a toothache, so she went to see a doctor. My brother was out all day on an errand that had something to do with village affairs. Ichisuke minded the shop today.

August 9, Friday

Sunny
Southwest wind
20°C at 7:00 a.m., 24°C at 2:00 p.m.

Breakfast:	cabbage and eggplant miso soup, salt-pickled cucumbers, *takuan*
Lunch:	*momi* cucumbers, roasted salted herring, salted cucumbers, *takuan,* etc.
Supper:	potato and cabbage miso soup, side dishes: same as lunch

I did the first thinning on the third-planted napa cabbages—just planted early this morning—finishing that before making breakfast. After eating, I weeded among the eggplants again. In the afternoon, my wife and I weeded around the eggplants together, carried away the weeds we'd picked, and weeded among the pumpkin plants after that. In the evening, I cut grass for the rabbits. I was able to pick an oriental melon for the first time this year, so we all ate that together.

Today's income:

104 eggplants	1 yen
67 tomatoes (2 *kan* 200 *momme* (8.25 kg))	60 sen
2 pumpkins (700 *momme* (2.6 kg))	15 sen
cabbages (1 *kan* (3.75 kg))	15 sen

At the main house: Kanezō cut hay early and then again after breakfast, and he overturned the compost and piled it up again with my brother—just as I explained before—in the afternoon. In the evening, he went out to cut more hay. My brother's wife went up into the hills in the morning to do what's called *makkō o kogu*. This means to pick green leaves of certain kinds of trees, lay them in the sun to dry, grind the crunchy dry leaves in a mortar, put it through a sifter (*kona-toshi*), and then burn the powder and offer it to the household ancestors (*hotoke-sama*) at the family Buddhist altar (*butsudan*). Here in the village, in local speech, this is called *shin-makkō o koede, hotoke-sama ni ageru* [to make new *makkō* and offer it to *hotoke-sama*]. A variety of leaves can be used: *sakura, agahira, gomagi, nemu no ki, kuwa* (mulberry), etc. Among these, silk tree (*nemu no ki*) leaves make the best *makkō*. Because of this, people sometimes call silk trees *makkō* trees. In the afternoon, my brother's wife weeded the rice seedling nursery paddies (*nashiro*) and Ichisuke minded the shop because my brother was dealing with the compost. My mother cooked, etc.

August 10, Saturday

Sunny—very fine weather
16°C at 6:00 a.m., 30°C at 1:00 p.m.

Breakfast:	cowpea miso soup with *kuromo* seaweed in it, salted cucumbers, *takuan*. We received the *kuromo* from my wife's brother's family. It was from Kita Iso.[4]
Lunch:	miso-preserved cucumbers and *takuan*
Supper:	whale-stock eggplant miso soup, salted cucumbers, *takuan*

I harvested cabbages and cut grass for the rabbits early, then had breakfast. My wife didn't go to Funagawa today; instead, she made breakfast and after eating went to the Wakimoto market to sell cabbages. I plowed the pumpkin fields and planted buckwheat seeds in those fields in the morning. I don't have any night soil or compost now, so I just planted the buckwheat without using anything. But anyway, people have said since long ago that buckwheat makes a field more fertile (*hatake o koyasu*), so I think that not using any fertilizer should be okay. In the afternoon I went down to the Wakimoto market to have a look around. My wife came back by noon with a bunch of things she'd bought, she boiled agar-agar (*tengusa*) and made *tokoroten* with it in the afternoon. I made lunch today. I came back home early in the evening and picked vegetables for selling tomorrow.

Today's income:

170 eggplants	1 yen 20 sen
24 tomatoes (800 *momme* (3 kg))	24 sen
2 pumpkins (700 *momme* (2.6 kg))	15 sen

Today's expenditures:

1 hat (*hitomoji shapo*)[5] for me	45 sen
spending money for my wife	2 yen
kimono cloth (1 bolt)	1 yen 50 sen
1 salted trout	50 sen
whale (blubber)	15 sen
summer *geta* for the kids (2 pairs)	30 sen
2 scrubbing brushes (*tawashi*)	5 sen
3 teacups	10 sen
agar-agar (100 *momme* (375 g))	25 sen

At the main house: Kanezō cut hay early, and then cut *bukkomi-gusa* after breakfast. In the afternoon, he went down to the Wakimoto market to have a look (see Fig. 6.1). But he also cut more hay. My brother's wife weeded the nursery paddies (*nashiro*) again in the morning, and in the afternoon, she went to the market, too. My mother cooked and such.

August 11, Sunday

(July 13 according to the old calendar)

Sunny and very nice
18°C at 6:00 a.m., 32°C at 2:00 p.m.

Breakfast:	cowpea and eggplant miso soup, salted cucumbers, *takuan*
Lunch:	*tokoroten* [explanation below] with soy sauce for flavor, roasted salted trout, salted cucumbers, etc.
Supper:	I'll write later.

I got up at three o'clock, strapped a basket (*kago*) of vegetables for selling on my back, and walked all the way to Funagawa with it. I left the basket at the home of someone I know and then came back. The vegetables I took were for my wife to sell. The kids were sleeping during that time. I got back at six o'clock and started making breakfast. After eating, I weeded more in the pumpkin patches, plowed a bit again, and planted more buckwheat—as I did yesterday. I also cut grass for the rabbits. We ate lunch at twelve-thirty. My wife sold two buckets full of vegetables this morning and returned by lunchtime. In the afternoon, I thinned out my first-planted napa cabbage crop for the third time, leaving only two sprouts in each spot. Next, I did the second thinning on my second-planted napa cabbage crop, leaving three of four sprouts per spot. After that, I poured some *rokueki* insecticide again. Once that was done, I cut more grass for the rabbits and then we had supper a

Figure 6.1. The crowded Wakimoto market. From Yoshida's personal collection: not included in 1938 book.

little on the early side this evening since it's *obon*—July 13—according to the old calendar. We prepared and ate as much food as we could. My wife spent the afternoon preparing the following dishes:

steamed eggplants
momi cucumbers
soup consisting of *konnyaku*, eggplant, and cowpeas, flavored with soy sauce
cabbage with soy sauce for flavor
red bean rice (*sekihan*)
tokoroten with soy sauce for flavor

To make *ten* [*tokoroten*], you take a piece of dried agar-agar (*tengusa*) and pound it thoroughly with a pipe or tube to remove any bits of rock or shells or sand, etc. Then, when it's all clean, you wash it very thoroughly, and put it in a pot—completely submerged in water—and simmer it. After the water starts to boil, you add 3 *shaku* (54.1 ml) or so of vinegar per 100 *momme* (375 g) of agar-agar to the pot. And you have to stir it some while it simmers. Once the agar-agar has dissolved, and the liquid has become nice and thick, you pour the liquid into a shallow container in which you've spread out a cloth bag. You strain it once like this, but some people do it twice, I hear. If you cool it down, it will firm up, and then you just chop this up into small pieces and eat it with soy sauce for flavor. It's very tasty, but not so easily digested, as a matter of fact.

Today's income:

280 eggplants	2 yen 3 sen
102 tomatoes (3 *kan* 400 *momme* (12.75 kg))	95 sen
10 pumpkins (2 *kan* 800 *momme* (10.5 kg))	50 sen
cabbages (10 *kan* 200 *momme* (38.25 kg))	1 yen 70 sen

Today's expenditures:

soy sauce (1 *shō* (1.8 L))	30 sen
white and brown sugar	10 sen

At the main house: Kanezō took the horse out to cut *bukkomi-gusa* very early—before sunrise. He returned around seven o'clock with one horse load, and after breakfast he went up into the hills to cut hay. So, by lunchtime he had cut two horse loads of grass and straw and such. In the afternoon he cleaned up around the house, finished a bit early, ate supper, and then went out to spend his free time as he pleased. My brother's wife weeded the nursery paddies (*nashiro*) in the morning, and helped my mother prepare food for this evening's feast in the afternoon.

August 12, Monday

[Saburō's wife, Katsue, wrote this entry and the next three as well.]
Sunny
18°C at 6:00 a.m., 28°C at 1:00 p.m.

Breakfast:	exactly the same as yesterday's supper
Lunch:	[*tokoro*]*ten* with soy sauce for flavor, *momi* cucumbers, salted cucumbers, etc.
Supper:	eggplant miso soup, *momi* cucumbers, fried cabbage, etc.
Snack:	*shiratama mochi* [small *mochi* balls] with sugar

Two friends of mine who live in a different village came over to visit.

My husband went to the Towada Lake area on an agriculture-related trip today.[6]

I did laundry and entertained the children during the morning, and cut grass for the rabbits in the afternoon. I took off the rest of the day.

Today's income:

72 eggplants	60 sen
16 tomatoes	15 sen
cabbages (1 *kan* 600 *momme* (6 kg))	32 sen

Today's expenditures:

cooking oil	5 sen
candies	5 sen

At the main house: Kanezō cut hay for the horse early in the morning, but he took the rest of the day off—he went to his natal home for a while in the afternoon. My husband's brother's wife did some sewing and cooking, etc. My husband's brother minded the shop and went out on some errands.

August 13, Tuesday

Sunny
18°C at 6:00 a.m., 15°C at 1:00 p.m.

Breakfast:	cowpea miso soup, *takuan,* fried eggplant with soy sauce for flavor
Lunch:	*mochi, ten* with soy sauce for flavor, salt-pickled cucumbers, etc.
Supper:	eggplant miso soup, *ten* flavored with soy sauce, *shishizuke* cucumbers[7]
Snack:	(late at night) *udon* noodles in kelp (*konbu*) soup

Four people (both male and female) came over to visit.

I made breakfast this morning. After breakfast, I did the first thinning of the daikon crop, and while doing that I sprinkled soil around each plant by moving my hand in a circular motion. I left four or five sprouts in each spot. I also weeded in the cabbage fields and cut some more grass for the rabbits. In the afternoon I took off from such work and did some sewing and entertained the children.

Today's income:

25 eggplants	20 sen
42 tomatoes	41 sen
cabbages (3 *kan* 500 *momme* (13.1 kg))	47 sen

Today's expenditures:

candies	5 sen
candles	1 sen

At the main house: Kanezō took the horse out in the morning and cut *bukkomi-gusa,* returning later with one horse load. He took off in the afternoon. My husband's brother watched the shop. My husband's brother's wife went out in the morning to pick a variety of vegetables for offering to the spirits of the ancestors. She took off from farmwork in the afternoon and did some sewing and cooking at home.

August 14, Wednesday

Cloudy
Southwest wind
17°C in the morning, 27°C at 1:00 p.m.

Breakfast:	I took the kids with me to my natal home to eat. We ate miso soup with *natsuna*[8] and eggplant in it, grated daikon, and chrysanthemum leaves with soy sauce.
Lunch:	roasted salted salmon, *takuan, ten* with soy sauce
Supper:	cowpea and cabbage miso soup, salted cucumbers, *takuan*, etc.

Five people came over to visit.

In the morning I weeded in the cabbage fields, cut grass for the rabbits, and picked vegetables for selling tomorrow. In the afternoon, I took off [from farmwork] and did some reading, sewing, cooking, etc.

Today's income:

64 eggplants	60 sen
79 tomatoes	55 sen

Today's expenditures:

candies	5 sen

At the main house: My husband's brother and Kanezō overturned the compost during the morning (see Fig. 6.2). Both took off in the afternoon, and each went out on his own. My husband's brother's wife went to the fields and thinned the daikon crop in the morning, leaving four or five sprouts in each spot, and spread some soil around them. She also picked some cowpeas before returning home. In the afternoon, she took off and sewed and cooked while minding the shop.

August 15, Thursday

Sunny
Southwest wind
19°C at 6:00 a.m., 26°C at 1:00 p.m.

Breakfast:	*nigiri-meshi* (*onigiri*) eaten with miso-preserved cucumbers
Lunch:	*ten* with soy sauce, *takuan*, etc.
Supper:	eggplant miso soup (stock made from little sardines we received from the main house), roasted salted salmon

Figure 6.2. Cutting and turning over the compost in front of the main house.

I got up quite early today and went out to sell vegetables. I returned by lunchtime and cut grass for the rabbits and prepared lunch. In the afternoon, I didn't do farmwork—I did some sewing and such. But I picked vegetables for selling, of course.

Today's income:

418 eggplants	2 yen 60 sen
6 tomatoes	5 sen
cabbages (1 *kan* 600 *momme* (6 kg))	30 sen

Today's expenditures:

apron (*maekake*)	15 sen
candies	5 sen

At the main house: Kanezō cut hay for the horse early, and after breakfast he cut *bukkomigusa*. He didn't work in the afternoon. My husband's brother's wife picked cucumbers in the fields.

August 16, Friday

Sunny
Southeast wind
12°C at 6:00 a.m., 28°C at 1:00 p.m.

Breakfast:	leftover soup from last night, salted cucumbers, etc.
Lunch:	roasted salted trout, pickled cucumbers, *ten* with soy sauce
Supper:	eggplant and potato miso soup (stock made from whale), [*tokoro*]*ten* with soy sauce, *takuan*

I was away on a trip until yesterday, so I slept in. After I got up, I made breakfast. I walked around my fields to have a good look at them in the morning, and cut grass for the rabbits. My wife got back by noon and both of us rested after lunch. But I did go out and pick vegetables in the evening for selling.

Today's income:

110 eggplants	65 sen
4 pumpkins (1 *kan* (3.75 kg))	20 sen
cabbages (800 *momme* (3 kg))	15 sen
70 tomatoes (2 *kan* (7.5 kg))	60 sen

Today's expenditures:

polished rice (1 *shō* (1.8 L))	27 sen
postage for mailing photo film	10 sen

At the main house: Kanezō cut hay early, and after breakfast he cut *bukkomi-gusa*. Later, he cut hay again and chopped it up. He rested in the afternoon. My brother and his wife did some work in the family sweets shop.

A new communal work shed for the hamlet (*buraku*) was built in the dancing area, so the young fellows of the village—all villagers, actually—are unable to dance for the Bon Festival (*obon*),[9] so it was called off this year. The young fellows seem especially sad about not being able to dance for the Bon Festival (see Figs. 6.3 and 6.4).

August 17, Saturday

Sunny
North wind
20°C at 6:00 a.m., 28°C at 1:00 p.m.

Breakfast:	eggplant and cabbage miso soup, pickled cucumbers and *takuan*, etc.
Lunch:	eggplant and whale soup (soy sauce flavored), roasted salted trout, pickled cucumbers, *takuan*
Supper:	cowpea and whale miso soup, roasted salted herring, pickled cucumbers, *takuan*

Figure 6.3. Ōkura hamlet Bon festivities (photo probably taken in 1936). From Yoshida's personal collection: not included in 1938 book.

Figure 6.4. Ōkura hamlet Bon Festival dancing (photo probably taken in 1936). Yoshida's mother is in front, fourth from right. His wife's mother is at far right. From Yoshida's personal collection (detail): not included in 1938 book.

I cut some grass early, and cleaned out the rabbit hutch and put fresh grass in it for them. Then I made breakfast. After eating, I cleaned out part of the cabbage fields and planted more buckwheat—without using any fertilizer, as before. In the afternoon, I took off. My wife came back by lunchtime, and she also took off in the afternoon. I picked vegetables for selling in the evening.

Today's income:

140 eggplants	82 sen
25 tomatoes (800 *momme* (3 kg))	20 sen
4 pumpkins (1 *kan* 500 *momme* (5.6 kg))	30 sen
cabbages (4 *kan* (15 kg))	60 sen

At the main house: Same as yesterday for Kanezō, and off in the afternoon. My brother's wife went out to the fields in the morning and picked cowpeas and cucumbers, and she took off in the afternoon. My mother was out, so my brother's wife took care of the cooking, etc.

August 18, Sunday

Sunny, with some clouds
North wind
17°C at 6:00 a.m., 28°C at 1:00 p.m.

Breakfast:	eggplant and cowpea miso soup, pickled cucumbers, *takuan*
Lunch:	red bean rice, fried cabbage with soy sauce, roasted salted herring, pickled cucumbers, *takuan*
Supper:	eggplant, cowpea, and whale miso soup, roasted salted herring, pickled cucumbers, *takuan*

Four people—old friends from elementary school, and even the teacher—came over to visit.

Early this morning, I prepared to make insecticide again: fetching water, measuring the soap, etc. After breakfast, I mixed up the *rokueki*. Later, I did the third thinning to the second-planted napa cabbages, and I did the second thinning to the third-planted napa cabbages. After that, I administered the insecticide. I took off in the afternoon. My wife picked vegetables for selling in the evening.

Today's income:

90 eggplants	55 sen
56 tomatoes (1 *kan* 600 *momme* (6 kg))	35 sen
6 pumpkins (1 *kan* 500 *momme* (5.6 kg))	30 sen
cabbages (2 *kan* 300 *momme* (8.6 kg))	40 sen

Today's expenditures:

(school) alumni association party fee	1 yen 50 sen

processed sugar	10 sen
candies	5 sen

At the main house: Kanezō cut hay for the horse early, and cut *bukkomi-gusa* after breakfast. In the afternoon, even though it would have been fine for him to take off and rest (because of *obon*), he went up into the hills to cut more hay. My brother's wife went out to the fields in the afternoon and did exactly the same things as yesterday. She also cooked, and watched the shop in the afternoon.

August 19, Monday

Sunny, with some clouds
Strong southwest wind
22°C at 6:00 a.m., 28°C at 1:00 p.m.

Breakfast:	eggplant miso soup with *kuromo* in it, pickled cucumbers. We got the *kuromo* from an acquaintance in Funagawa.
Lunch:	miso-preserved cucumbers, roasted salted trout, *takuan,* etc.
Supper:	cucumber and cowpea miso soup (stock made from whale fat) with *kuromo* in it, roasted salted trout, *takuan*

Six village women came over to visit.

I thinned my Aokubi Hōryō daikon[10] crop for the first time early, before making breakfast. After eating, I picked cucumbers, and pulled up the plants that were done producing. Later, I went down to the main part of Wakimoto Village (*honson*) on an errand. In the afternoon, I had to run another errand—this time to a different village. My wife made it back by lunchtime. She prepared for preserving cucumbers in the afternoon.

Today's income:

111 eggplants	75 sen
42 tomatoes (1 *kan* (3.75 kg))	25 sen
6 pumpkins (1 *kan* 700 *momme* (6.4 kg))	33 sen
cabbages (1 *kan* 400 *momme* (5.25 kg))	25 sen

Today's expenditures:

1 book	2 yen

At the main house: Same as yesterday for Kanezō this morning, but he took off in the afternoon, since he worked all day yesterday without a rest. He went to a different village to watch a sumo tournament. My brother's wife did some weeding and such in the fields. My mother cooked, etc.

August 20, Tuesday

Sunny, with some clouds
Strong southwest wind
24°C in the morning, 28°C in the afternoon

Breakfast:	cowpea and eggplant miso soup with *kuromo* seaweed in it, *takuan*
Lunch:	fish I received in Funagawa cooked with soy sauce and eggplants, *takuan.* I have no idea what kind of fish it was.
Supper:	eggplant miso soup (stock made from whale) with *kuromo* in it, roasted salted herring, *takuan,* etc.

I cleaned up the poles I'd used for growing cucumbers and cut grass for the rabbits early, and started making breakfast at six o'clock. After eating, I did *kusakezuri*[11] in the cucumber fields, and then weeded. I did the same in the afternoon. After weeding I stuffed the weeds in a *goshin* and strapped it on my back and carried it to the compost area and dumped it. My wife did the same work from the afternoon. In the evening, I picked more vegetables for selling.

Today's income:

170 eggplants	88 sen
90 tomatoes (2 *kan* 400 *momme* (9 kg))	52 sen
3 pumpkins (1 *kan* 400 *momme* (5.25 kg))	30 sen
cabbages (1 *kan* 800 *momme* (6.75 kg))	30 sen

Today's expenditures:

salt (1 *shō* (1.8 L))	10 sen
fish	10 sen
candies	5 sen

At the main house: Kanezō cut hay early. After breakfast, he went up to Mt. Kanpū to participate in "dividing the mountain" (*higusa yama o wakeru*). Mt. Kanpū is divided into multiple areas—each with its own name, such as Sabugaze,[12] Nashi no Kidai,[13] Yosobe Yama, Hachimori Tonke, Udo, Urata Iri, etc. The divisions are made according to the size of each hill—not per household, but per horse—and decided every year according to the number of (eligible) horses, which fluctuates. For example, this year the mountain was divided in the following way:

Sabugaze	8 horses
Nashi no Kidai	13 horses
Yosobe Yama	2 horses
Hachimori Tonke	10 horses
Udo	16 horses
Urata Iri	6 horses

In this way, the mountain is divided into large sections, and each horse is assigned to one. Then, they further divide each area by the predicted number of horses. The way they decide the divisions is like this: At the first stage (*ō-wake*), for example, they prepare a bunch of small pieces of paper, equal to the number of horses. Then they assign a number to each area of the mountain, such as one for Sabugaze, two for Nashi no Kidai, three for Hachimori Tonke, four for Yosobe Yama, five for Udo, and six for Urata Iri. Then, they write a number on each paper like this: if, say, Sabugaze is to be divided among six horses, they write "1" on six pieces of paper, and so on. Then, they put all the papers in the hat and mix them up. After that, they divide the village into four areas: Iwakura, Higashiguchi, Nishiguchi, and Sawa. One representative from each area participates in a drawing that has been prepared separately—picking a number from one to four. Whoever gets number one . . . well . . . for example, if the Higashiguchi representative gets number one, then the Higashiguchi [horse-owning] residents will begin drawing lots from the hat first. Then, whoever draws a "1" goes to Sabugaze, whoever gets a "3" will go to Hachimori Tonke, whoever ends up with a "5" will be assigned to Udo, etc.—divided up just like that. And, to make the boundaries between these territories clear, for example, they'll build up a little row of soil between Hachimori Tonke and Udo along the dividing line. And once the general dividing (*kowake*) is done, they have to decide on how to share each area. It works like this: For example, if eight horses are to use a certain area, they'll generally split it up into eight parts, or maybe into sixteen smaller parts. Then they cut grass along the boundaries to make them clear. Next, as one might expect, they number the separate parts, just like the way that the first divisions were decided, and they number each area—one, two, three, etc.—and then make lots. So, in this example case, they'd prepare eight thin ropes and use these as lots. Then they'd weave them together, as shown in the diagram (see Fig. 6.5).

After weaving the ropes together as shown, they'd roll it all up so no one can tell which rope is which. Each man would grab a rope and hold on to it. Then, they'd unravel the whole thing, and the man holding the first one to come out, would get part number one, the second, part number two, and so on. In this way, each house's territory on the mountain is decided. The man who prepares the lots for this final stage gets the last rope left at the end.

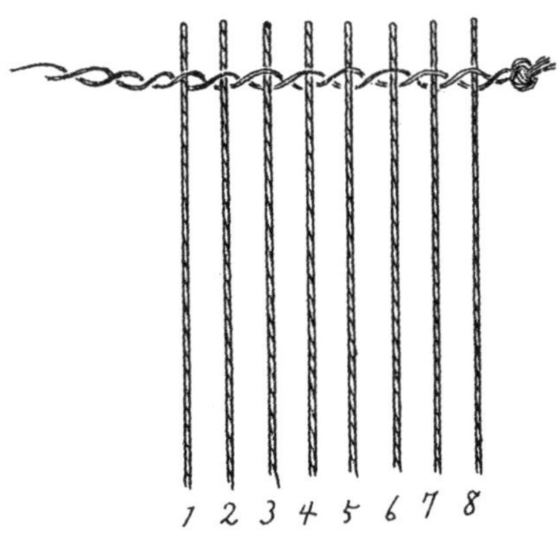

Figure 6.5. Ropes used as lots for deciding how to divide the mountain.

This is how Mt. Kanpū is divided. Today, it was decided that the main house can use part of the Udo area. Kanezō came back down from the mountain by lunchtime, and in the afternoon he and my brother made compost. They really have quite a pile of it at the main house now. My brother's wife went out to the fields in the morning and removed bugs from the daikon crop, etc. She minded the shop in the afternoon. My mother cooked and such.

August 21, Wednesday

Intermittent rain
Southwest wind
24°C in the morning, 27°C in the afternoon

Breakfast:	cowpea and eggplant miso soup, *takuan*
Lunch:	roasted salted trout, *takuan*
Supper:	cowpea and potato miso soup with *kuromo* in it, roasted salted trout, *takuan*

Two people came from a different village to see my farm.

Early this morning I re-plowed my cucumber field and did the fourth thinning on my first-planted napa cabbage crop, leaving only one plant standing in each spot. Then I dug a little hole on one side of each plant and poured three full buckets of a 1:1 mixture of night soil and water, in which I'd dissolved 300 *momme* (1.1 kg) of ammonium sulfate, into all the holes. I was able to do all of this by six o'clock because I started working so early this morning. Then, I made breakfast. After eating, I planted more buckwheat in the cucumber field—without using any fertilizer, as before. Then I filled in all the fertilizer holes in the napa cabbage patch. At ten o'clock I went over to my old school to hear Obara Kuniyoshi[14] speak about his philosophy on education. But I left at one-thirty, while he was still speaking, and came back home because I had so much work to do. In the afternoon, I planted more buckwheat, harvested vegetables for selling, etc. I finished all my buckwheat planting today. All in all, I planted 5 *shō* (9 L) of buckwheat seeds. In the afternoon, my wife weeded in the napa cabbage fields. When I fertilized the napa cabbages this morning, that was the first time to do it (this year).

Today's income:

80 eggplants	45 sen
124 tomatoes (2 *kan* 300 *momme* (8.6 kg))	55 sen
4 pumpkins (1 *kan* 300 *momme* (4.9 kg))	25 sen
cabbages (2 *kan* 500 *momme* (9.4 kg))	40 sen

Today's expenditures:

agricultural chemicals	65 sen

At the main house: Kanezō cut hay early, and he and my brother's wife took lunch with them up into the foothills of Mt. Kanpū to cut hay there. My mother cooked.

August 22, Thursday

Cloudy, with rain
Strong southwest wind
24°C in the morning, 26°C in the afternoon

Breakfast:	cabbage miso soup, pickled cucumbers with soy sauce, *takuan,* etc.
Lunch:	roasted salted trout, pickled cucumbers, *takuan,* etc.
Supper:	We ate a one-pot soup dish containing whale and daikon leaves that we'd gotten from thinning the daikon crops, flavored with miso. Side dishes: same as lunch

Early this morning, I hauled a bunch of weeds and such that we had pulled out of the fields before on my back over to the compost heap, and made breakfast after that. After eating, I had to go out and do more of the same work. Once all that was done, I did the second thinning on my daikon crop, leaving only two plants standing in each spot (*nihondachi*). In the afternoon, I prepared 2 *to* (36.1 L) of four-*to*-style Bordeaux mixture for my fields, and I used it on my daikon, napa cabbages, and cabbages, etc. But rain fell in the evening, which nullified my efforts. I also harvested vegetables for selling, among other things, in the afternoon. My wife helped me administer the fungicide [Bordeaux mixture] in the afternoon, and after that she weeded in the napa cabbage fields.

Today's income:

85 eggplants	45 sen
3 pumpkins (1 *kan* 200 *momme* (4.5 kg))	25 sen
1 watermelon (900 *momme* (3.4 kg))	20 sen
cabbages (2 *kan* 500 *momme* (9.4 kg))	40 sen

At the main house: Same as always in the early morning for Kanezō. Every day the horse must be taken out and bathed in water at least two times—maybe even three times—without fail. Again today, Kanezō took his lunch out with him and spent the whole day cutting hay up in the foothills of the mountain. My brother's wife didn't go today—she went out to the fields and picked bugs off the daikon, and picked cucumbers, before going home. In the afternoon, she sewed while minding the store. Ichisuke took care of bathing the horse. My brother's wife and my mother cooked together today.

August 23, Friday

Cloudy, with rain
Strong southwest wind
21°C in the morning, 26°C in the afternoon

Breakfast:	eggplant and cabbage miso soup, *takuan*
Lunch and supper:	basically, the same as yesterday

I cut grass for the rabbits early. Then I made breakfast. I finished eating by seven o'clock, and then went up into the hills (*higusa yama*)[15] to cut hay for the main house. My wife weeded around the cabbages and such in the afternoon.

Today's income:

80 eggplants	40 sen
37 tomatoes (1 *kan* (3.75 kg))	25 sen
2 pumpkins (800 *momme* (3 kg))	20 sen
cabbages (2 *kan* 300 *momme* (8.6 kg))	35 sen

At the main house: Kanezō did his usual early morning work, and then (after breakfast) he took his lunch with him up to the mountainside to cut hay. He spent the whole day on that again. My brother's wife went out to cut hay today, too.

The *higusa yama* is kind of steep, so when you cut hay there you just work your way up the incline from down below, moving across as you slowly advance, and then let the hay you've cut dry in the sun a bit. The way to cut is as follows: You cut a path about 2 *shaku* (60.6 cm) to 2 *shaku* 5 *sun* (75.8 cm) wide, with the blade angled upward, for a distance of about 4–5 or maybe 6 *ken* (7.2–10.8 m), moving across the mountainside. Then you go back across, cutting downward, so you get a long row of cut hay all piled up. At this point, you have cut one row, or *ippon*. But you've got to sharpen your scythe after cutting each row, because if the blade doesn't cut well the hay won't pile up right. There's no point in cutting if the hay doesn't pile up properly. That's why my father always used to tell me to spend more time sharpening the scythe than actually cutting hay. Even now, when I think about it, he really was right on the mark. My mother cooked and such today.

August 24, Saturday

Rain, later just cloudy
Southwest wind
20°C in the morning, 24°C in the afternoon

Breakfast:	eggplant and cowpea miso soup with dried kelp shavings (*tororo konbu*) in it, also: fried eggplant with soy sauce for flavor and *takuan*
Lunch:	same as yesterday
Supper:	eggplant miso soup (stock made from whale), side dishes: same as lunch

I didn't work outside this morning because of the heavy rain. My wife stayed home and made breakfast, etc. During the morning, I went down into the main part of the village on an errand. My wife went out and bought some salt, which she used to pickle cucumbers, and she also did laundry and such. In the afternoon, she went out to sell vegetables. As for me, I just took it easy for a while, but I did harvest some vegetables and such in the evening.

Today's income:

135 eggplants	90 sen
20 tomatoes (500 *momme* (1.9 kg))	10 sen

Today's expenditures:

salt	20 sen
candies	5 sen

At the main house: Kanezō gave up on cutting hay for the horse early this morning because of the rain, and instead he worked on fixing the saddle. But he finished that pretty early, so he went out to cut some hay among the paddies and the dry fields. In the afternoon, he took the horse out with him and cut *bukkomi-gusa*. My brother's wife sewed, and she and my mother cooked together, etc.

August 25, Sunday

Rain
Southwest wind
19°C in the morning, 20°C in the afternoon

Breakfast	eggplant and cabbage miso soup, *takuan*, pickled cucumbers
Lunch:	potatoes boiled in salt water, roasted salted herring, *takuan*, etc.
Supper:	eggplant miso soup (stock made from whale), roasted salted trout, pickled cucumbers, etc.

I cut grass for the rabbits early, and cut weeds and grass for composting. Later, I prepared breakfast. I didn't do any outside work after eating because of the rain. In the end, I was just idle for pretty much the whole day. My wife didn't make it back until evening. She didn't do much else today other than pick vegetables for selling tomorrow.

Today's income:

84 eggplants	50 sen
50 tomatoes (1 *kan* 200 *momme* (4.5 kg))	30 sen
watermelons (2 *kan* 600 *momme* (9.75 kg))	45 sen
cabbages (1 *kan* 800 *momme* (6.75 kg))	20 sen

Today's expenditures:

polished rice (1 *shō* (1.8 L))	25 sen

At the main house: Kanezō cut hay early, and after breakfast he took the horse up into the hills to cut more. He cut *bukkomi-gusa* in the afternoon. My brother's wife stayed home and sewed and cooked and such, but she went to the fields in the evening and picked cowpeas and cucumbers, etc.

Today I'll explain how to bundle *bukkomi-gusa.* Everybody has their own way, but there really isn't much difference. I'll write about my own experience. First, when you go up into the hills to cut *bukkomi-gusa,* you want to cut the longest stuff you can find. Generally, to make for one *maro,* you need two *sanba tsunagi* bunches. You have to set the two bunches down about 2 *shaku* (60.6 cm) apart. Then, you overlap them so the ends of the straw you cut are sticking out on both sides. Then, you lay the ends of the straws over the "ropes." Then, you arrange all the ends of the straws of each bunch evenly and tie them all up as tightly as possible. If you pull them together tightly, the center part will bunch up nice and round (see Fig. 6.6). If you pile these up on a horse, it looks good. Some young guys even break their bones trying to tie up bundles like this.

August 26, Monday

Rain, later cloudy
Southwest wind
20°C in the morning, 22°C in the afternoon

Breakfast:	eggplant and cabbage miso soup, pickled cucumbers, *takuan,* etc.
Lunch:	*soba nereke* with soy sauce, *takuan,* etc.
Supper:	eggplant and cowpea miso soup (stock made from whale), roasted salted herring, *takuan*

I cut grass for the rabbits, and also cut compost straw and stuff like that, in the early morning. Then, at seven o'clock I started making breakfast. At around nine o'clock heavy rain started falling, so I had to quit working outside. In the afternoon I went out on errands, and I harvested vegetables for selling in the evening. My wife got home before lunchtime, and she did laundry and sewing and such in the afternoon.

Today's income:
70 eggplants 35 sen

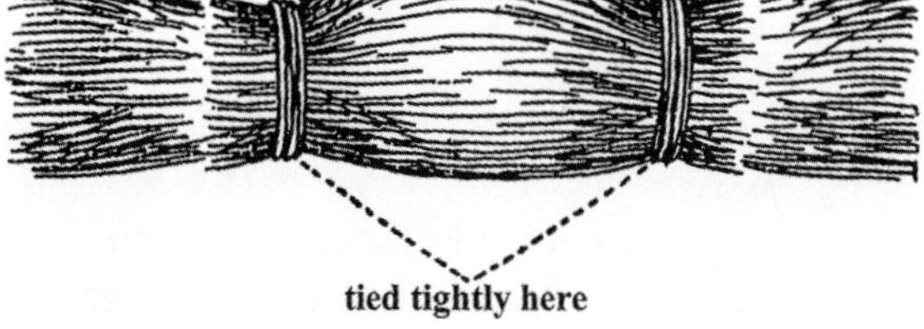

Figure 6.6. A *bukkomi-gusa* bundle.

31 watermelons (3 *kan* 600 *momme* (13.5 kg))	62 sen
cabbages (1 *kan* 700 *momme* (6.4 kg))	30 sen

Today's expenditures:

cotton cloth for pillows	20 sen
mochi	5 sen

At the main house: Kanezō cut hay for the horse early, and cut *bukkomi-gusa* later. In the afternoon, he went out to cut more hay for the horse among the dry fields, and in the rice paddies. Ichisuke went out in the evening to get the horseshoes taken care of. My brother's wife did some sewing, and she cooked with my mother, etc. In the evening, she washed potatoes. This is for having starch extracted from them by machine tomorrow.

August 27, Tuesday

Sunny
West wind
20°C in the morning, 28°C in the afternoon

Breakfast:	eggplant miso soup, *takuan*
Lunch:	potatoes boiled with salt, roasted salted trout, *takuan,* etc.
Supper:	eggplant and cowpea miso soup, boiled cabbage flavored with soy sauce, side dishes: same as lunch

Early this morning, I cut grass for the rabbits. Then I ran into the village on an errand, and then I made breakfast. After breakfast, I retied a number of my eggplant plants to their poles because many of the knots I'd already tied came apart in the strong wind and rain we had recently. After that, I cut compost straw. Then I made lunch. My wife got back by lunchtime, and she weeded in the green onion fields in the afternoon—by hand. I cut more compost straw in the afternoon. In the evening, I harvested vegetables for selling, as I do every day.

Today's income:

120 eggplants	60 sen
40 tomatoes	20 sen
4 watermelons (5 *kan* 500 *momme* (20.6 kg))	1 yen 5 sen
cabbages (3 *kan* (11.25 kg))	40 sen

At the main house: Kanezō cut hay for the horse in early morning, and after breakfast he took his lunch with him up into the hills (*higusa yama*) to cut more hay. That was the last of the hay-cutting for this year (up in the hills). Ichisuke took the horse and washed it off. My brother's wife got the potatoes ground up by machine, and then had starch extracted from them. She also did laundry, and she cooked with my mother, etc.

August 28, Wednesday

Sunny
Light southeast wind
22°C in the morning, 30°C in the afternoon

Breakfast:	eggplant miso soup, pickled cucumbers, *takuan,* etc.
Lunch:	roasted salted trout, pickled cucumbers, *takuan,* pickled eggplants, etc.
Supper:	cowpea and eggplant miso soup, side dishes: same as lunch
Snack:	Lately, our daily snacks consist of oriental melon (*amauri*), watermelon, and the like.

I weeded the green onion field from early morning. Then, I made breakfast at about seven o'clock. After eating, I did the first follow-up fertilizing to the green onions. And for the ones I planted in the plot that had been used for peas, I sprinkled 1 *kan* 800 *momme* (6.8 kg) of ammonium sulfate around their bases. Right after that, I pulled soil up around their roots for the first time this season. I didn't pull soil up from both sides of the onions—only from one side. In other words, I scooped soil up from the trough I'd dug for planting them [did *tsuchiyose*]. In the afternoon, I went to the station area to buy rolled oats. After I got back, I picked vegetables for selling and cut grass for the rabbits. My wife came back by lunchtime. She weeded in the napa cabbage and green onion fields in the afternoon.

Today's income:

125 eggplants	63 sen
15 tomatoes (300 *momme* (1.1 kg))	5 sen
3 pumpkins (1 *kan* 600 *momme* (6 kg))	25 sen
cabbages (2 *kan* 400 *momme* (9 kg))	35 sen

Today's expenditures:

rolled oats (1 bag)	2 yen 40 sen

At the main house: Kanezō went out to cut hay for the horse early, and after breakfast he hauled two barrels of night soil out to the fields using the horse. With my brother's wife's help he spread it on the green onion fields—follow-up fertilizer. In the afternoon, my brother's wife went back out and picked bugs off the daikon and picked the dry leaves off the daikon as well. Kanezō took the horse up into the hills—to the *higusa yama.* By himself, he used a *komazari*[16] to collect hay, bundled it, and brought one horse load—seven big bundles—back

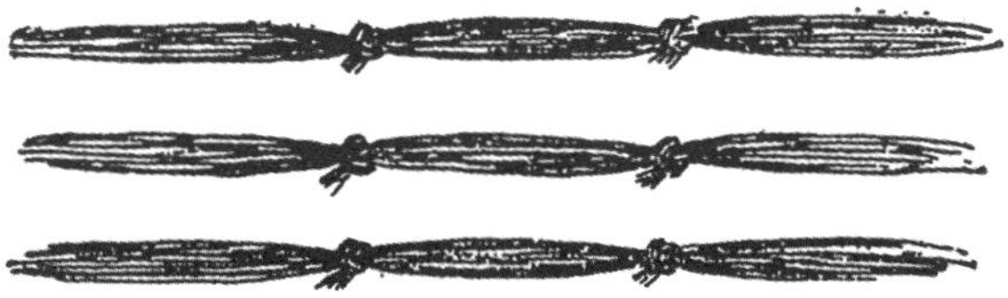

Figure 6.7. Straw *tsunagi* ropes for tying *marogi* bundles.

with him. Each of these *marogi* bundles is *midoko tabake.*[17] You tie these up with three *sanba tsunagi* or *yonba tsunagi*,[18] by laying the hay down on top of the straw *tsunagi* ropes, as shown in the diagram here, and tying it all up (see Fig. 6.7).

In the old days, people used to weave two rice plant straws together—especially weaving the tips—to make these *tsunagi,* so they were called *futabatsunagi* (two-straw *tsunagi*). But today folks feel that it is too much trouble to weave everything, so they simply make *tsunagi* and use them without weaving.

Once cut hay that has been lying in the sun for a while gets nice and dry, you gather it all up using a fine-toothed wooden rake (*konohasarai,* or *komajari* in local speech) and then bunch the hay together to make lengths of about 3 *shaku* (90.9 cm). Three of these make one *marogi* bundle.

August 29, Thursday

(August 1 according to the old calendar)

Cloudy, later rain
Southeast wind
22°C in the morning, 22°C in the afternoon

Breakfast:	eggplant and cowpea miso soup, pickled eggplants, *takuan*
Lunch:	red bean rice (*sekihan*), bonito (*katsuo*) bought at Funagawa cooked with eggplant in soy sauce, pickled eggplants, *takuan*
Supper:	*sekihan,* eggplant miso soup (stock made from whale), roasted salted trout, pickled eggplants, *takuan*
Snacks:	same as yesterday

Early this morning, I gave the good cabbage leaves I picked off the plants to the rabbits and I piled the bad ones on the compost heap. Then, I made breakfast. After eating, I did more of the same work. Also, I pulled rotten rice straws out of my hotboxes and prepared them for using as compost. In the afternoon it was raining, for one thing, but it was also August 1 according to the old lunar calendar, so all the villagers took off and rested. I did the same. In the evening, I picked vegetables for selling. My wife got back by lunchtime, and she sewed and such during the afternoon.

Today's income:

110 eggplants	53 sen
15 tomatoes (300 *momme* (1.1 kg))	7 sen
4 watermelons (5 *kan* 400 *momme* (20.25 kg))	95 sen
cabbages (2 *kan* 600 *momme* (9.75 kg))	35 sen

Today's expenditures:

1 conical hat (*karakasa*)	60 sen
fish-shaped candies	15 sen

At the main house: Kanezō cut hay in the early morning, as usual, and then later in the morning he and my brother's wife took the horse up into the hills with them to gather more hay. She collected it with a wooden fine-toothed rake (*komajari*) and he tied and bundled the hay. My brother brought the horse back down from the mountain. Ichisuke minded the shop. But they got rained on up in the hills, so that messed everything up. Yesterday they got one horse load. Today, (they got) two. Nobody did any farmwork in the afternoon, but my mother and my brother's wife did some sewing, cooking, and such.

August 30, Friday

Sunny
Strong west wind
20°C in the morning, 25°C in the afternoon

Breakfast:	*yujige meshi*,[19] miso-preserved daikon, pickled eggplants, *takuan*, etc.
Lunch:	*yujige meshi*, pumpkin miso soup, side dishes: same as morning, roasted salted herring
Supper:	eggplant and cowpea miso soup (stock made from whale), *myōga* (Japanese ginger) boiled and eaten with soy sauce, pickled eggplants, *takuan*

Yoshida Yonekichi's mother died today.

I didn't work outside early; I just made breakfast first. After eating, I cut grass for compost and for the rabbits. Then, I cut the seed cucumbers in half, removed the seeds, washed them thoroughly, and put them in the sunlight to dry, etc.

I got about 8 *shaku* (144.3 ml) of seeds from 12 cucumbers. I prepared lunch, and planted green pea seeds I'd prepared before in the afternoon. The peas I planted in the spring have insect eggs on them, so they won't be any good for seeds. I'll have to specially prepare seeds to grow beans for seed and plant them in the summer or fall. Up until last year I planted in the summer, but the result wasn't so great, so I'm planning to experiment with planting in the fall. What I planted today was just four rows in part of one onion field—2 *shaku* 5 *sun* (75.8 cm) wide—about 6 *tsubo* (20 m^2). For fertilizer, I used 10 *kan* (37.5 kg), consisting of a bucket of pee, 200 *momme* (750 g) of superphosphate of lime, and 4 *shō* (7.2 L) of wood ash—this for about 1 *gō* (180 ml) of seeds. I planted them the same way as I planted the beans in the spring. After that, I picked vegetables for selling. My wife weeded in the napa cabbage fields in the afternoon.

Today's income:

144 eggplants	75 sen
15 tomatoes (300 *momme* (1.1 kg))	6 sen
1 watermelon (1 *kan* 300 *momme* (4.9 kg))	25 sen
onions (3 *kan* 800 *momme* (14.25 kg))	60 sen

At the main house: Kanezō went out early to cut hay. Later in the morning, he took the horse up into the hills to cut more hay. In the afternoon, he tossed the manure out of the horse stall and hauled dry grass up to the hut in the hills. My brother's wife sewed and such all day while minding the shop. My mother cooked.

August 31, Saturday

Thin clouds in the sky
Wind out of the southwest
14°C in the morning, 22°C in the afternoon

Breakfast:	eggplant miso soup with pickled eggplants, pickled cucumbers, *takuan*, etc.
Lunch:	roasted salted herring, pickled eggplants, pickled cucumbers, *takuan*, etc.
Supper:	miso soup with eggplants (stock made from whale), *tsukudani* rice topping consisting of various little fish caught in Lake Hachirō, pickled eggplants, etc.

In the early morning, I did the fourth crop-thinning in the second-planted napa cabbage field, and then I did the third thinning in the third-planted napa cabbage field. Then I made breakfast. After eating, I washed the napa cabbage seedlings I'd picked for feeding to the rabbits, and I dug small holes along one side of each row of the second-planted napa cabbages, to give them their first follow-up fertilizer. Into the holes, I poured a mixture of one bucket of night soil combined with a bucket of water, and about 100 *momme* (370 g) of ammonium sulfate. After that, I used a hoe to weed some fields where we need to walk, and I dumped what I'd cut into the compost pile. I made lunch and then I dug more holes for fertilizer, and worked some on fixing up the little pond. After that, I picked vegetables for selling. My wife weeded the daikon fields, among other things, in the afternoon.

Today's income:

105 eggplants	50 sen
18 tomatoes (600 *momme* (2.25 kg))	10 sen
4 watermelons (3 *kan* (11.25 kg))	53 sen
onions (1 *kan* 400 *momme* (5.25 kg))	27 sen

Today's expenditures:

whale (blubber)	10 sen
tsukudani rice topping	50 sen

At the main house: Kanezō did the same as always in the early morning. Today, he and my brother's wife took their lunches with them up into the hills, hauling dry grass with them. Ichisuke did *uma no shireboi* four times today. In other words, he hauled four horse loads. My mother cooked and such.

Figure 0.23. Pouring hot water down a pipe to warm a hotbox (April 24). [Saburō's wife pours water while her daughter Yaeko watches.]

Figure 0.24. Handmade tools: A basket (*tekago*) and a bamboo colander/sifter (*koezaru*). The latter is necessary for spreading compost in dry fields.

Figure 0.25. Making miso: Cooking soybeans (May 19).

Figure 0.26. Making miso: Stomping on cooked soybeans (May 20).

Figure 0.27. Paddy preparation: Plowing (*ta-uchi*) using a horse (April 25).

Figure 0.28. Paddy preparation: Plowing (*takaki—aragurekaki*) with a horse (May 22). [The boy is Saburō's nephew, Ichisuke.]

Figure 0.29. Paddy preparation: Dragging a big pole across a paddy (doing *ta-narashi*) (June 3).

Figure 0.30. Paddy preparation: Rolling a *kata* across a paddy's surface (June 3).

Figure 0.31. Paddy preparation: Pulling seedlings from the nursery paddies (doing *naetori*) (June 7).

Figure 0.32. Paddy preparation: *Sodome* (*saotome*) hand-planting the paddies. (The boy [Ichisuke] standing on the ridge on the left is throwing little bunches of seedlings to the women as they move along).

Figure 0.33. Weeding in the paddies (July 13).

Figure 0.34. Cutting *higusa* on the slopes of Mt. Kanpū (August 21).

Figure 0.35. A row of rice sheaves standing in a harvested paddy (front view) (October 8).

Figure 0.36. A row of rice sheaves standing in a harvested paddy (side view) (October 8).

Figure 0.37. Stacking sheaves of rice (doing *inekeshi*): Stacking sheaves on a post after moving them from another post, for further drying (October 24).

Figure 0.38. Stacking sheaves of rice (doing *inekeshi*): A completed *honiyo* (October 8).

Figure 0.39. Soybean plant *tsuboke* (harvested plants arranged for drying) (November 1).

Figure 0.40. Harvested rice plant bundles (*ine-marogi*) (November 1).

Figure 0.41. Millet (*awa*) plant *tsuboke* (harvested plants arranged for drying) (October 24).

Figure 0.42. Buckwheat (*soba*) plant *tsuboke* (harvested plants arranged for drying) (November 1).

Figure 0.43. *Waraniyo* (November 1).

Figure 0.44. Pounding soybeans (November 19). [Saburō's wife pounds while her daughters watch.]

Figure 0.45. Making a *senaka-ade*: Early stage.

Figure 0.46. A *senaka-ade* (front side).

Figure 0.47. A *senaka-ade* (back side).

Figure 0.48. A *toshina* and dried *mochi* hanging over the front door of the main house (February 8).

Figure 0.49. Children holding *white shiratama mochi* on willow branches and large hearth chopsticks made of *tara* tree wood (February 8). [The children are Saburō's daughter Fusae and his nephew Tsugio.]

Figure 0.50. The straw *kedashi* (*kede*) worn by Namahage have been tied around the neck of a stone dog statue in front of the village shrine (February 8). [*Kedashi* appear to have also been tied around a post. The statue would have been a *komainu* ("lion-dog") shrine guardian.]

Chapter 7: September 1935

September 1, Sunday

Sunny
Wind out of the southwest
12°C in the morning, 24°C in the afternoon

Breakfast:	eggplant miso soup, boiled Japanese ginger (*myōga*)—with soy sauce for flavor, pickled eggplants, and cucumbers and such
Lunch:	roasted salted herring, *tsukudani* [rice topping] pickled eggplants, *takuan*
Supper:	miso soup with potatoes, *konnyaku* and eggplants (stock made from whale), *tsukudani,* pickled eggplants, etc.

Early this morning I went to the Yokoyama household—in this village—to collect night soil. After that, I made breakfast, and after eating I did the same work as yesterday. My wife got back before noon and prepared lunch. In the afternoon she gathered weeds that had been pulled from the field and dumped them in the compost pile. I applied the first follow-up fertilizer to the third-planted napa cabbage crop, and the first follow-up fertilizer to five rows of green onions—all with one and a half barrels of night soil with the same amount of water and 200 *momme* (750 g) of ammonium sulfate. Later, I filled in the holes and picked vegetables for selling.

Today's income:

100 eggplants	40 sen
1 watermelon (2 *kan* 400 *momme* (9 kg))	30 sen
onions (1 *kan* (3.75 kg))	20 sen

Today's expenditures:

candies	5 sen

At the main house: Kanezō spent the early morning doing his usual work. Later in the morning he cut *bukkomi-gusa,* and in the afternoon he hauled the last horse load of dried straw up into the hills. That makes for a total of eight horse loads. My brother's wife had to go to another village on an errand in the morning, and in the afternoon, she took care of the daikon in the fields—picking bugs and such. She cooked breakfast today, and my mother prepared lunch and supper. It's the usual daily arrangement.

September 2, Monday

Sunny
14°C in the morning, 27°C in the afternoon

Breakfast:	miso soup with *konnyaku* and eggplant, boiled onions with soy sauce for flavor, pickled eggplants
Lunch:	roasted salted herring, *tsukudani,* pickled eggplants, onions with soy sauce, *takuan*
Supper:	miso soup with eggplants (stock made from whale), the rest: same as lunch

A child of Yoshida Tadaharu's died today.

I cut straw for feeding the rabbits early. Then I prepared breakfast. After eating, I dug little holes on one side of each of my first-planted napa cabbages and did the second follow-up fertilizing. I used a load of night soil diluted 1:1 with water, with 200 *momme* (750 g) of ammonium sulfate dissolved in it. After that, I dug more little holes—between each autumn cabbage plant—and gave them their third follow-up fertilizing: a load of night soil mixed 1:1 with water and 200 *momme* (750 g) of ammonium sulfate mixed in. Once that was done, I dug holes between the taro plants and poured in two loads of night soil mixed 1:1 with water. My wife returned before noon and did some weeding with the hoe and then made lunch. After lunch, she knocked the soil off the roots of everything she'd pulled up with the hoe and hauled all the weeds to the compost pile. I pulled earth up over the roots of the taros for the second time, and later I picked vegetables for selling.

Today's income:

198 eggplants	80 sen
20 tomatoes (350 *momme* (1.3 kg))	7 sen
1 watermelon (1 *kan* 500 *momme* (5.6 kg))	20 sen
cabbages (1 *kan* 400 *momme* (5.25 kg))	25 sen

Today's expenditures:

rice	35 sen

At the main house: Kanezō did the usual early morning work. After breakfast he went out to the Bishamon paddies to dig drainage ditches. In the afternoon he went to cut *bukkomi-gusa* as always, and then he chopped it up with a big knife and tossed it in the horse stall. He also washed the horse and cut hay to feed it. My brother's wife was in the fields all day, picking cowpeas (*sasage*), and red beans (*tenko* in local talk). My mother helped her cook and such.

September 3, Tuesday

Cloudy, later rain
South wind
19°C in the morning, 22°C in the afternoon

Breakfast:	miso soup with eggplants in it, pickled eggplants, *takuan*
Lunch:	roasted salted herring, *tsukudani*
Supper:	exactly the same as last night

Early this morning I cut grass for the rabbits, and then made breakfast. After eating, I cut grass and such for making compost, and then did some weeding. I didn't work outside in the afternoon due to the rain. Instead, I went down to the main part of the village on an errand. But I did pick vegetables for selling later. My wife came back before noon, and she sewed and such in the afternoon.

Today's income:

122 eggplants	60 sen
2 watermelons (2 *kan* 500 *momme* (9.4 kg))	35 sen

Today's expenditures:

summer cabbage seeds	1 yen 40 sen
postcards	15 sen

At the main house: Kanezō did the same work as usual early. After breakfast, he hauled more dried grass up to the shack on the mountainside and dumped it. In the afternoon he went out among the paddies and fields to cut hay for the horse. In the morning my brother's wife went out to the fields and thinned the daikon crop and picked bugs off the daikon. She sewed and such in the afternoon. And she and my mother cooked together, etc.

September 4, Wednesday

Rain, later sunny
Southwest wind
20°C in the morning, 26°C in the afternoon

Breakfast:	*exactly* the same as every day
Lunch:	roasted salted herring, *takuan*, pickled eggplants
Supper:	miso soup (stock made from whale) with eggplants and cowpeas in it, roasted salted herring, *takuan*, pickled eggplants

Early this morning I cut grass for the rabbits. Then I made breakfast. I cut grass and straw for composting both in the morning and in the afternoon, and picked vegetables for selling in the evening. My wife came back before noon, and she sewed and such in the afternoon.

Today's income:

190 eggplants (97 small)	73 sen
2 watermelons (2 *kan* 600 *momme* (9.75 kg))	35 sen

Today's expenditures:

kerosene (*sekiyu*)	12 sen

At the main house: Kanezō did the same work as usual before breakfast. He spent the rest of the day hauling *kamaki*—you break tree parts up and boil them in a cauldron. While doing that, after lunch, he applied chemicals to the daikon crop together with my brother's wife. They mixed lime casein with arsenate (*hisaen*) to make about 2 *to* (36.1 L) of the stuff. My brother's wife also picked red and green beans in the fields and debugged the daikon. She did some cooking and such as well.

September 5, Thursday

(August 8 according to the old calendar)
Sunny
14°C in the morning, 30°C at 3:00 p.m.

Breakfast:	same as every day
Lunch:	roasted salted herring and miso soup made with whale stock, with eggplant in it, pickled eggplants, *takuan*
Supper:	exactly the same as lunch

Yoshida Sanjirō's wife gave birth to a baby girl.

I accidentally slept in this morning! After getting up, I made breakfast and then cut grasses for composting all day—cutting various weeds and such all around my fields, and inside them. My wife returned before lunch, and she went up into the hills to enjoy the scenery with a friend in the afternoon. In the evening, I picked vegetables for selling, and cut grass for the rabbits, and such.

Today's income:

200 eggplants (110 small)	95 sen
4 watermelons (4 *kan* (15 kg))	55 sen

Today's expenditures:

rice polishing	15 sen

At the main house: Kanezō did the same work as yesterday, but he took the afternoon off (and rested) because today was the eighth of August according to the old lunar calendar—it has always been a day to rest. They made *mochi* and ate them (at the main house). My brother's wife did the preparations for that during the morning. Then, she went up into the hills for the scenery in the afternoon. My mother went to another village yesterday on a shrine-related matter and hasn't returned, so my brother's wife took care of all the food preparations alone.

September 6, Friday

Sunny
13°C in the morning, 31°C at 2:00 p.m.

Breakfast:	same as yesterday
Lunch:	roasted salted herring, pickled cucumbers and eggplants, *takuan*, etc.
Supper:	miso soup with potatoes and eggplants in it, other dishes: same as lunch

Early this morning I thinned the autumn daikon crop for the third time, and while doing that I dug little holes between each plant for applying follow-up fertilizer. Then I made breakfast. After eating I poured one *ka* of water into 1.5 *ka* of night soil and fertilized the daikon crop. Then I collected more night soil from other houses and chatted a bit with a local fellow. In the afternoon I went down into the village on an errand, and I picked vegetables for selling in the evening. My wife weeded the cucumber fields in the afternoon and hauled weeds to the compost pile (and piled them up).

Today's income:

160 eggplants	90 sen
2 watermelons (2 *kan* (7.5 kg))	28 sen

At the main house: Kanezō did the same things he always does early. After breakfast, he took brown unpolished rice (*genmai*) to the mill to have it polished. My brother did that, too. His wife minded the shop. In the afternoon Kanezō weeded the nursery paddies (*nashiro*). My brother's wife went out in the afternoon and debugged the daikon and picked green and red beans before going home. My mother cooked.

September 7, Saturday

Cloudy
20°C in the morning, 28°C in the afternoon

Breakfast:	miso soup with onion pieces in it, pickled cucumbers, *takuan*, etc.
Lunch:	ray (*kashibe*) [standard: *kasube*] cooked with eggplants in soy sauce, *takuan*, pickled cucumbers
Supper:	eggplant miso soup, others same as lunch

Early in the morning I did the fourth thinning on my third-planted cabbage crop, leaving only single plants standing, and did the third thinning on the Aokubi Hōryō daikon. Then I prepared breakfast. In the evening, I waited for my wife to return before taking Fusae up to

Mt. Kanpū to play. After getting back I picked vegetables for selling. My wife sewed and such in the afternoon.

Today's income:

57 eggplants	35 sen
27 pumpkins (800 *momme* (3 kg))	10 sen
2 watermelons (2 *kan* 200 *momme* (8.25 kg))	25 sen

Today' expenditures:

fish	10 sen
sweets	5 sen

At the main house: Kanezō went out to cut hay for the horse early. Later, he took his lunch with him up to Mt. Kanpū to join in the Seinen Gakkō[1] Oga Branch training session. My brother's wife minded the shop all day. My mother cooked and such.

September 8, Sunday

Sunny, sometimes cloudy
West wind
23°C in the morning, 28°C in the afternoon

Breakfast:	miso soup with eggplant in it, pickled cucumbers, *takuan,* etc.
Lunch:	roasted salted herring, pickled eggplants, *takuan,* etc.
Supper:	miso soup with potato and eggplant in it, side dishes: same as lunch

I cut grass for the rabbits early, and then prepared breakfast. After breakfast I cut grass and weeds for making compost. In the afternoon, I went to watch the South Akita County Young Men's Association sports day events. My wife came back before noon. She picked bugs off the napa cabbage plants, and picked vegetables for selling tomorrow, etc., in the afternoon.

Today's income:

110 eggplants (35 small)	65 sen
3 watermelons (3 *kan* 200 *momme* (12 kg))	45 sen

Today's expenditures:

train fare	18 sen

At the main house: Kanezō did the usual early morning work. After breakfast he cut *bukkomi-gusa,* and in the afternoon he went up into the hills to cut hay for the horse. My brother's wife minded the shop all day, and she cooked, with my mother.

September 9, Monday

Sometimes sunny, sometimes cloudy
West wind
22°C in the morning, 28°C in the afternoon

Breakfast:	same as yesterday
Lunch:	same as yesterday
Supper:	same as yesterday

Today each household in the village had to send one person to help work on the roads. Someone came to see me from a different village early this morning, so I didn't get anything done. I made breakfast and then cut grass for the rabbits and fixed up the rabbit hutch during the rest of the morning. My wife cut grass and weeds for making compost in the morning, and picked vegetables for selling. She did the laundry in the afternoon.

Today's income:

65 eggplants	55 sen

At the main house: Kanezō cut hay for the horse early, and after breakfast he went out to repair the roads. In the afternoon he hauled tinder for boiling (*kamaki*). My brother's wife sewed while minding the shop. My mother cooked.

September 10, Tuesday

Rain in the morning, just cloudy in the afternoon, but more rain later
North wind
19°C in the morning, 17°C in the afternoon

Breakfast:	miso soup with onions in it, pickled cucumbers, *takuan*, etc.
Lunch:	steamed eggplants, *takuan*, etc.
Supper:	eggplant miso soup, roasted salted herring, daikon miso soup, *takuan*, etc.

Yamazaki Daikichi, the beekeeping pioneer, came to visit today.

I couldn't do any work in the early morning because of the rain, so I just made breakfast after getting up. Then I talked with Yamazaki Daikichi—the beekeeping pioneer—until nine-thirty. After that, I cut grass and such for composting. Rain fell again in the afternoon, so I had to give up on outdoor work. My wife returned before lunchtime, and she sewed in the afternoon. In the evening, my wife and I picked vegetables for selling.

Today's income:

93 eggplants	45 sen
5 pumpkins (6 *kan* (22.5 kg))	72 sen

At the main house: Kanezō did his usual early morning work. After breakfast he hauled *kamaki,* but he took the rest of the morning off because of the rain. My mother cooked. My brother's wife went to Shichiga Village[2] to pray at the shrine there because there has been a lot of disease lately, but mainly because of the sleeping sickness that's been going around. People say that if you pray at the shrine in that village, you won't get sick. You're supposed to go with candles and rice balls (*onigiri*). I don't believe such superstitions, so I don't pray there. People in my village laugh at me for that.

September 11, Wednesday

Sunny
North wind
15°C in the morning, 24°C in the afternoon

Breakfast:	eggplant miso soup, miso daikon, *takuan,* etc.
Lunch:	roasted salted herring, *takuan,* miso daikon, etc.
Supper:	potato and eggplant miso soup, boiled onions with soy sauce poured on them for flavor, *takuan,* etc.

Tonight, at my old school, there was a political speech by Sekiyama Shigenosuke, a candidate for the prefectural assembly. Sekiyama-kun[3] is a native of this village.

I cut grass for the rabbits early and then made breakfast. After eating, I collected urine and prepared for mixing fungicide—washing out the container (*oke*), fetching water, and measuring the ingredients. In the afternoon, my wife and I applied the fungicide together. When I applied four-*to*-style Bordeaux mixture to the napa cabbages previously, the chemicals harmed the plants, so this time I made six-*to*-style Bordeaux mixture. To do this, I made four-*to*-style mixture and dissolved 22 *momme* (82.5 g) of casein lime and 100 *momme* (375 g) of arsenate in it. I applied this to the napa cabbages and the daikon and the autumn cabbages. In the evening, I picked vegetables for selling.

Today's income:	
150 eggplants	85 sen
2 watermelons (3 *kan* 200 *momme* (12 kg))	42 sen
cabbages (1 *kan* (3.75 kg))	20 sen
Today's expenditures:	
cotton	30 sen

At the main house: Kanezō cut hay for the horse early, and after breakfast he cut *bukkomi-gusa.* In the afternoon he pulled dung and straw out of the stable and made compost with it. My brother's wife went out to the fields in the morning and picked the dry leaves from the daikon and did some crop-thinning. Her way of doing this is different from mine. She always does it many times, at any time of the day, but I do it three times. Most farmers eat what they pick when thinning crops, so they do it whenever they feel a need. My mother cooked today. My brother's wife stayed home in the afternoon and sewed and such, and she also cooked supper.

September 12, Thursday

Sunny, sometimes cloudy
Strong southeast wind
11°C in the morning, 23°C in the afternoon

Breakfast:	eggplant miso soup, pickled cucumbers, *takuan*, etc.
Lunch:	roasted salted herring, pickled cucumbers, *takuan*, etc.
Supper:	We ate *yama mochi*[4] we received from the main house with *azuki* beans, and boiled cabbage and eggplant with soy sauce for flavor, and Japanese ginger (*myōga*) miso soup, etc.

I prepared breakfast early. After eating, I planted seeds for next year's summer cabbages in a field that I used for cucumbers. I prepared rows, and used about 20 *kan* (75 kg) of compost soil I made by mixing in 5 *gō* (900 ml) each of superphosphate of lime and oilcake, and 1 *shō* (1.8 L) of powdered dried anchovies. Then I covered all that up with about 1 *sun* (30 mm) of soil and poured on about two buckets of night soil diluted 1:1 with water. Last, I applied about 6 *shō* (10.8 L) of wood ash to the surface. Finally, I planted the seeds, and covered them with a layer of soil about 5 *bu* (1.5 cm) thick. Then I put a good layer of rice straws over the soil and laid wooden poles down over them so they wouldn't blow away. I planted ¼ *shaku* (4.5 ml) of king cabbage seeds, ⅓ *shaku* (6 ml) of Nozaki Kyokuwasei [very early ripening] cabbage seeds, 1 *shaku* (18 ml) of Nozaki Wasei [early ripening] cabbage seeds, ½ *shaku* (9 ml) of Hōmaru cabbage seeds, and ½ *shaku* (9 ml) of Sadaya cabbage seeds. In the afternoon, I cut grass for the rabbits, and ran some errands, etc. My wife worked on remaking the *futon* in the afternoon. She also prepared *mochi* for eating, among other things.

Today's income:

25 eggplants	20 sen
watermelons (1 *kan* 400 *momme* (5.25 kg))	65 sen
cabbages (1 *kan* 700 *momme* (6.4 kg))	30 sen

Today's expenditures:

coarse (*zaramen*) brown sugar	10 sen
pears	10 sen
sweets	5 sen

At the main house: Kanezō did his usual early morning work, and after breakfast he hauled tinder for boiling (*kamaki*). He rested in the afternoon. There are five days off from today. My brother's wife minded the shop in the morning, and helped my mother prepare to pound *mochi*, etc., in the afternoon. Kanezō did the pounding. Today is August 15th according to the old lunar calendar—the day of the moon festival.[5]

September 13, Friday

Cloudy with some sunshine
Southeast wind
20°C in the morning, 26°C in the afternoon

Breakfast:	*yama mochi,* miso soup with eggplant in it, Japanese ginger (*myōga*) preserved in miso, cabbage with soy sauce for flavor
Lunch:	*yama mochi* and roasted dried herring, pickled cucumbers, *takuan,* etc.
Supper:	*yama mochi,* eggplants preserved in miso, other side dishes: same as lunch

Just now there was a fire in Funagawa, so I went to check it out. This morning I did the second follow-up fertilizing to the second-planted cabbage crop. As usual, I dug little holes on one side of each plant, and into the holes I poured two *ka* of night soil and water (equal parts) into which I had dissolved 200 *momme* (750 g) of ammonium sulfate. After that I did the first follow-up fertilizing on the Aokubi [Hōryō] daikon. I dug little holes between the plants and poured in night soil—one bucket diluted with the same amount of water. In the afternoon I rested. My wife didn't go out to sell vegetables this morning. Instead, she weeded and helped with the fertilizing by filling in all the little holes I'd dug. She also rested in the afternoon.

Today's income:

20 eggplants	10 sen

At the main house: Kanezō did the usual early morning work and then after breakfast he went into the hills to cut hay for the horse. He rested in the afternoon. My brother's wife went out to the fields in the morning and picked bugs off the daikon, etc. She also rested in the afternoon. The cooking was done by my brother's wife and my mother.

September 14, Saturday

Rain, later just cloudy
Northwest wind
14°C in the morning, 16°C in the afternoon

Breakfast:	miso soup with potatoes and eggplant in it, miso-preserved daikon, *takuan,* etc.
Lunch:	eggplant cooked with salted krill, roasted dried herring, pickled cucumbers, *takuan*
Supper:	same as breakfast

Oikawa Shūjirō's wife gave birth to a baby boy.

Again today, my wife did not go out and sell vegetables. She sewed most of the day. I cut grass for the rabbits in the morning and repaired the rabbit hutch, etc. I didn't work in the afternoon. Yesterday evening, the hutch broke open and two young rabbits got killed by a dog. I'm beside myself with frustration and disappointment.

At the main house: Kanezō did the usual early work. After breakfast he cut *bukkomi-gusa* and in the afternoon he hauled firewood. My brother's wife did some sewing and cooking while minding the shop.

September 15, Sunday

Sunny
Northwest wind
10°C in the morning, 22°C in the afternoon

Breakfast:	same as yesterday
Lunch:	same as yesterday
Supper:	same as yesterday

Early this morning I applied the third follow-up fertilizer to the first crop of cabbages. I dug holes between the plants and poured in 1 *ka* of night soil diluted with the same amount of water. Then I applied the second follow-up fertilizer to the green onions I planted where peas were before, and in part of the summer cabbage space. As usual, I used a 1:1 mix of night soil and water—1.5 *ka* of each this time—and I poured that around the bases of the onion plants (see Fig. 7.1). After breakfast, I scraped soil up over the roots of the green onions for the second time this season, pulling soil up from both sides. In the afternoon, I repaired the rabbit hutch again, and picked vegetables for selling. Again, my wife didn't go out to sell vegetables. Instead, she weeded the cabbage fields and filled in the holes I'd dug for doing the follow-up fertilizing, etc.

Today's income:

40 eggplants (small)	15 sen

Today's expenditures:

rice (4 *shō* (7.2 L))	1 yen

Figure 7.1. The green onion crop (note the tall ridges). From Yoshida's personal collection: not included in 1938 book.

At the main house: Kanezō did the usual early morning work before breakfast, and later he went out to the paddies to cut millet (*hie*) among the rice plants. After lunch he went out to the fields and spread arsenate and lime casein on the daikon. He took off in the afternoon. My brother's wife made sweets with my brother all day. My mother cooked.

September 16, Monday

Sunny
8°C in the morning, 24°C in the afternoon

Breakfast:	miso soup with cabbage in it, miso-preserved daikon, *takuan*, etc.
Lunch:	same as yesterday
Supper:	miso soup made with dried roasted fish with potatoes and eggplants in it, pickled cucumbers, *takuan*, etc.

I didn't do any work early this morning—just made breakfast. After eating I cut grass for the rabbits, and ran some errands later. My wife returned (from selling vegetables) before noon, and she rested in the afternoon.

Today's income:

123 eggplants:	60 sen
3 watermelons (3 *kan* 500 *momme* (13.1 kg))	50 sen

At the main house: Kanezō cut hay for the horse early, and after breakfast he went out to cut more millet in the rice paddies. He took it easy in the afternoon. My brother's wife sewed and such while minding the shop all day.

September 17, Tuesday

[Saburō's wife, Katsue, wrote the entry for this day.]
Cloudy
Southeast wind
12°C in the morning, 20°C at 1:00 p.m.

Breakfast:	same as yesterday
Lunch:	miso soup and *takuan*
Supper:	miso soup with eggplant in it (stock made from dried roasted anchovies), pickled eggplants

My husband was out all day today on errands and didn't return. I prepared breakfast. During the morning I weeded in the autumn cabbage fields. In the afternoon I rested and sewed, did some reading, and cut some grass for the rabbits.

At the main house: Kanezō cut hay for the horse early. After breakfast he took his lunch with him to Nakamura in Oga to post notices and hand out flyers—notices about candidacy in the upcoming elections for the prefectural branch of the National Farmers' Association. He works for the association. He was paid 80 sen for the day. My husband's brother's wife minded the shop. My husband's brother attended a *yore*[6]—a gathering of all members of the household heads' association (*koshukai*). My husband's brother's wife and mother cooked together.

September 18, Wednesday

Rainy, later cloudy with some sunshine
Northwest wind
14°C in the morning, 22°C at 4:00 p.m.

Breakfast:	eggplant and potato miso soup, pickled eggplants, pickled cucumbers, *takuan*, etc.
Lunch:	salted krill (cooked), otherwise same as breakfast
Supper:	miso soup with eggplant and cowpeas in it (stock made from dried roasted anchovies), *shijigo*,[7] pickled eggplants, *takuan*

We finished eating the last of the watermelons today.

Again, today, I did no early work—just made breakfast. I cut grass for the rabbits in the afternoon. My wife weeded the watermelon fields.

Today's income:

60 eggplants	35 sen

Today's expenditures:

matches	5 sen
shijigo (fish)	10 sen
mochi	5 sen

At the main house: Kanezō rested in the early morning, and after breakfast he went up into the hills to cut hay for the horse. In the afternoon he did exactly the same work as yesterday. My brother's wife sewed. In the afternoon she went out into the fields and picked bugs off the daikon and thinned the crops.

September 19, Thursday

Sunny
Northwest wind
11°C in the morning, 22°C at 1:00 p.m.

Breakfast:	potato and eggplant miso soup, pickled eggplants, *takuan*
Lunch:	fish and eggplant cooked in soy sauce, pickled eggplants, *takuan* (I don't know the name of the fish—they were caught off the Wakimoto shore)
Supper:	same soup as last night, other dishes: same as lunch today

After breakfast, I cut grass for the rabbits. Then I applied the second fertilizer to the third-planted cabbage crop. I did it just like last time—a bucket of night soil mixed with a bucket of water and 700 *momme* (2.6 kg) of ammonium sulfate dissolved in that. I rested in the afternoon. In the evening, I went out again and filled in all the little holes I'd dug for fertilizing. My wife was busy caring for our eldest daughter [Fusae] because she got sick.

Today's income:

20 eggplants	10 sen

Today's expenditures:

white sugar	5 sen
soy sauce	6 sen
pears	10 sen
fish	10 sen

At the main house: Kanezō cut hay for the horse early, and after breakfast he went out to distribute flyers again. My brother's wife took her lunch with her to Funakoshi, to look for materials for making *mushiro* mats. My mother went to the reservoir to catch shrimp to eat.

September 20, Friday

Sunny
8°C in the morning, 26°C in the afternoon

Breakfast:	potato and cabbage miso soup, shrimp we got from the main house, *takuan*, pickled eggplants
Lunch:	exactly the same as yesterday
Supper:	I ate *udon* at the hospital

Fusae's illness got worse overnight, so my wife took her to a hospital in Akita City first thing this morning. I cut grass for the rabbits and such early. I got a telegram telling me that Fusae had been hospitalized, so I got ready quickly and headed for Akita City. It was a big relief (upon arriving) to hear that her illness is actually a fairly mild one.[8]

Today's expenditures:

hospital fee	16 yen 75 sen[9]
fertilizer (1 *kan* 500 *momme* (5.6 kg))	70 sen (anchovy powder)

At the main house: Kanezō did exactly the same work as yesterday. My brother's wife minded the shop, did laundry, sewed, and cooked and such all day. My mother wove straw *mushiro* mats and cooked, etc.

September 21, Saturday

Sunny, some clouds
10°C in the morning, 26°C in the afternoon

Breakfast:	leftover soup from yesterday morning
Lunch:	same as breakfast, with some *shijigo* fish
Supper:	eggplant and cabbage miso soup, miso-preserved daikon, *shijigo* and such

Early this morning I cut straw for the rabbits, and then I made breakfast. After breakfast I did some tilling in the area where I'm growing peas for seed and poured 5 *shō* (9 L) of night soil mixed 1:1 with water around the roots. After that, I picked some seed eggplants and scraped soil up over the roots of the cowpeas. In the afternoon I went to my old school to hear Murayama Yōichi, candidate for the National Farmers' Association (Zenkoku Nōmin Kumiai), give a political speech. Okada Sōji[10] came from Tokyo for that.

At the main house: Kanezō did exactly what he did yesterday. My brother's wife made sweets with my brother in the morning, and she went out and picked dry leaves off the daikon and searched for *aomame* in the afternoon. *Aomame* are immature soybeans. My mother cooked.

September 22, Sunday

Cloudy
13°C in the morning, 21°C in the afternoon

Breakfast:	same as yesterday's supper, but without the *shijigo*
Lunch:	fish, bought yesterday, that we packed in salt overnight and roasted, Japanese ginger (*myōga*) miso soup
Supper:	cabbage and eggplant miso soup, the rest: same as yesterday's lunch

First this morning, I made breakfast. Then, after eating, I went out on an errand. In the afternoon I weeded among the watermelon plants, and then hauled away the weeds, cut grass for the rabbits, and peeled chestnuts.

Today's income:

36 eggplants	20 sen

Today's expenditures:

candies	2 sen

At the main house: Kanezō cut hay for the horse early, and he cut millet in the rice paddies later in the morning and after lunch. My brother's wife did this, too. My mother cooked. My brother went up to Mt. Kanpū on village business.

September 23, Monday

Cloudy
13°C in the morning, 24°C in the afternoon

Breakfast:	cabbage miso soup, miso-preserved daikon
Lunch:	We ate bonito that we received from the main house, and daikon miso soup and such. The bonito was caught off the coast of Minami Iso.
Supper:	canned whale cooked with eggplant in soy sauce, *takuan,* chestnuts

This morning I prepared breakfast first, and after eating I cut grass for the rabbits, and shelled and cooked *aomame* (immature soybeans). In the afternoon I took Yaeko[11] with me to the hospital in Akita City to visit Fusae.

Today's expenditures:

train fare	1 yen 5 sen
canned whale and sweets	40 sen
book	1 yen 60 sen

At the main house: Kanezō did the usual early morning work, and he cut millet in the paddies along with my brother's wife again. My mother cooked and such.

September 24, Tuesday

Cloudy, later rainy
South wind
12°C in the morning, 16°C in the afternoon

Breakfast:	eggplant miso soup, miso-preserved daikon, *takuan,* etc.
Lunch:	*shijigo,* daikon miso soup, etc.
Supper:	cabbage and eggplant miso soup, roasted salted fish, daikon miso soup, etc.

I made breakfast this morning, and after eating I did the third fertilizing to the second crop of cabbages—dug little holes between the plants and poured in 1 *ka* of night soil mixed with the same amount of water. I also picked seed pumpkins, and eggplants to sell. In the afternoon, I filled in all the little holes and cut grass for the rabbits.

Today's income:

190 eggplants	95 sen
3 young rabbits	45 sen

Today's expenditures:

postage stamps	3 sen

At the main house: Kanezō did as usual in the early morning. After breakfast he did the same as yesterday. In the afternoon, he took the horse up into the hills to cut hay. My brother's wife did the same as yesterday. My mother cooked. My brother pulled hay and manure from the stable and made compost with it.

September 25, Wednesday

Rain, later just cloudy
Strong south wind
12°C in the morning, 26°C in the afternoon

Breakfast:	rice pilaf (*gomokuhan*) with soybeans (*aomame*) mixed in, miso-preserved daikon, etc.
Lunch:	same as breakfast
Supper:	daikon miso soup, *takuan, tsurumo* [sea lace: *Chorda filum*] eaten with grated daikon, etc. We ate supper at the main house. The *tsurumo* was from Minami Iso.

First this morning, I made breakfast. After eating I cleaned the droppings and grass and such—compost—out of the rabbit hutch (and put in fresh stuff). Then I went to the school to serve as a monitor for the prefectural assembly election. I got home from that at eight o'clock in the evening.

Today's expenditures:

tobacco [cigarettes or powder]	10 sen

At the main house: Kanezō cut hay for the horse during the morning, and he did *godosarai*[12] with my brother's wife in the afternoon. The way was the same as in spring. In the afternoon they cut more millet in the paddies. My mother cooked and such.

September 26, Thursday

Rain, later just cloudy
Strong south wind
12°C in the morning, 20°C in the afternoon

Breakfast:	leftover rice from yesterday's breakfast, miso-preserved daikon, *shijigo* and such
Lunch:	basically, the same as breakfast
Supper:	eggplant and cabbage miso soup, all else: same as morning

First this morning, I made breakfast. After eating, I just stayed in and rested because of the rain. In the afternoon I cut grass for the rabbits and ran some errands. I also made supper.

Today's income:

75 eggplants	40 sen
2 young rabbits	40 sen
"get well soon" (*mimai*) cash for Fusae	50 sen

Today's expenditures:

paper	5 sen
sweets	5 sen

At the main house: Kanezō did the same as always early today. He couldn't do any (outdoor) work after breakfast, so he wove *noma* mats with rice plant straws (*wara*). In the afternoon he took the horse up into the hills to cut hay. My brother's wife did some sewing in the morning. In the afternoon, she went out to thin the daikon crop and picked bugs off the daikon before going back home. My mother wove *mushiro* mats while cooking. Lately my mother has been doing this every day while cooking.

September 27, Friday

Sunny
Southwest wind
14°C in the morning, 22°C in the afternoon

Breakfast:	eggplant miso soup, miso-preserved daikon
Lunch and supper:	ate in Akita City

We prepared breakfast early this morning, and after eating we went to Akita City to pick Fusae up (at the hospital) because she was discharged today.

Today's expenditures:

hospital fee	2 yen 76 sen[13]

At the main house: Kanezō cut hay for the horse early, and after breakfast he went up into the hills to cut red hay. My brother's wife went out to the fields in the morning to pick dried leaves off the daikon plants. In the afternoon, Kanezō and my brother's wife pressed red beans. To do this, first you must remove all the beans from their shells (which we call doing *komamefugi*). After that, you lay a *mushiro* mat down on the floor and then mash them with a *joge*.[14] You then scrape up the unwanted parts with a piece of bamboo and pick up the mashed beans. It has been decided that the second son of the big main house[15] will move out and establish a branch household, so there was a moving ceremony for that. And at the same time, they had a *mago-ue* ceremony[16] for the birth of a grandchild. My mother and my brother went next door for those events. My brother's wife took care of the cooking.

September 28, Saturday

Cloudy with some sun

Southwest wind
16°C in the morning, 24°C in the afternoon

Breakfast:	daikon and potato miso soup, pickled eggplants and cucumbers, *takuan*
Lunch:	roasted salted herring—otherwise, same as breakfast
Supper:	*udon* noodles boiled and eaten in soup flavored with soy sauce (stock made from roasted dried anchovies), side dishes: same as breakfast
Snack:	chestnuts

My wife made breakfast, and after eating she went out shopping. She returned before noon, and did laundry and wove *noma* mats and such in the afternoon. I harvested vegetables for selling in the early morning, and did the first thinning of the summer cabbage crop—just picked sprouts from areas where too many seeds had fallen in clusters when I scattered them there. Then I picked withered leaves off some of the daikon. In the afternoon, I peeled chestnuts and picked more vegetables for selling.

Today's income:

200 eggplants	77 sen
daikon	10 sen

Today's expenditures:

soy sauce (1 *shō* (1.8 L))	15 sen

At the main house: Kanezō started preparing to harvest rice early by weaving ropes with rice straws. After breakfast, he took his lunch with him up into the hills to cut hay for the horse. He brought back a total of six bundles (*maro*). My brother's wife did laundry and sewed while minding the shop all day. My brother, as a member of the agricultural support association (Shinōkai) went out to inspect rice paddies around the village. My mother wove *mushiro* mats—usual work for her these days—and cooked.

September 29, Sunday

Cloudy, sometimes sunny
Northwest wind
17°C in the morning, 23°C in the afternoon

Breakfast:	cowpea and daikon miso soup, pickled eggplants and pickled cucumbers, *takuan*
Lunch:	*shijigo,* pickled eggplants, *takuan,* etc.
Supper:	eggplant and *gunji* soup—flavored with soy sauce, pickled eggplants and cucumbers, *takuan,* etc. *Gunji* are fish that come from the waters of Lake Hachirō.[17] We received them from someone as a condolence gift (*mimai*) because of Fusae's illness.

Yoshida Tamesuke's wife gave birth to a baby girl today.

After breakfast, I went out to harvest rice all day. This year's rice harvest started today. My wife went out this morning to sell vegetables, and she picked withered leaves off the daikon in the afternoon.

Today's income:

180 eggplants	80 sen
daikon	15 sen

At the main house: Kanezō went out early as usual to cut hay for the horse, and then later he began harvesting rice. My brother's wife and my brother also harvested rice (with Kanezō)—all three of them. In the morning they harvested rice at Shinden, and in the afternoon, they started harvesting rice in the paddies at Sanbonmatsu. The type of rice is generally known as Aigame, but its scientific name is Riku'u No. 132.[18] My mother wove *mushiro* mats and cooked.

September 30, Monday

Sunny
Northeast wind
14°C in the morning, 24°C in the afternoon

Breakfast:	potato and cowpea miso soup, side dishes: same as yesterday morning
Lunch:	roasted salted *gunji*, pickled eggplants and cucumbers, *takuan*
Supper:	eggplant miso soup (stock made from roasted anchovies), salted krill, side dishes: same as lunch
Snack:	roasted chestnuts

Today I went out to help another family harvest their rice again. My wife and I both went. We took our lunches with us. I was able to cut 70 bundles of rice—which I hanged on tall poles (*kui*) to dry—by myself. For rice, ten bunches make one bundle, and one bunch is two handfuls. Each bundle is split into four parts and hung to dry.

Today's expenditures:

salted trout	40 sen

At the main house: Early this morning, Kanezō pulled the wooden poles for drying rice out from under the house and tied ten of them together. After breakfast he did the same work as yesterday, and my brother's wife did so as well. My brother went out to the paddies in the evening. Ichisuke minded the shop. My mother did the same things as yesterday.

Chapter 8: October 1935

October 1, Tuesday

Sunny
Northwest wind
16°C in the morning, 24°C in the afternoon

Breakfast:	potato and eggplant miso soup, pickled eggplants, pickled cucumbers, *takuan*
Lunch:	roasted salted trout, pickled eggplants and cucumbers, *takuan*
Supper:	soy sauce–based soup with eggplant in it, other dishes: same as lunch
Snack:	cooked immature soybeans (*aomame*)

Today I spent the day harvesting rice again. But first, I cut grass for the rabbits—before breakfast. I finished the rice harvesting at about two-thirty, and then I went back home and picked up chestnuts. Then I made supper. My wife was out harvesting rice all day.

At the main house: Kanezō cut hay for the horse early, ate breakfast, and spent the rest of the day harvesting rice and hanging all the big bundles up on stakes to dry [see Figs. 0.37 and 0.38]. My brother's wife harvested rice all morning, and she minded the shop and cooked in the afternoon. My brother went out on an errand in the afternoon, and he went to the paddies and hanged rice plant bundles to dry in the evening. My mother went out to harvest rice in the afternoon.

October 2, Wednesday

Cloudy
Southwest wind
12°C in the morning, 22°C in the afternoon

Breakfast:	napa cabbage miso soup, boiled Japanese ginger (*myōga*) with soy sauce for flavor, pickled eggplants, pickled cucumbers, *takuan*
Lunch:	roasted salted trout, pickled eggplants and cucumbers, *takuan*
Supper:	same soup as yesterday evening, side dishes: same as this morning
Snack:	soybeans (*aomame*) and chestnuts

Today my wife and I both spent the day harvesting rice—pretty much the same as yesterday.

Today's income:

toothbrushing powder	10 sen
kerosene (*sekiyu*)	10 sen
sweets	10 sen

At the main house: Kanezō pulled more stakes out from under the house and bundled them early this morning. He spent the rest of the day harvesting rice and stacking the sheaves on stakes to dry. Today my brother's wife and my mother harvested rice all day. My brother didn't go out to the paddies until evening.

The grandmother at Yoshida Yonekichi's household died some days ago [on Aug. 30] at the age of sixty-something. I heard that just before she died, she told her children that there was a bag containing about 280 yen[1] in a drawer of her dresser, as a last will and testament. Sure enough, according to the story, the bag was there. They said it even contained coins from the tenth year of the Meiji Era [1877]. This family has been very poor for a long time, so people of this village were duly impressed by her ability to amass a fortune like that under such circumstances.

October 3, Thursday

Rain, stopping later, then just cloudy
Strong southwest wind
16°C in the morning, 18°C in the afternoon

Breakfast:	pumpkin miso soup, miso-preserved daikon, pickled eggplant, *takuan*, pickled tomatoes, etc.
Lunch:	roasted salted trout, all else: same as morning
Supper:	green onion miso soup, all else: same as lunch
Snack:	chestnuts

I cut grass for the rabbits early this morning, and after breakfast my wife and I both went out to help harvest rice again. But rain started falling heavily around ten o'clock, making it impossible to do the work, so we came back home. We didn't work in the afternoon.

Today's expenditures:

4 *shō* (7.2 L) of rice	1 yen 20 sen
white sugar	10 sen

At the main house: Kanezō pulled more stakes out from beneath the house and bundled them before breakfast. He also returned from the paddies early because of the rain. So did my brother's wife. In the afternoon, Kanezō took the horse up into the hills to cut hay. My brother's wife went out to the fields to pick *aomame*. She also cooked. My mother went to a different village on an errand.

Today I'll write some about harvesting rice. You must always cut rice with the wind direction in mind, to make it easier. In the past, we used to harvest rice using a method called "triple-handful chopping," but today we use the "double-handful chopping" method. The individual bundled sheaves are smaller now than before because of the greater reliance on machines and such. To do triple-handful chopping, you grab three fistfuls of rice plant straws and cut them and make one bunch [sheave] with that. Of course, to do double-handful chopping you grab two fistfuls of plants and cut them. Every year it's a little different because of the quality of the rice crop, but this year it seems to be about normal, and we've been making one bunch with six individual plants—or two with twelve—and then tying the bunches with one more. But in some cases, with a great crop, we've been able to make one fistful with three plants, or two with six. As you go, you simply place the second fistful atop the first, and then later you tie them together. They say that the reason for doing it this way dates from back when people used to lay the harvested sheaves of rice plants down on the ridges between the rice paddies—standing them up so that the tips would drape down over the sides of the ridges—instead of stacking them on big stakes immediately, as we do today. It seems to have made the job easier back then. But today we stack the bundled sheaves on stakes right away, so it's not really necessary to tie them together.

Standing the sheaves on the ridges and draping the tips down like that is called *sogudate* here in the village. To do *sogudate*, you point the stems upward and let the rice grains touch the ground. The thing is, if you leave them like that for some weeks and don't stack them on stakes, the rice may start to sprout. I hear that this happened a lot in the past. But the methods have improved since then, and now we stack them on stakes to dry. We shove the stakes into the ridges in rows and hang four or five bundled sheaves on the stakes. The first two are tied to the stakes with rice straws, and after that more sheaves are stacked up over them one by one, so the rice grains hang outward and down (the double-handfuls of rice plants are tied to the center of each stake, separately). Four or five of them are laid down, and then two more are tied to the stake. Finally, at the end you split some single handfuls and drape them over the top of each stake, at opposing angles, so the tips hang downward. There are two ways of tying up the handfuls: bending the end of the straw you're using for tying sharply and sticking it under the other part, and simply wrapping the straw over itself and sticking it under the other part without bending it sharply. The former way is the older way, and the latter way is how we've been doing it lately, and this is more convenient when machines are involved (see Fig. 8.01).

There have also been many kinds of scythes used to harvest rice (see Fig. 8.02). Way back, people fashioned them out of weed-cutting blades they already had. Then, special scythes made especially for harvesting rice began to appear. You'd have to take a whetstone out to the paddies with you and sharpen your scythe many times in one day when you used any of these kinds of blades. But recently, saw-toothed scythes (*nokogiri gama*) have begun to appear. They say you don't even have to sharpen them and that you can use them for years. It's great that such useful tools have become available.

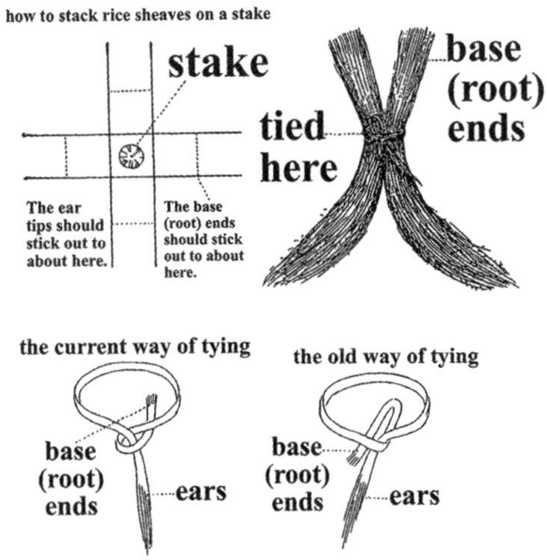

Figure 8.01. How to stack and tie sheaves of harvested rice.

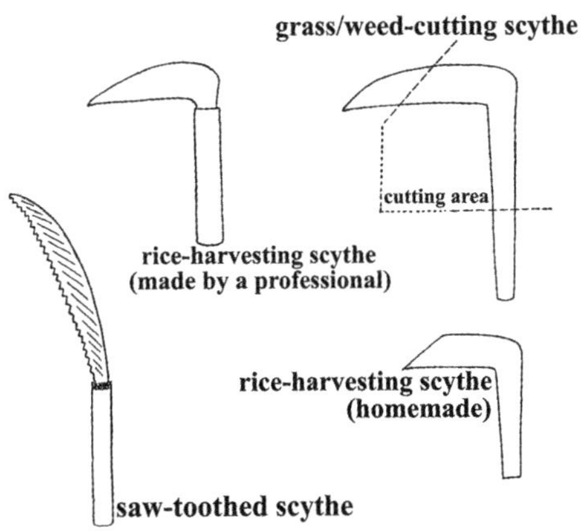

Figure 8.02. Changes in rice-harvesting scythes over time.

October 4, Friday

Occasional rain
Southwest wind
13°C in the morning, 22°C in the afternoon

Breakfast:	potato and cowpea miso soup, pickled eggplant, *takuan*
Lunch:	exactly the same as yesterday's lunch
Supper:	eggplant miso soup, all else: same as breakfast

Early this morning, I collected and hauled piss, and I picked up chestnuts. I spent the rest of the day helping to harvest rice. By myself, I did the work of one and a half people. Of course, my wife also helped harvest rice all day.

At the main house: Kanezō pulled out more big stakes early this morning and bundled them for taking them out to the paddies, and he cut hay for the horse. After breakfast, he and my brother's wife took their lunches with them out to the paddies at Bishamon and harvested rice there all day. My brother went out, too, to help stack rice plant sheaves on stakes to dry. My mother cooked and wove straw *mushiro* mats all day.

October 5, Saturday

Sunny, later cloudy
Southwest wind
10°C in the morning, 20°C in the afternoon

Breakfast:	potato and eggplant miso soup, side dishes: same as every morning
Lunch:	small octopi cooked in soy sauce, pickled eggplants and cucumbers, *takuan*. The octopi were caught off the Oga coast.
Supper:	*mochi* with sweet red beans (*azuki*), pickled eggplants, *takuan*, etc.

I cut grass for the rabbits early, then ate breakfast. I spent the rest of the day helping to harvest rice—again doing the work of one and a half people. I cut 3 *shō* (5.41 L) worth of unpolished rice. My wife also helped another family harvest rice all day.

Today I checked to see exactly how many individual rice plant stalks there are in a single bundle, and I came up with a figure of about 160. I also checked to see how many stalks are used for tying each bundle. These are called *tabanetsura*. I counted about 10. That's for a mid-sized bundle. Some are bigger and some are smaller.

Today's expenditures:

fish	10 sen
work gloves	12 sen
brown sugar	8 sen

At the main house: Last night the *bakurō* brought a new horse to the main house and took away the one they'd been using.[2] Early this morning, Kanezō took that new horse up into the hills and hauled logs with it. After breakfast he did more of the same. In the afternoon, he went out to harvest rice, but he returned early in the evening and made *mochi* and cut hay for the horse. My brother's wife took her lunch with her and harvested rice all day and my brother prepared stakes during the morning. He also went out to harvest rice in the afternoon. My mother stayed home and wove more *mushiro* mats and made *mochi* and such.

October 6, Sunday

Cloudy
West wind
13°C in the morning, 20°C in the afternoon

Breakfast:	leftover *mochi* from last night, eggplant miso soup, pickled eggplants, *takuan*
Lunch:	roasted salted trout, pickled eggplants, pickled cucumbers, *takuan*
Supper:	miso soup (stock made from dried roasted green onions), pickled eggplants, pickled cucumbers, *takuan*
Snack:	*aomame* that we cooked

I went out and collected chestnuts early this morning, and after breakfast my wife and I spent the better part of the day harvesting rice. My wife cut 70 bundles and got home around two o'clock. Again, I did the work of one and a half people. I got home around three-thirty. After that I cut grass and such for the rabbits.

At the main house: Kanezō cut hay for the horse early, and wove ropes for bundling, and spent most of the day harvesting rice. My brother's wife did the same. My brother set up stakes and made more for stacking rice sheaves to dry. Ichisuke minded the shop. My mother wove *mushiro* mats with straws and cooked and such, as usual.

October 7, Monday

Cloudy, some rain
Strong west wind
14°C in the morning, 16°C in the afternoon

Breakfast:	miso soup with pumpkin in it, pickled eggplants and *takuan*
Lunch and supper:	same as yesterday
Snack:	cooked *aomame*

The wind and rain were too strong in the early morning to do anything outside, so I went out to work later, after it had settled down. I spent the day stacking harvested rice bundles on big stakes to dry. My wife went out to harvest rice again, and I also helped the same household. In the afternoon I went over to my younger sister's house to drink celebratory saké because her husband had been promoted to the rank of corporal in the army cavalry.

Today's expenditures:

postage stamps	10 sen

At the main house: Kanezō and my brother's wife did exactly the same work as yesterday. Even my mother went out to help harvest rice today. All three took their lunches out with them. My brother went out in the evening to hang rice bundles on stakes. After getting back they went to my sister's house, but my mother went home early and cooked.

October 8, Tuesday

Sunny
North wind
5°C in the morning, 16°C in the afternoon

Breakfast:	eggplant miso soup, pickled eggplants, miso-preserved daikon, *takuan,* etc.
Lunch:	strips of daikon and eggplants cooked in soy sauce, roasted salted trout, pickled eggplants, *takuan*
Supper:	eggplant miso soup (stock made from dried roasted anchovies), salted napa cabbage with soy sauce, pickled eggplants

I cut grass for the rabbits and collected chestnuts, etc., in the early morning. Later, I took my lunch with me and went out to help harvest rice again. My hand was sore, so I couldn't do as much as I normally can. My wife went out to help harvest rice again, too.

At the main house: Everyone did the same work as yesterday. They finished harvesting the paddies at Sakura today.

October 9, Wednesday

Sunny, occasional light rain
Southwest wind
10°C in the morning, 20°C in the afternoon

Breakfast:	eggplant miso soup, side dishes: same as yesterday evening
Lunch:	same as yesterday's lunch
Supper:	same as yesterday evening

Early this morning I cut grass for the rabbits. And once again, I took my lunch with me and spent the day stacking rice bundles up to dry. And in the evening, I went out to re-stack rice bundles on the stakes, to dry them. My wife stayed home—our second-born daughter, Yaeko, developed a fever during the night so my wife stayed with her. I think she caught a cold. The rice harvesting of the family I've been helping—Sangobē's household—finished today, so they celebrated tonight. I was invited over for that. We ate *udon* noodles in broth with chicken in it, and the soup was flavored with strong miso from the bottom of the miso vat (*miso no tamari*). We also ate dried whitebait with grated daikon and soy sauce. There also was dried herring cooked with cowpeas in soy sauce, summer vegetables with grated daikon and soy sauce, and more. We also drank saké.

Today's income:
60 small eggplants 20 sen

At the main house: Kanezō did exactly the same work as yesterday. My brother and my mother didn't go out to harvest rice today. My mother cooked and such instead. My elder sister came over to help with the harvesting, instead of my mother doing it. Today the harvesting of the paddies at Bishamon was completed.

October 10, Thursday

Sunny
13°C in the morning, 23°C in the afternoon

Breakfast:	eggplant miso soup, miso-preserved Japanese ginger (*myōga*), pickled eggplants
Lunch:	pumpkin cooked with salt, roasted salted herring, pickled eggplants, *takuan*
Supper:	miso soup with green onion pieces in it, side dishes: same as lunch

Lately we're consuming a 1 *shō* (1.8 L) to 5 *gō* (902 ml) rice/oat mixture.

Early this morning, I went out and did some light hoeing along the south sides of the rows in my green onion fields. Then, I dug little holes between all the plants in the third crop of napa cabbages, to give them their third follow-up fertilizing. After breakfast, I did the third follow-up fertilizing to the napa cabbages and the green onions—using 5 *ka* of a 3:2 night soil/water mixture. For the green onions I just poured the stuff on the surfaces of the sides of the rows that I'd hoed earlier in the morning. After this, I picked eggplant seeds—plucking them from the insides of the fruits, then washing them, and then putting them in cloth bags for hanging in places where the airflow is good, to dry them. After that, I made lunch. After lunch, I filled in all the little holes I'd dug between the napa cabbage plants earlier, and I scraped soil up over the roots of the green onion plants for the third time. After doing all that, I prepared supper. My wife was helping her natal household harvest rice all day.

Today's income:
20 eggplants 10 sen

At the main house: Kanezō cut hay for the horse and such early. After breakfast he hauled firewood using the horse, and he helped my mother clean up both inside and outside the house. There was a cleanliness inspection, and he had to clean out the little shed and the rice storeroom (*inabe*) as well. My brother's wife minded the shop in the morning. My brother spent the day helping to harvest rice in the local agricultural association's experimental paddies. In the afternoon, Kanezō used the horse to carry loads of rice plants (did *shireboi*). My brother's wife bundled rice. They finished doing the Shinden paddies today. My mother put rice plant stalks into the rice storeroom (*inabe*) and cooked.

Figure 8.03. Harvesting rice beneath the main house (1) (October 10).

Figure 8.04. Harvesting rice beneath the main house (2) (October 10).

October 11, Friday

Cloudy, rain in the evening
Southeast wind
8°C in the morning, 19°C in the afternoon

Breakfast:	miso soup with pumpkin, boiled onions with soy sauce for flavor, pickled eggplants, *takuan*
Lunch:	*shijigo* fish, pickled eggplants, Japanese ginger (*myōga*) preserved with plums (*umezuke*)
Supper:	eggplant and onion miso soup, side dishes: same as lunch

In the early morning, I cut grass for the rabbits and collected chestnuts, etc. And again today, I took my lunch with me and went out to help another family by tying up their rice plants. I did that—bundling sheaves—during the morning, and in the afternoon, I stacked the sheaves up on big stakes to dry. All these rice plants had already been cut but had just been left lying around due to a lack of stakes. From today, my wife started heading over to a different village[3] to help harvest rice, because it has pretty much finished in our village.

Today's expenditures:

shijigo (fish)	10 sen
brown sugar	5 sen

At the main house: Kanezō and my brother's wife got up at three o'clock to take care of the newly harvested rice plants. They broke for breakfast at about six o'clock, and then after eating both went out to harvest more rice. It was looking like rain in the afternoon, so they stopped harvesting; my brother's wife bundled sheaves and Kanezō hauled them using the horse (did *shireboi*). They finished the Sanbonmatsu paddies today. My brother went over to Shinden to stack rice sheaves up on stakes. Kieko minded the shop. My mother put bundles into the storeroom and cooked and such.

Now I'll explain a bit about how to tie rice plants up in sheaves. Normally, you tie together fifteen bunches that can each be grabbed with one hand. First, you grab bunches of rice plants from a *honiyo* stack (see Fig. 0.38) with your right hand and then lay them on your left hand. Once you've laid out five, you reach back and grab a piece of cord (*tsunagi*)[4] from your back and take one end, with your left hand, and lay it across the plants on your left hand and then turn them all over and lay them on the ground at your feet. So, then you'll have five bunches laid out with a cord. Then, you put four more down, going the other way. Then you put six more on top of those, in the same way as the first five, and then wrap the cord around them and tie them all together with the cord, pulling very tightly. Then, you align the tips of the cord and wrap them around one more time with both hands, and tie again, and you're done. In this way, you can make the final sheaves long or short. If you're going to put them straight into the rice storeroom (*inabe*), then you'll want to make them short, but if you're going to stack them high to make a [new] *honiyo,* you'll want them to be as long as you can make them. There's also another way to bundle rice plants for making sheaves to stack on stakes called *futataba-marogi*. This way is usually used when the rice plants droop down—which happens in certain years—or when you're going to haul rice plants on a horse-drawn cart, etc.

To do this, you take handfuls of rice plant stalks one by one and lay them down so that the rice grains are on the inside, with five on top and five on bottom, or ten in total, in your left hand. Then you tie them just as I explained above while gripping them tightly, so that the rice grains are on the inside of the sheaf. This way, you'll have *futataba-marogi* sheaf bundles. If you use a horse, you can normally haul twelve big ones (*maro*). It depends on the strength of the horse. A strong one can manage over twelve—even up to twenty-two perhaps—while a weaker horse might only be able to carry six or ten.

October 12, Saturday

Rain from yesterday evening till this morning, later cloudy
Southwest wind, later out of the northeast
12°C in the morning, 16°C at 5:00 p.m.

Breakfast:	miso soup with eggplant in it, pickled eggplants, *takuan,* etc.
Lunch:	pumpkin miso soup, roasted salted herring, miso-preserved daikon, *takuan*
Supper:	miso soup with cabbage and onion in it, *shijigo* (fish), pickled eggplants, *takuan,* shrimp we received from the main house (my mother caught these in a drainage channel)

I cut grass for the rabbits early. After breakfast, my wife and I both went out to help another family harvest their rice—in another village.

> At the main house: It was exactly the same as yesterday—up early, dealing with harvested rice plants. After breakfast, Kanezō and my brother did that again together. My brother's wife minded the shop. My mother went out to catch shrimp. In the afternoon, my brother's wife and Kanezō went out to harvest rice, and my mother cooked. Today they finished their rice harvesting, so they celebrated by eating *tororo meshi*.[5]

The main house's total rice harvest amounts for this year are as follows:

Sanbonmatsu paddies	855 bundles (*taba*)
Sakura paddies	695 bundles (*taba*)
Shinden paddies	315 bundles (*taba*)
Bishamon paddies	570 bundles (*taba*)

They harvested even more rice than last year, which was an especially good year.[6] In local talk, people joyously shout *kabekitta!*

October 13, Sunday

Sunny, some clouds
Southwest wind
13°C in the morning, 18°C at 4:00 p.m.

Breakfast:	daikon and potato miso soup, pickled eggplants, *takuan*
Lunch:	potato miso soup, a horsehead fish (*guji*) and eggplant dish, *takuan*
Supper:	cabbage and eggplant miso soup, pickled eggplants, *takuan*, roasted salted herring

Early this morning, I cut grass for the rabbits, and again my wife and I went to help harvest rice in another village. I was served lunch at the home of the landowner. We sent the kids to the main house to play for the day. They took their lunch with them. Today I harvested 80 bunches of rice plants—a good day's work—and headed back home at two-thirty. After that, I made supper.

Today's expenditures:
candies and apples 5 sen

At the main house: They got up at two o'clock this morning. Kanezō, my brother's wife, and my brother did *momi toshi*. I'll explain exactly how to do that, step-by-step. First, you put whole rice grains that have been cleaned by machine into a *toshi* called a *momi toshi*. This is a round container made of bamboo strips—about 2 *shaku* (60.6 cm) tall and 3 *sun* (9.1 cm) deep. There are two of them at the main house. My elder brother and his wife held the *toshi* and Kanezō fed rice grains into it constantly, using a scoop. When you do this, once the rice grains have been sifted the rice must be put through a winnowing machine (*tōmi*). My brother's wife's job is to pour the un-hulled rice grains into the winnower, and my brother's task is to put all the "number one" (best quality) grains into a bag.

All the "number two" grains are put through the winnowing machine again, and the *karumada* are removed. These are rice grains that have remained stuck to the rice plants. Once this is done, the "number two" grains are put through the machine again. Then, both the "number one" and the "number two" grains are removed, and the former are put into a sack. The "number two" grains are put through the winnower once more, and all the "number one" grains that come out are set aside and saved as *hito no shida*. These will be ground with a millstone and made into a powder and used to make *nereke* and other kinds of *mochi*. The "number two" grains from this final processing are set aside as *ume no shida*. This refers to horse (*uma*) feed. *Shida* is a local word for *shiina* [empty grain husks, in this case].

Once this work is finished, the next job is to do what we call *shikkuri tobashi*. *Shikkuri* refers to the stuff that remains mixed with the rice grains after they've gone through the

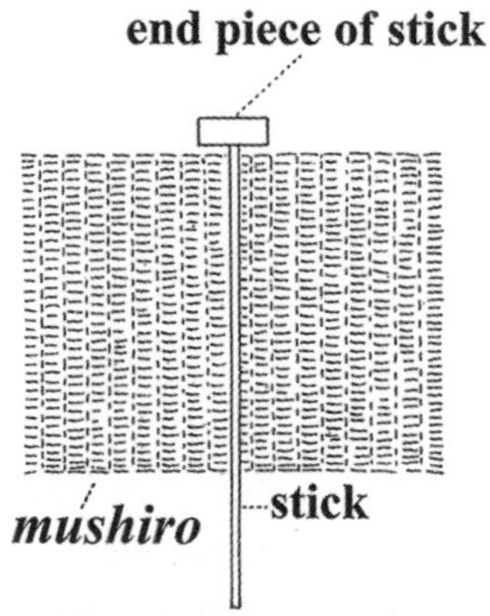

Figure 8.05. *Mushiro* mat and stick device used for *shikkuri tobashi.*

winnower—stuff like pieces of rice plants that weren't adequately processed, and stamens and such. Basically, this must be separated from the grains, and this process is called *shikkuri tobashi.* First, Kanezō takes a large *mushiro* mat and folds it in half. Then, he opens it up and sets it down on the ground and lays a long stick along the fold (see Fig. 8.05). Next, he stands with one end of the stick pointing at him and grabs the part of the mat farthest away from himself and pulls the two sides upward, leaving the center—held down by the stick—on the ground. Then, he puts about one-third of the mat between his legs and holds it tightly. And finally, if he bats the loose two-thirds of the mat hard with both hands, he can create a lot of wind.[7]

So, at this point, my brother's wife drops all that *shikkuri* over the folded mat while Kanezō bats it, and the stamens and stuff blow away, all the pieces of rice plant remain (see Fig. 8.06). My brother then beats the remaining rice plant pieces (ears) and knocks off the grains that are still clinging on. Then, the grains are sifted and put through the winnower again, and the leftover residue is set aside as horse fodder. All the stuff produced by doing this is called *shikkuri kashi,* and it's fed to horses. Other stuff that blows out of the rice winnower is called *hashika,* and it's collected and dumped in the compost pile to be used as fertilizer. This was the big work for them this morning. After breakfast, Kanezō cleaned out the stable and turned over the compost. My brother's wife went out to the fields and picked vegetables for eating. In the afternoon they stacked up more harvested rice plants for drying, and Kanezō used the horse to haul more rice plants (did *uma no shireboi*). My brother re-stacked cut rice plants to dry (did *keshi*).[8] My brother's wife bundled rice plants, and she headed back early in the evening and put more rice plant bundles into the rice storeroom. My mother helped with that, and cooked and such.

October 14, Monday

Cloudy, later sunny
Southwest wind
14°C in the morning, 21°C at 2:00 p.m.

Figure 8.06. Kanezō doing *shikkuri tobashi* in the main house's stable. From Yoshida's personal collection: not included in 1938 book.

Breakfast:	miso soup with napa cabbage and potato in it, miso-preserved *myōga*, pickled eggplants, *takuan*
Lunch:	*fukuran mochi,* roasted salted trout, pickled eggplants, *takuan* (I'll explain how to make *fukuran mochi* later.)
Supper:	miso soup made with roasted dried fish stock and eggplants and sea lace (*tsurumo*) in it, side dishes: same as lunch

Yoshida Jōsuke's eldest daughter left the village to work in a factory, but eventually she came back because she got sick. Today, she died.

Today I took off from work during the morning. In the afternoon, I picked dried leaves off the Hōryō daikon, and cleaned out the rabbit hutch (that stuff will be used as fertilizer) and picked eggplants and collected chestnuts. My wife made *fukuran mochi* during the morning, to have for lunch. You just peel potatoes and grate them, and some daikon, and squeeze all the juice through a cloth. And you cook up a pot of *azuki* beans and pumpkin pieces, flavored with salt. Then, you pinch off bits of the stuff left over from squeezing through a cloth and drop them into the pot and cook them in that for a bit. This is what we call *fukuran mochi.* In the afternoon, my wife did laundry and sewed, and pickled some eggplants. In the evening, we received some sea lace (*tsurumo*) that was harvested at Minami Iso in Oga from my wife's brother's household, so we chopped it up and put it in our miso soup and ate that. It was absolutely delicious.

Today's income:

18 eggplants	10 sen

At the main house: They got up at three o'clock again today, and my brother and his wife and Kanezō went out to thresh rice (do *inekoki*). My brother and Kanezō took turns operating the machine and processing the rice. They also took turns removing the stamens from among the sifted grains, and my brother's wife hauled bundles of rice plants from the storeroom and put them in the box attached to the machine. Kanezō grabbed those one by one with his right hand and fed them into the machine. My brother then took them from Kanezō and put them into another machine that removes the stamens and such, and then bundled them. He bundled parts from every five or so bunches of rice plants into one bunch, and those from every ten bundles into one. My brother's wife put a rice sieve under the machine's spout and collected the stamens and stuff and sifted everything constantly. This stuff is called *shikkuri,* as I mentioned above, and after sifting everything, the little stamen parts must be separated from everything else by fanning a *mushiro* mat. Once this is done, people will generally either bundle ten, or twelve, or sixteen bunches into bundles and then haul all of these to their rice storeroom. This is the same for all the stamens—these must be done separately because they'll be further separated afterward. Later, this stuff might be put into the horse stall or may be burned in the garden. But this is basically how the rice grains are separated from the plants and how the process works.

In the afternoon, Kanezō cleaned off some of the big wooden stakes and put them back under the house for storage, and he prepared a place for storing the rice bran. My brother's wife, as usual, went out to the fields and dug up burdock roots and taros, and pulled up some daikon, and then returned to the house. My mother went to a different village to see their festival. In the afternoon, they stacked up more rice plants on stakes to dry. Kanezō bundled rice plants (did *marogi*) and Ichisuke hauled the bundles using the horse (did *shireboi*). My brother sifted rice grains and his wife put bundles away for storage and cooked and such. Kieko minded the shop. When one hauls rice plant bundles using a horse (does *uma no shireboi*), one must bring stakes back from the rice paddies when one returns. A kid can manage one or two stakes, while an adult (male) can manage to walk back with five or six.

October 15, Tuesday

Sunny, later cloudy
Southwest wind
12°C in the morning, 20°C at 2:00 p.m.

Breakfast:	daikon miso soup with sea lace in it, pickled eggplants, *takuan*
Lunch:	*shijiko*,[9] pickled eggplants, *takuan*, raw eggplants with some miso on them
Supper:	eggplant miso soup with sea lace in it, roasted salted salmon, pickled eggplants, *takuan*

Yoshida Kiyomatsu's eldest daughter, Mii, gave birth to an illegitimate child (today), but the baby died.

Early this morning, I cut grass for the rabbits. It takes one big basketful of grass each day to feed more than ten rabbits.

My wife went to another village to help a family harvest rice again today. During the morning, I bundled sheaves of harvested rice plants (did *marogi*), and hauled them on my shoulders and piled them on the horse's back for transporting, etc. In the afternoon, I went out to stack rice plant sheaves on stakes for drying. I stacked about 50 stakes' worth and finished early, so I hauled more rice plant bundles and stacked them on the horse's back again, etc., before coming home. Then I made supper. The daylight is getting shorter day by day, so it's already pretty dark if I head home around five o'clock, and it's totally dark by the time supper is ready.

Today's expenditures:

salt	10 sen
salted salmon	50 sen
brown sugar	5 sen

At the main house: They get up at three o'clock every morning and thresh rice. It's always the three of them—my brother, his wife, and Kanezō—doing that work. After breakfast, Kanezō and my brother went out to the paddies at Sakura to stack rice sheaves to dry. My brother's wife minded the shop and cooked and such. Kanezō used the horse to haul rice plants (did *shireboi*) until Ichisuke got home from school—around three o'clock. My brother bundled cut rice plants. After school, Ichisuke took over guiding the horse, and Kanezō switched to bundling rice plants and stacking them on the horse's back. My brother then went out to stack cut rice plants up on stakes to dry. His wife stayed home and put rice plants in the storeroom and cooked and such. Kieko minded the shop.

I'll write about how to re-stack cut rice plants on stakes to dry (how to do *keshi*). Basically, you put a stake in the ground near rice plants that have already been hung up (or stacked) to dry and then move them to the new stake. In other words, you re-stack them, this time putting the plants that were on top at the bottom. It's the opposite of the way it was done before. The first time, the objective was to dry the full ears of the rice plants, but the second time the goal is to dry both the stems of the plants and the ears. So, the first time, the point was to keep the ears on the outside, away from the stake. But the second time, the point is to do it so that the stems stick out, away from the stake. And you want to stack them so the stems are sticking out in four directions. Then you adjust the stems and fluff out the ears a bit. I've drawn a diagram to show what I mean (see Fig. 8.07).

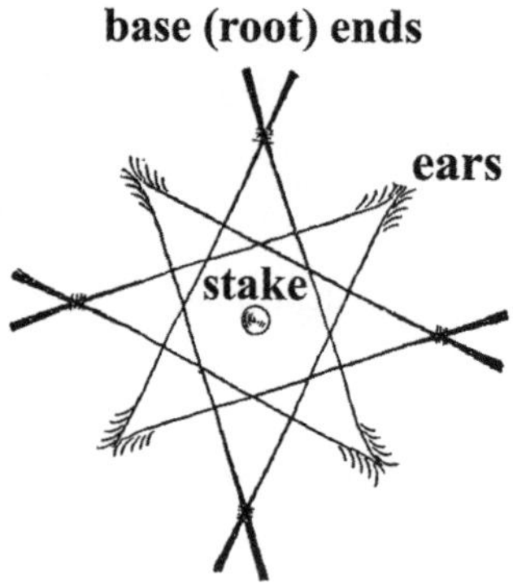

Figure 8.07. How to stack rice plants on a stake to dry.

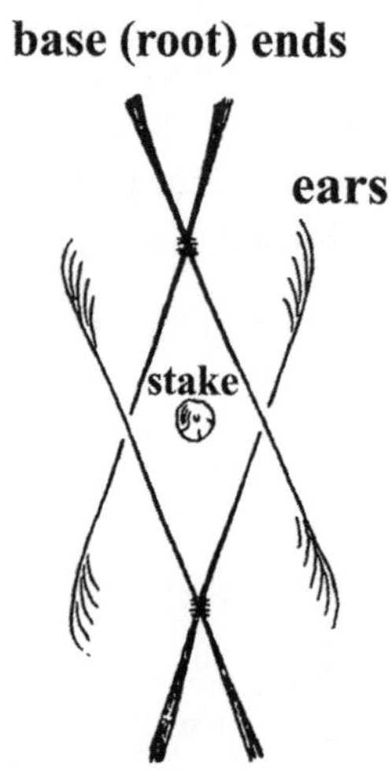

Figure 8.08. How to do the second stacking (rice plants on a stake).

And when you do the second stacking (*keshi*)—just like before—you want to keep everything nice and straight and centered. And then, the last four you put down, you should make sure that the ears are sticking way out. And the final two bunches, called *sadaba,* should be tightly interlocked. That's how it's done. I've drawn another diagram to show how it all works (see Fig. 8.08). Both diagrams are drawn to show the view from directly above.

October 16, Wednesday

Sunny
10°C in the morning, 22°C at 2:00 p.m.

Breakfast:	daikon miso soup, pickled eggplants, *takuan,* etc.
Lunch:	roasted salted salmon, pickled eggplants, *takuan*
Supper:	miso soup with green onions, side dishes: same as lunch

I cut grass for the rabbits early this morning, and then fastened new hoops to the barrels for collecting water and night soil. Then I applied follow-up fertilizer to the summer cabbages—dissolved 100 *momme* (375 g) of ammonium sulfate in 1 *to* 5 *shō* (27.1 L) of water and poured that over the beds from above. I also picked dried leaves off the daikon. After that, I prepared lunch. In the afternoon, I went over to (Yoshida) Sangobē's place to help them re-stack their rice plants. In three hours—from one-twenty to four-twenty—I stacked more than 111 five-bunch stakes with rice plant bunches. This is something that can't be achieved without working at full throttle ["full horsepower"]. I was just determined to see exactly how much I could accomplish if I pushed as hard as I could. People have said since the old days that one person can only manage to fill 200 stakes in a single day. After finishing, I went home and started preparing supper. Again today, my wife went to another village to help harvest rice.

Today's expenditures:

candies for the kids	5 sen

At the main house: They got up at two o'clock this morning, and the usual group of three went out to thresh rice. After breakfast they used the horse to haul threshed rice to the communal work shed to hull it. Then they cleaned many of the stakes for stacking rice sheaves and put them back under the house. My brother's wife cooked and minded the shop and such. In the afternoon, sure enough, my brother and Kanezō went out to hull more rice. They filled straw containers (*tawara*) with rice, and sold some, and the rest of the rice grains and bran and such they brought back to the house for eating. They finished that pretty early, so Kanezō threshed more rice alone.

October 17, Thursday

Cloudy, rain later, then just cloudy after the rain
South wind, later out of the southwest
11°C in the morning, 15°C at 2:00 p.m.

Breakfast:	miso soup with pumpkin in it, pickled eggplants, *takuan*, etc.
Lunch:	*shijigo*, pickled eggplants
Supper:	eggplant miso soup, roasted salted salmon, pickled eggplants

I cut grass for the rabbits early, and I picked dried leaves off the daikon. After breakfast, I went over to help Sangobē's family re-stack rice plants again. But I came back home around eleven o'clock because of the rain. I rested a bit in the afternoon, but I did prepare lunch and supper. Once again, my wife went to a different village to help harvest rice.

Today's expenditures:

shijigo	10 sen
white sugar	5 sen

At the main house: They got up at three o'clock and threshed rice again. They also filled sacks with rice grains. Kanezō went out to re-stack rice plants after breakfast, but he had to quit and return to the house (because of the rain) around ten o'clock. Then, he and my brother and my brother's wife did more threshing together. They did more in the afternoon, too.

October 18, Friday

Rain
Strong south wind
12°C in the morning, 18°C in the afternoon

Breakfast:	cabbage, napa cabbage, and potato miso soup, pickled eggplants, *takuan*
Lunch:	roasted salted salmon, pickled eggplants, *takuan*
Supper:	miso soup with taro roots in it, *shijigo,* pickled eggplants, *takuan*

I didn't work outside early because of the rain, but later in the morning I dug up some taros—15 plants, which made for about 6 *shō* (10.8 L). I was a little concerned because the stems aren't thickening up as much as last year, so I harvested those plants to check, and the result was reassuring—not bad. I knocked the dirt off the tubers and picked those off, etc. After that, I went into the village to buy some rolled oats (*tsubushi mugi*). In the afternoon, I picked dried leaves off the daikon and fed them to the rabbits, and washed the taro plant stems. Up until yesterday my wife was helping others harvest rice, but that's over now so today she was at home. She sewed during the morning, and she scraped taro stems in the afternoon. She did this in the night, too—up until going to bed. The surface scrapings from the taro plants are tied up with straws and hung out to dry in the sun, and during the winter they're put into soup.

Today's expenditures:

1 bag of rolled oats	2 yen 30 sen
sweets	5 sen

At the main house: They got up at three o'clock, and the usual team of three threshed rice. After breakfast, Kanezō made *tawara* in preparation for hulling rice. When you make *tawara,* you need to make lids (*sadara*) for them so they'll be ready to hold rice once it has been hulled. My brother filled sacks with rice, cut hay for the horse, and went out to negotiate for rice hulling.[10] In the afternoon, he and Kanezō used the horse to haul rice to the communal work shed, hulled rice, filled *tawara* with hulled rice grains and tied them up, and hauled bran (*nuka*). My brother's wife minded the shop all day, and my mother cooked and such.

Figure 8.09. The Ōkura hamlet communal work shed. From Yoshida's personal collection: not included 1938 book.

Using government subsidies, the village finally prepared a communal work shed (see Fig. 8.09). It seems to be organized like a cooperative association. Rice polishing is free, and it costs 12 sen per *tawara* to hull it. I hear they've asked a member of the association to serve as manager of the machines for 20 yen per month, just for this autumn. This compares with the prices charged by the business owners in the village in the following way.

The capitalists in the village have been charging 10 sen per *tawara* for rice polishing and 20 to 25 sen for hulling. But if someone processes his rice at the communal work shed, he'll save about 13 yen. There's something important about this that people have been talking about, and that I should make note of here.

The main capitalist is Yoshida Saburobē—a plutocrat—and the thing is, not a single plutocrat is a member of this new (communal work shed) association (*kumiai*). I mentioned this before, but since the communal work shed was built on the village dancing ground, they've decided not to hold the annual *bon-odori* dance, or the *saku-odori* dance,[11] from this year.

October 19, Saturday

Cloudy and sunny
Strong south wind
21°C in the morning

Breakfast:	miso soup with taro roots in it, pickled eggplants, *takuan*
Lunch:	*bego* cooked in soy sauce with eggplants, pickled eggplants, pickled cucumbers (*Bego* is a fish caught off the Minami Iso coast in Oga.[12])

Supper:	miso soup with taro roots in it, boiled cabbage with soy sauce for flavor, pickled eggplants, *shijigo*

I picked eggplants early this morning. From today, my wife started going out to sell vegetables again, so I had to prepare for that. After breakfast I went to the main house to help stack a dried rice plant *waraniyo*.[13] You must stack the straw bunches in a circular pattern to make one of these (see Fig. 8.10). I did this together with Kanezō. He tossed two bunches of rice plants at a time to me, and I arranged and stacked them. To do it, you just place the bunches down with the bottom ends of the plants pointing outward and the ears (which held rice grains before the threshing) pointing inward and stack them going around in a circle. You go all the way around twice and then stack them in the center once, and then go around again. If you don't stack them in the center as you go, then a hole will appear in the center later, and all the surrounding stacked plant bunches will easily slip apart and the whole thing will collapse, so it's crucial to go around twice and then stack one in the center, again and again. The height and diameter of a *waraniyo* varies from person to person. The one I made today is about 9 *shaku* (2.7 m) in diameter and 7 *shaku* (2.1 m) tall, but that's only the distance from the bottom to the lower edge of the "roof" on top. To make this, you go around and around and stack the bunches so that you get a pointy raised center. Specifically, with each layer of bunches, you put down 3–4 *sun* (approx. 10.5 cm) toward the center. After that, you lay down *noma* mats over the top and then a cone-shaped woven straw cover (*geboshi*), and then tie that up from the bottom with ropes in six directions, and you're done. Highly skilled people can make *waraniyo* that are very big—both in diameter and in height. It really is very difficult.

To make that *waraniyo* today I used about 1,500 bunches of dried rice plants.

I ate lunch at the main house today [probably with both daughters]. We ate *mochi* made from this year's rice. In the afternoon, I dug up taros, harvested eggplants, and scraped the surfaces of taro plant stems. My wife headed out early this morning to sell vegetables and came back around two o'clock. After returning, she scraped taro plant stems.

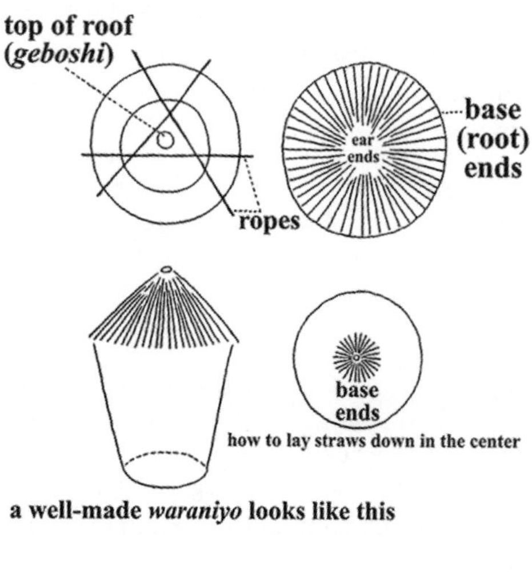

Figure 8.10. How to make a *waraniyo* rice plant stack.

Today's income:

170 eggplants (30 big, 140 small)	50 sen
taros (6 *shō* (10.8 L))	75 sen
cabbages (800 *momme* (3 kg))	15 sen

Today's expenditures:

soy sauce	10 sen
sweets	5 sen
salt	15 sen

At the main house: They rested in the early morning. They had already finished dealing with the harvested rice for the day—because of the rain—so there wasn't any of that to deal with. But Kanezō was out cutting hay for the horse at the crack of dawn. During the morning, he and I made that *waraniyo*. In the afternoon, he went out to the paddies at Sanbonmatsu and dug out the drainage ditches. My brother's wife minded the shop in the morning, and went to her natal household to help harvest millet in the afternoon. There's no special skill or technique for harvesting millet. You just tie a straw belt around your waist and attach a big scythe to that and cut with it. You hold the scythe with your right hand and grab the millet plants with your left hand, and then cut them as close to the roots as you can. Then you tie single handfuls up in sheaves, or maybe double handfuls. Then, as you go, you put a dozen sheaves together and tie them tightly up near the ears and stand them up to dry. Each of these is called one *tsuboke* [see Figs. 0.39, 0.41, and 0.42]. My mother cooked and such today.

October 20, Sunday

Sunny
Strong west wind
12°C in the morning, 18°C in the afternoon

Breakfast:	eggplant miso soup, cabbage [probably boiled] with soy sauce for flavor, pickled eggplants, *takuan*, etc.
Lunch:	roasted salted salmon, pickled eggplants, *takuan*
Supper:	miso soup with taro roots in it, *shijigo* (fish), pickled eggplants, *takuan*

In the early morning, I cut grass for the rabbits, and then later in the morning I harvested more taros, and then I scraped off the surfaces of the stems and tied them up with straws for hanging. Then I made lunch. I dug up more taros in the afternoon, and picked eggplants, and went into the village to collect night soil, and I also scraped the surfaces of more taro plant stalks. My wife came back by lunchtime, and she scraped taro plants and picked dried leaves off the daikon during the afternoon.

Today's income:

190 eggplants (70 big, 120 small)	50 sen
taros (5 *shō* 5 *gō* (9.9 L))	60 sen
cabbages (1 *kan* (3.75 kg))	20 sen

Today's expenditures:

pears	5 sen

At the main house: They rested in the early morning, but Kanezō didn't—he cut hay for the horse. They hauled rice plants in from the paddies all day. Kanezō loaded it and guided the horse. My brother bundled rice plants. His wife stayed at home and put bundles of rice plants into the storeroom and cooked, and my mother went out to the fields to harvest buckwheat.

October 21, Monday

Occasional rain, some sleet
Strong west wind
10°C in the morning, 16°C in the afternoon

Breakfast:	same soup as yesterday evening, miso-preserved eggplants, *takuan*
Lunch:	same as yesterday
Supper:	miso soup with taro roots and stems in it (stock made from crucian carp), pickled eggplants, and a little fish cooked in a pot. The carp, which was caught in the waters of Lake Hachirō, we received from my wife's natal household. The little fish were from the Funagawa coast—I received them from a friend.
Snack:	some cooked crabs

I didn't work outside in the early morning because of the rain—I just cleaned up and made breakfast. After eating, I tied taro stems up with straw ropes, and dug up more taros, and then made lunch. In the afternoon, I scraped more taro plants and tied them up. From the afternoon, my wife did the same work as me, and she picked eggplants and such in the evening.

Today's income:

190 small eggplants	40 sen
taros (8 *shō* (14.4 L))	80 sen

At the main house: They got up at two-thirty, and the usual team of three threshed rice. After breakfast, they did more of the same, but my brother didn't participate. Kanezō and my brother's wife also brought dried rice plants in from the paddies, but they were only able to bring in two horse loads by the time they had to quit, because of the rain. After that, they immediately started threshing rice. My mother cooked and such.

October 22, Tuesday

Sunny
Strong west wind
10°C in the morning, 13°C in the afternoon

Breakfast:	eggplant miso soup with sea oak (*arame*) in it, pickled eggplants. The sea oak was gathered on the Funagawa shoreline.
Lunch:	last night's leftover fish, pickled eggplants, etc.
Supper:	miso soup made with carp stock, with taro pieces and sea oak in it, *shijigo,* pickled eggplants, etc.

I didn't work early this morning because I felt a bit sick—like I'd caught a cold. I cleaned up after breakfast, and then I tied up napa cabbages—just gathered all the leaves of each and tied them up tightly with (rice plant) straw ropes. This took until twelve-thirty, and after that I ate lunch. After eating, I went to the Agricultural Cooperative (Nōkai) warehouse to buy some discounted rice. Then I took that to the communal work shed to get it polished. After getting home, I cut grass for the rabbits, and I dug up one of my Chinese yams to see how they're doing. I was happy to find that they've grown better than expected. My wife got back around two o'clock, and after eating lunch she picked especially tiny eggplants. We'll sell some of these, and the rest we'll preserve for later. In the evening, she picked the tops off all those little eggplants.

Today's income:

cabbages (2 *kan* (7.5 kg))	30 sen
taros (7 *shō* (12.6 L))	72 sen

Today's expenditures:

rice (one *tawara*)	11 yen 15 sen
candies	5 sen
soap	5 sen

At the main house: The usual three-person team threshed rice from the early morning. After breakfast, Kanezō went out to the rice paddies to dig out drainage ditches, and in the afternoon my brother tied dried rice plants up in bunches and Kanezō stacked them on the horse and led it, to haul them. Ichisuke got home from school around three o'clock and took over leading the horse, and Kanezō bundled rice plant bunches and my brother re-stacked rice plants on stakes for drying. My brother's wife stayed at home during the morning and did laundry and minded the shop. In the afternoon, she put rice bundles into the storeroom and cooked and such. My mother went out to the fields in the afternoon and harvested daikon, etc.

October 23, Wednesday

Sunny
4°C in the morning, 18°C in the afternoon

Breakfast:	eggplant miso soup with sea oak (*arame*) in it, pickled napa cabbage with soy sauce for flavor, pickled eggplants and such
Lunch:	roasted salted salmon, all else: same as morning
Supper:	miso soup made with halfbeak (*sayori*) [*Hyporhamphus sajori*] stock, with taro pieces in it, other dishes: same as lunch. The halfbeaks came from the waters of Lake Hachirō—I received them from an acquaintance in Tennō Village.

I didn't go out and work in the early morning—just made breakfast right away. After cleaning up, I dug up taros and removed little baby taros from the bases of the plants. I cut the plants at the bases and washed the lower parts to remove all the little taros. Then I scraped the surface parts of the stems and made lunch. After eating, I went back to removing the surface parts of the taro plant stems and then tied them up with straw ropes. Once all that was done, I went into the village—to the communal work shed—to get the rice I left there yesterday for polishing, and I walked back home with it on my back. My wife got home around one o'clock, and she helped me remove the surface parts of the taro plant stems, and she picked eggplants and pulled off their tops, etc.

Today's income:

tiny eggplants (5 *shō* (9 L))	40 sen
taros (7 *shō* (12.6 L))	72 sen
napa cabbages (1 *kan* 300 *momme* (4.9 kg))	15 sen

Today's expenditures:

candies for the kids	3 sen

At the main house: The usual group of three got up at two-thirty this morning and did the usual thing—thresh rice. During the (early) morning, after breakfast, and in the afternoon, Kanezō and my brother hauled dried rice plants in from the paddies. They finished the Sakura paddies today. My brother's wife minded the shop, put rice plant bundles into the storeroom, and cooked and such. My mother took her lunch out to the fields and picked pea pods and pounded them to release the peas. She also pounded sesame seed pods: laid straw *mushiro* mats on the ground and pounded them on top of those with a stick. Then, after pounding the pods, she laid sesame stems down on top of some pebbles and pounded them. After that, she used a woven sifter for rice grains and sifted them to remove all the unwanted stuff. Then she stacked the sesame and pea plant stems in the fields and left them there.

October 24, Thursday

First frost this morning, but a very sunny day
Weak west wind
1°C in the morning, 20°C in the afternoon

Breakfast:	eggplant miso soup, side dishes: same as yesterday morning
Lunch:	same as yesterday's lunch
Supper:	same as yesterday's supper

I cut grass for the rabbits in the early morning, and picked and washed daikon, and after breakfast I cleaned up and then went straight over to the communal work shed and got two (free) bags of rice bran and carried them home with me. Then I went over to Sangobē's place to bundle rice plants. I stacked these on a horse and hauled them over to the roadside, where I stacked them on (the ground). My wife got home around one o'clock, and then she picked soybeans. She tied up each big handful in a bunch with straws. After finishing the soybeans, she harvested eggplants, and in the evening, she picked off their tops.

Today's income:

eggplants (5 *shō* (9 L))	50 sen
taros (7 *shō* 5 gō (13.5 L))	81 sen
8 daikon	12 sen

Today's expenditures:

brown sugar	5 sen

At the main house: They got up at two o'clock and the three of them immediately started threshing rice. Kanezō continued threshing rice together with my brother's wife after breakfast until around ten o'clock, and then he went out to the Bishamon paddies to re-stack rice plants (on stakes) for drying. My brother's wife continued threshing rice the rest of the day until evening, and then cleaned up. My brother went out to re-stack rice plants in the Bishamon paddies in the afternoon, and Kanezō went to the fields to pound soybeans. My mother asked my aunt next door [the big main house] to accompany her out to the fields—with their lunches—to pound soybeans. Normally, you'd pick these and then dry the pods in the sunshine and then pound them on a different sunny day, but there's also a way known as *fugioshi,* which is the way they did it today over at the main house. During the morning, my mother and my aunt picked soybeans and laid them out on mats to dry in the sunshine. In the afternoon, Kanezō and my aunt pounded them with sticks, and then my mother pounded the remaining shells and such with a small stick. In the evening, they sifted the beans with rice sifters and removed all the junk from among them. Then they piled the plant parts and shells and such in the fields.

October 25, Friday

(September 28 according to the old calendar)

Very sunny
5°C in the morning, 22°C in the afternoon

Breakfast:	eggplant miso soup, pickled eggplants, pickled napa cabbage, etc.
Lunch:	roasted salted salmon, pickled eggplants, pickled napa cabbage, etc.
Supper:	as explained below

In the early morning, I cut grass for the rabbits, and then after the post-breakfast cleanup I went back to Sangobē's place to bundle dried rice plants and haul and stack them. But we finished this *ine-age* work early, so I collected the stakes (used for stacking rice to dry) from the paddies and tied them up in sets of ten. My wife got home at one-thirty and washed green onions and tied them up. She also harvested buckwheat.

Today is September 28 according to the old lunar calendar—the day we celebrate as the second seasonal festival (*sekku*)—so we pounded *mochi* and ate them. I received *mochi* from the main house. We ate some with sweet *azuki* beans, put some in miso soup made with halfbeak stock and ate them that way, and had some with grated daikon mixed with sea lace (*tsurumo*) from Minami Iso in Oga, and with miso, and with sea lace and vinegar, etc.

Today's income:

eggplants (5 *shō* (9 L))	50 sen
6 daikon	10 sen
taros (7 *shō* (12.6 L))	70 sen

Today's expenditures:

brown and white sugar	20 sen (10 sen each)

At the main house: The usual group of three got up at two o'clock and immediately began threshing rice. After breakfast, they piled rice straw bunches up in circular stacks (*wara-niyo*). They stacked about forty bundles' (*maro*) worth. At around nine o'clock they went out to haul dried rice plants in from the paddies (to do *ine-age*). My brother tied up the bundles, and Kanezō managed the horse—loading and guiding. My brother ate lunch in the paddies and finished all the bundling work by three-thirty, then headed home. Then he pounded *mochi*. Kanezō's work took the rest of the day—until evening—to finish. My brother's wife spent the day at home, minding the shop, putting rice plant bundles in the storeroom, and filling up and heating the bath, etc. My mother was busy all day with the *mochi* preparations.

October 26, Saturday

Sunny, later cloudy
Strong south wind
5°C in the morning, 16°C in the afternoon

Breakfast:	eggplant miso soup, leftover *yama mochi* from last night, pickled eggplants, vinegar-flavored sea lace (*tsurumo no suiko*)
Lunch:	roasted salted salmon, pickled eggplants, vinegar-flavored sea lace, etc.
Supper:	halfbeak stock miso soup with taro pieces in it, pickled napa cabbage, pickled eggplants, etc.

I cut grass for the rabbits early. Then I pulled up soybean plants and stood them up to make *tsuboke* clusters in the fields, to dry them.[14] After cleaning up following breakfast, I dug up more taros, and scraped off the surfaces of the stems, and then prepared lunch. After eating, I tied up taro stem scrapings with straw strings, and harvested buckwheat and stood the plants up in *tsuboke* to dry. My wife came home around one-thirty and got right to work harvesting buckwheat. We finished harvesting our buckwheat today.

Today's income:

green onions (6 bunches)	10 sen
daikon	10 sen
taros (8 *shō* (14.4 L))	70 sen

At the main house: They got up at two o'clock again this morning. The usual group of three started threshing rice right away. After breakfast, my brother and Kanezō headed out to the paddies to bring back more dried rice plants, just like yesterday. My brother's wife also repeated her activities from yesterday. My mother went out to the fields and picked dry leaves off the daikon.

October 27, Sunday

Cloudy, with rain later
Strong southeast wind
7°C in the morning, 16°C in the afternoon

Breakfast:	eggplant miso soup, pickled eggplants, pickled napa cabbage, etc.
Lunch:	roasted salted salmon, pickled eggplants, etc.
Supper:	crucian carp stock miso soup with taro pieces in it, *shijigo*, pickled eggplants, etc.
Snack:	pears

I harvested cabbages and such early this morning, and then I made breakfast. After eating, I cleaned up and then picked and cleaned some leftover soybeans from yesterday. After that, since the wind was very strong, I went into the cedar groves and collected a bunch of dead leaves, and then I dug up some taros and made lunch. The afternoon was rainy, so I just cleaned up indoors. My wife got back around one o'clock and did laundry and removed the stains and such in the big cauldron (*kama*) we use for cooking rice, etc. She also picked lice from the girls' heads.

Today's income:

cabbages (1 *kan* 800 *momme* (6.75 kg))	25 sen
10 daikon	15 sen
taros (8 *shō* (14.4 L))	70 sen

Today's expenditures:

shijigo (fish)	10 sen
pears (2 *kan* (7.5 kg))	60 sen

At the main house: It was exactly the same situation as yesterday morning. These days, my mother always cooks breakfast. After breakfast, my brother and Kanezō went out again to re-stack rice plants in the paddies (did *inekeshi*). My brother's wife minded the shop and put fresh rice bundles into the storeroom and such. In the afternoon, they threshed rice because the rain prevented them from working outside—the three of them threshed. The kids minded the shop. Today, they finished bringing in all the dried rice from the paddies at Bishamon. During the morning, my mother went into the woods with the kids to collect dried cedar leaves, and in the afternoon, she stayed at home and did some food preparations.

October 28, Monday

Rain
Strong west wind
12°C in the morning, 21°C in the afternoon

Breakfast:	eggplant miso soup, pickled eggplants, miso-preserved eggplants
Lunch:	roasted salted salmon, pickled eggplants
Supper:	exactly the same as lunch
Snack:	sweet potatoes and pears

I didn't work outside early this morning because of the rain. But the days are getting shorter, so I can't do much early work anymore, anyway. During the morning I scraped the surfaces of taro plant stems together with my wife. In the afternoon, I had to go into the village on an errand, and I cut grass for the rabbits. My wife sewed and picked eggplants this afternoon.

Today's expenditures:

postage stamps	6 sen
sweets	4 sen

At the main house: The usual team of three did rice threshing in the early morning and again after breakfast, and sifted rice grains (did *momi toshi*) in the afternoon. The kids minded the shop. They finished threshing all the rice from the Bishamon paddies today. My mother cooked and such.

Today, I'll write about how to make an un-threshed rice plant straw stack (*inaniyo*). This year, at the main house, they didn't stack the rice plants up like this—they simply threshed what they'd brought in as they went along. But up to this year they always stacked the rice plants up in such a way. It's not all that different from making a straw stack (*waraniyo*), which is basically always done in the way I explained before, but there are people who stack large *maro* bundles of plants tightly together, going round and round, and there are those who lay them down in *ichi maro ogi* style.[15] I think the latter way is correct. What is meant by *ichi maro ogi* style is just putting one (*ichi*) bundle (*maro*) down and then leaving a space for one more between it and the next one you lay down, and going all the way around like that, and then going back around later and filling in all those empty spaces in order. I'll include an illustration to show what I mean (see Fig. 8.11). And just like when stacking straw bundles, you must stack the bundles once around the center every two times you go around. And you must try to make the stack bigger, little by little, as you move upward. It's not really so difficult to achieve this (if you do it right).

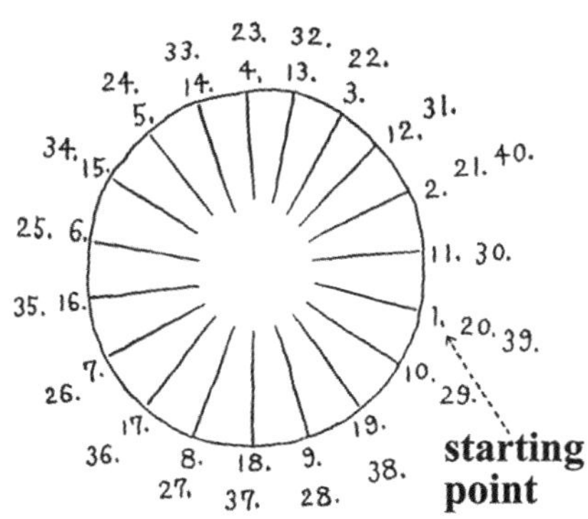

Figure 8.11. How to make an *inaniyo* rice plant stack.

October 29, Tuesday

Occasional rain
West wind
12°C in the morning, 16°C in the afternoon

Breakfast:	eggplant miso soup, pickled eggplants, etc.
Lunch:	*shijigo*, pickled eggplants, etc.
Supper:	halfbeak stock miso soup with pieces of taro in it, side dishes: same as lunch

This morning, I washed daikon and picked cabbages and such to prepare for selling vegetables. After eating breakfast, I cleaned up and worked on building a new outhouse to cover the latrine (see Fig. 8.12). It's just like what I built before. I used chestnut wood, bamboo, wheat straws, rice straws, and rope and such.

I was short on materials, so I went into the landlord's forest and disrespectfully helped myself to what I needed and came back. I surrounded it with wheat straws to make the walls, and the roof I made with straw *noma* mats. Then I made lunch, and we ate, and I did this and that, which took until three o'clock. Then, I put some new *noma* mats on the awnings of the house and harvested some daikon and cabbages and such. My wife came back by lunchtime, and she wove *noma* mats in the afternoon.

Today's income:

eggplants (2 *shō* (3.6 L))	20 sen
7 daikon	10 sen
cabbages (2 *kan* (7.5 kg))	30 sen
taros (7 *shō* (12.6 L))	65 sen

Today's expenditures:

a scale	33 sen

At the main house: There was no more rice to thresh, so they didn't do any early morning work. Kanezō went out to Sanbonmatsu and spent the day digging drainage ditches in the

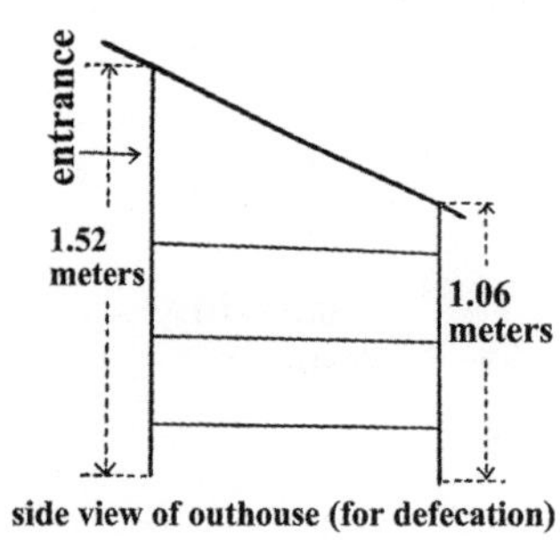

Figure 8.12. The outhouse.

paddies. My brother's wife was just hanging around in the shop all day. My mother cooked and such.

October 30, Wednesday

Sunny, later cloudy with occasional light rain
Strong west wind
10°C in the morning, 16°C in the afternoon

Breakfast:	same as yesterday
Lunch:	roasted salted salmon, pickled eggplants, etc.
Supper:	*udon* noodles in soup made with boiled dried anchovies and soy sauce, boiled napa cabbage with soy sauce for flavor, pickled eggplants, etc.
Snack:	dried persimmons we received from the main house

After eating breakfast, I cleaned up and started preparing compost soil for next year's hotboxes. I thoroughly mixed 50 *kan* (187.5 kg) of fully decomposed compost with 100 *momme* (375 g) of soil, dried anchovy powder (2 *kan* (7.5 kg)), two *ka* of night soil, 2 *to* (36.1 L) of ash from burned wood and brush, etc., and piled it all up. Specifically, I started with the soil and over that I poured the night soil, then the anchovy powder, and then the rich compost over that, and then finally the ash, in that order. Later, I'll turn it all over again and pile it up (do *honzumi*) and leave it like that over winter. This took me until two o'clock. But I made lunch while doing that. After eating, I cut grass for the rabbits, and went over to someone else's fields to collect daikon leaves that they had just thrown there. I also picked vegetables for selling. My wife came back at two o'clock and wove *noma* mats.

Today's income:

napa cabbages (1 *kan* 500 *momme* (5.6 kg))	25 sen
10 daikon	15 sen
taros (7 *shō* (12.6 L))	70 sen

Today's expenditures:

udon	22 sen
soy sauce	15 sen
dried anchovies	5 sen
sweets	3 sen

At the main house: Kanezō cut hay for the horse early, and spent the rest of the day digging drainage ditches in the Sakura paddies. My brother went out on errands, and his wife was in the shop. My mother cooked and such.

October 31, Thursday

Occasional rain
Strong west wind
12°C in the morning, 16°C in the afternoon

Breakfast:	daikon miso soup, pickled eggplants, boiled napa cabbage with soy sauce for flavor
Lunch:	*shijigo*, pickled eggplants, etc.
Supper:	miso soup with pieces of taro and some sea lace in it, boiled *seri*[16] with soy sauce for flavor, pickled eggplants and such. We received the *seri* and the sea lace from my wife's brother's family.
Snack:	pears

First this morning, I went over to another person's fields to collect dried daikon leaves to feed to the rabbits. After that, I harvested more taros. To sell the stems of the plants I dug up today, I washed them carefully with water and tied them up. After breakfast, I cleaned up and then started preparing lunch. In the afternoon I yanked the tubers off the taro plants I'd harvested in the morning, and I also washed green onions and harvested napa cabbages, etc. My wife returned at one o'clock, and she took taro plant stems to someone who had ordered some from us, and then went to her brother's home to deliver a *futon* stuffed with dried rice plant pieces (*shibebuton*).

Today's income:

4 bulb daikon (radishes)	10 sen
napa cabbages (1 *kan* 500 *momme* (5.6 kg))	20 sen
cabbages (1 *kan* 200 *momme* (4.5 kg))	20 sen
taros (5 *shō* (9 L))	50 sen

Today's expenditures:

geta (sandals—two pairs, for the kids)	10 sen

At the main house: Kanezō cut hay for the horse early, and again today he went out to dig drainage ditches in the paddies—the Sakura paddies in the morning and Shinden in the afternoon. My brother's wife was in the shop during the daytime, but she went out to the fields and picked some daikon in the evening. My mother cooked and such.

Chapter 9: November 1935

November 1, Friday

Sunny
Strong west wind
10°C in the morning, 18°C in the afternoon

Breakfast:	daikon miso soup with sea lace (*tsurumo*) in it, pickled napa cabbage, pickled eggplants, etc.
Lunch:	*shijigo,* pickled napa cabbage with soy sauce for flavor, pickled eggplants, etc.
Supper:	miso soup with taro pieces and sea lace in it, roasted salted trout, pickled napa cabbage, pickled eggplants

After the breakfast cleanup, I removed the surfaces of taro plant stems. I tied them up with straw ropes, and made lunch. In the afternoon, I went into the village to buy some straw, and after getting home, I wove *noma* mats and picked vegetables for selling. My wife got back at three-thirty, then wove a *noma* mat and made supper.

I'll explain, as best I can, how to make a *noma* mat. The length depends on the person who makes it. Also, whether to place the bases (bottom ends) of the rice plant straws (*wara*) pointing to the left or the right side is up to the maker. I always point the bases of the straws to my right when I weave a *noma* mat. And the thickness of the mat also depends on the maker. I usually use about 24 or 25 handfuls of straws. To weave a mat, you just weave the straws in turn. Specifically, you take some straw and wrap it tightly around some other straw, and then you tie that other straw around more straw, and so on. When you weave, you use the base ends of the straws. You use a rope to tie up the first straw, and then you tie the second straw with the end of the rope, and then the next straw, etc. Then you grab four or five of the first straws and take the end of the rope, keeping the second straw in the center, and bring the rope under from the left to the right, and then bring the rope back over to the left, under the second straw. Then you tie the third straw and repeat the process, and when you finish you just make sure that the rope doesn't come loose.

Today's income:

2 radishes (*kabudaikon*)	10 sen
napa cabbages (1 *kan* 600 *momme* (6 kg))	25 sen
taros (8 *shō* (14.4 L))	85 sen

Today's expenditures:

salt	10 sen
salted trout	10 sen
stamps	6 sen

At the main house: Kanezō cut hay for the horse early, and then after breakfast he went out to the Sakura paddies to dig drainage ditches. My brother's wife was at home during the morning, but in the afternoon she went out to help bring dried rice plants in from the paddies. She bundled them, and Kanezō carried the bundles from the paddies out to the horse path, and Ichisuke guided the horse. My brother stayed at the house and put rice plant bundles into the storeroom. At around three o'clock the horse got startled by a car and Ichisuke fell off the horse. He wasn't badly hurt, but that ended his participation in that activity for today. My brother took over guiding the horse, and my mother took over putting rice bundles in the storeroom, while cooking. Today they finished bringing all the dried rice plants in from the paddies. In the past we always celebrated such an occasion at least by eating rice with grated Chinese yam on it but they didn't eat anything special at the main house today.

November 2, Saturday

Sunny
Northwest wind
5°C in the morning, 18°C in the afternoon

Breakfast:	potato miso soup, pickled napa cabbage, pickled eggplants, etc.
Lunch:	*shijigo,* pickled napa cabbage, pickled eggplants
Supper:	same as yesterday evening

No early morning work—just made breakfast. After cleaning up, I hauled loads of straw I'd arranged to buy a few days ago. It's for making hotboxes. I borrowed the horse from the main house and used it. I made five trips during the morning, even with bundles on my own back. I paid 1 *rin* 5 *bu* for each small bundle—bought 3,500 of them. After all that, I made lunch. In the afternoon, I wove *noma* mats, and then I stacked *waraniyo.* Then I picked vegetables for selling. My wife returned at two o'clock. After that, she also wove *noma* mats, and she helped me stack *waraniyo,* etc.

Today's income:

10 daikon	10 sen
napa cabbages (1 *kan* 700 *momme* (6.4 kg))	20 sen
taros (7 *shō* (12.6 L))	70 sen

Today's expenditures:

straw	5 yen 25 sen
fish	10 sen

At the main house: They got up at twelve-thirty in the morning, and the usual team of three threshed rice right away. This morning they finished threshing all the *mochi* rice they brought in from the paddies yesterday. In the afternoon, the three of them pounded rice grains. *Mochi* rice grains have little hairs on them, so they must be pounded with special sticks. In the past, all that had to be done by hand, but now it's done by machine, so there's no need to do all that manual pounding anymore. So, at least that aspect of the process of dealing with *mochi* rice has become very convenient. People used to have to pound *mochi* rice in groups of three or five (in the old days) so it didn't work well if they weren't all in sync with one another—they really had a hard time with that. At the main house today, they sifted and cleaned all the threshed and pounded rice, so they completed their rice work for the year. In the afternoon they rested indoors for a bit and cleaned up outside and heated the bath, and in the evening, they celebrated the completion of the threshing—called *kokige* in local speech. They ate *udon* in soup with chicken meat in it.

November 3, Sunday

Sunny
2°C in the morning, 17°C in the afternoon

Breakfast:	miso soup with taro pieces in it, boiled *seri* with soy sauce, pickled napa cabbage, pickled eggplants
Lunch:	ray (*kashibe*) cooked in soy sauce, pickled napa cabbage, pickled eggplants
Supper:	same soup as morning, side dishes: same as lunch. The ray came from the coast of Oga.

First this morning, I went down to the pond and pulled out some rabbit pelts I'd been allowing to soak, and took them home and prepared to tan them. I was asked by someone to do this. I started the tanning work immediately after cleaning up following breakfast. I finished two pelts by ten-thirty, then I washed them with soap and rinsed them thoroughly with water and put them in an alum-soap solution to soak. After that, I made lunch. After eating, I scraped soil up over the roots of the green onions for the fourth time (did the fourth *tsuchiyose*), and I picked vegetables for selling. My wife came back at two o'clock and then got to work weaving *noma* mats again—as usual.

Today's income:

4 radishes (*kabudaikon*)	8 sen
napa cabbages (1 *kan* 500 *momme* (5.6 kg))	20 sen
cabbages (1 *kan* (3.75 kg))	14 sen
taros (6 *shō* (10.8 L))	60 sen

Today's expenditures:

sweets	10 sen

At the main house: Kanezō rested in the early morning. After breakfast, he and my brother's wife took the horse to her natal home to help them bring all their rice plants in from the paddies. My mother went out to the fields in the afternoon and dug up daikon and burdock roots and then went home.

November 4, Monday

Cloudy, later sunny
4°C in the morning, 15°C in the afternoon

Breakfast:	daikon miso soup, pickled napa cabbage, pickled eggplants
Lunch:	roasted salt-preserved ray, pickled napa cabbage, pickled eggplants, etc.
Supper:	*yama mochi* eaten with sweet *azuki* beans and in soup made with burdock root and soy sauce, also boiled napa cabbage with soy sauce and pickled eggplants

Early this morning I went out to cut grass for the rabbits. After the post-breakfast cleanup, I dug up some taros. Then I cooked some *azuki* beans for eating with *yama mochi* tonight, and then I made lunch. In the afternoon, I removed the surfaces of taro plant stems and tied them up for drying, and then went over to the main house to help pound *mochi.* My wife came back at two o'clock and prepared for pounding *mochi.*

Today's income:

10 daikon	15 sen
napa cabbages (1 *kan* 600 *momme* (6 kg))	20 sen
cabbages (700 *momme* (2.6 kg))	15 sen
taros (7 *shō* (12.6 L))	75 sen

Today's expenditures:

granulated sugar	10 sen

At the main house: Kanezō went out early and cut hay for the horse. He spent the rest of the day weaving *tawara* because some he wove last year got rained on and rotted. He pulled *migo*[1] out of rice plant straws and pounded them and wove rope first and then made *tawara.* He made five in total, and in the afternoon, he pounded *mochi.* My brother's wife was minding the shop all day—she and my brother made sweets together. My mother spent the day preparing to pound *mochi.*

Today is *kigitsune,* or the day to make *karasu mochi* (crow *mochi*).[2] Apparently, it's because they used to really feed them to the crows. Even inside the village of Wakimoto, there are hamlets that engage in this tradition and those that don't. Down in the old central part of the village (*honson*), they don't follow it anymore. People say that when Wakimoto's castle—Oibana Castle—fell, the enemy soldiers came by boat. And there were a whole lot of crows on the beach. And when the enemy soldiers saw that, they disembarked, and advanced unnoticed because of all the crows, and then the castle fell.[3] So, in the main part of Wakimoto they think of crows as enemies and don't participate in the tradition now. But we do make such *mochi* and offer them to crows here in Ōkura hamlet.

November 5, Tuesday

Cloudy, later rainy

Breakfast:	leftover *mochi* from last night, napa cabbage with soy sauce for flavor, pickled eggplants
Lunch:	roasted ray, pickled napa cabbage, pickled eggplants, leftover *mochi* from last night
Supper:	miso soup with taro pieces inside, side dishes: same as lunch

In the early morning I cut grass for the rabbits. Also, since the roof part of the *waraniyo* I made the other day blew off, I had to go over and repair that, etc. After breakfast I cleaned up and then I harvested some more taros. I peeled taro plant stems, and then made lunch. In the afternoon I peeled more stems and made things with straw. Then I peeled and hanged up some persimmons we got from the main house for drying and gathered up the skins and put those up to dry as well—they say that if you include some when you make preserved daikon (*takuan*), the color will be better. My wife got back by lunchtime. In the afternoon she also peeled taro plant stems, and she cleaned out the cooking pots and cooked and such.

Today's income:

8 radishes (*kabudaikon*)	15 sen
napa cabbages (1 *kan* 800 *momme* (6.75 kg))	25 sen
taros (6 *shō* (10.8 L))	70 sen

At the main house: Kanezō put the end parts (*sadara*) on *tawara* and prepared to fill and seal them. During the morning, he polished rice with my brother. In the afternoon, he hauled a bunch of rice bran, and made a new rope for the horse's saddle. My brother's wife and Ichisuke went out early to the fields and gathered up the buckwheat plants that had been left to dry—buckwheat *tsuboke* [see Fig. 0.42], in other words—and carried them back home on their backs. She was in the little shop in the morning and in the afternoon. My mother cooked and such.

November 6, Wednesday

Occasional rain

Strong northwest wind
7°C in the morning, 10°C in the afternoon

Breakfast:	miso soup with burdock root and napa cabbage, pickled napa cabbage, pickled eggplants and such
Lunch:	cooked ray, pickled napa cabbage, pickled daikon
Supper:	miso soup with taro pieces and *kuromo* seaweed in it, side dishes: same as lunch. We received the *kuromo* from the main house—they got it from relatives at Minami Iso.

In the early morning I picked vegetables for selling. After breakfast I cleaned out the rabbit hutch and cut new straws and put them in there, collected night soil from our own toilet, and washed green onions. Then I made lunch. In the afternoon I tied up the onions in bundles, picked napa cabbages, and cut grass for the rabbits to eat, among a few other things. My wife got back at one o'clock. In the afternoon she took the kids to get their hair cut.

Today's income:

5 daikon	5 sen
napa cabbages (1 *kan* 800 *momme* (6.75 kg))	25 sen
cabbages (1 *kan* 300 *momme* (4.9 kg))	20 sen
taros (6 *shō* (10.8 L))	65 sen

Today's expenditures:

spending money for my wife	25 sen

At the main house: Kanezō and my brother's wife threshed buckwheat by machine from the early morning, and then they sifted and cleaned it, just like they did with rice grains before. Kanezō also went out to dig drainage ditches in the paddies—both in the morning and in the afternoon. He finished the Sakura paddies in the morning, and he worked on the Bishamon paddies in the afternoon. My brother's wife was at home in the morning. She went to see a doctor in the afternoon because she was having some kind of stomach trouble. My brother was in the shop and my mother cooked.

November 7, Thursday

Rain
West wind
6°C in the morning, 8°C in the afternoon

Breakfast:	daikon miso soup with *kuromo* in it, pickled napa cabbage with soy sauce for flavor, pickled eggplants

Lunch:	roasted salted pumpkin, roasted salted salmon, pickled napa cabbage, pickled eggplants
Supper:	same soup as last night, side dishes: same as lunch

In the early morning I went to have rice processed. I bought two *tawara* of unpolished rice (*genmai*), so I got it polished—had it made into white table rice. Later, I braved the rain and hauled those two *tawara* home. Then I rested a bit. In the afternoon, I went to Tarusawa Elementary School for the village Nōkai agricultural product competition. My wife stayed home because of the rain; she sewed and such.

Today's expenditures:

2 *tawara* of rice	22 yen
hulling and polishing	10 sen
rice bran	40 sen
kerosene (*sekiyu*)	15 sen

At the main house: Kanezō cut hay for the horse early. He spent the day weaving *noma* mats for the roof of the hay storeroom. In the evening he took the horse to wash it, and my brother's wife was in the shop all day. My brother went to the agricultural product competition. My mother cooked and such.

November 8, Friday

Cloudy, sometimes sunny
West wind
1°C in the morning, 11°C in the afternoon

Breakfast:	daikon miso soup, pickled napa cabbage and daikon, pickled eggplants
Lunch:	roasted salted trout, pickled cabbage, pickled eggplants
Supper:	miso soup with green onion slices in it, side dishes: same as lunch

In the early morning I repaired the rabbit hutch. After breakfast, I cleaned up and then I dug holes in a field and buried some seed potatoes, to store them over the winter. Then I cut grass for the rabbits. After that, I went into the woods and dug up some tree stumps and roots. Then I made lunch. Once I'd finished cleaning up, I went over to the main house and borrowed the horse and used it to haul a load of straw back home. This was left over from what I hauled the other day. After that, I stacked the straw and made a nice, firm *waraniyo.* Then I picked vegetables for selling—picked napa cabbages and washed and tied up green onions. My wife returned at two o'clock and then she pulled out a bunch of *mushiro* and *komo* mats that had been stored for a while and beat them outside to clean them off and laid them out on the floor of the house's main room. She also helped me make the *waraniyo.*

Today's income:

green onions	20 sen
napa cabbages (1 *kan* 800 *momme* (6.75 kg))	25 sen
taros (6 *shō* (10.8 L))	65 sen

At the main house: Kanezō cut hay for the horse in the early morning, and after breakfast he went out to the Bishamon paddies to dig drainage ditches. He finished that work today. In the afternoon, he pulled the *migo* out of rice plant straws and pounded them. He's going to make ropes (*migonawa*) with them. My brother's wife went to see a doctor again and was also in the shop for a while. My mother cooked. My brother just goofed around (*burabura asobi*).

November 9, Saturday

Cloudy, sunny sometimes, some light rain
West wind
6°C in the morning, 13°C in the afternoon

Breakfast:	miso soup with taro pieces and napa cabbage in it, pickled napa cabbage with soy sauce, pickled eggplants
Lunch:	roasted salted trout, pickled napa cabbage, pickled eggplants, etc.
Supper:	miso soup with taro pieces in it, roasted flounder with soy sauce for flavor, pickled napa cabbage, pickled eggplants

After breakfast I cleaned up and then I put the rice into the storage box and such. If I just leave it in bags, rats will eat it. Then I dug up more tree stumps and roots and cut grass for the rabbits. Then I made lunch, and after eating I cleaned up and harvested more taros and picked vegetables for selling. My wife came back around three o'clock—she said her feet[4] hurt so she couldn't walk fast. After getting home she pickled napa cabbages and prepared the fish she'd bought for cooking, and cooked supper.

Today's income:

green onions (12 bundles)	20 sen (one bundle averages 10 onions)
napa cabbages (2 *kan* 55 *momme* (7.7 kg))	25 sen
taros (4 *shō* (7.2 L))	35 sen

Today's expenditures:

fish	10 sen
sweets	5 sen

At the main house: In the early morning Kanezō used the horse to haul a load of firewood from the pile. Later, he harvested daikon. My brother's wife tied them up in bundles. Then

Kanezō used the horse to carry them. Normally, there are 12–13 daikon per bundle, and you can bundle them together whether they are thick or thin. You just line them up evenly and tie them tightly by their green stems and leaves. This year's daikon crop wasn't so great, so they were able to haul all of them in only one day. This afternoon, my mother went to the school to hear Shioya Yōyū, the Christian, speak.[5]

November 10, Sunday

Sunny, later cloudy, then rainy
Strong southwest wind
2°C in the morning, 12°C in the afternoon

Breakfast:	daikon and potato miso soup, pickled napa cabbage, pickled eggplants
Lunch:	scrod[6] cooked in a pot, pickled eggplants. The fish was from Minami Iso in Oga; I bought it and brought it back home.
Supper:	miso soup with potato, napa cabbage and scrod in it, side dishes: same as morning

In the early morning I collected urine. After breakfast I cleaned up, and pulled up eggplant plants, as well as their support poles. Then I harvested Chinese yams. I'm disappointed because the Chinese yams didn't do well at all this year. I made lunch, and then rain fell in the afternoon so I didn't work outside from that point. My wife got back by lunchtime. She sewed and such in the afternoon.

Today's income:

napa cabbages (4 *kan* 200 *momme* (15.75 kg))	40 sen
taros (5 *shō* (9 L))	50 sen

Today's expenditures:

writing paper	10 sen

At the main house: Kanezō cut hay for the horse early. After breakfast he went out to the fields and harvested burdock roots. They're very good this year. In the afternoon he stayed indoors because of the rain—he wove ropes using *migo* (*migonawa*). My brother and his wife were in the shop, but she also sewed and such. My mother cooked, etc.

November 11, Monday

Rain, changing to snow
Powerful southwest wind
2°C in the morning, 4°C in the afternoon

Breakfast:	miso soup with eggplants and napa cabbage in it, pickled napa cabbage, pickled eggplants, etc.
Lunch:	pumpkin cooked with salt, pickled napa cabbage, pickled eggplants
Supper:	daikon miso soup, *takuan* we received from the main house, pickled napa cabbage, pickled eggplants

In the morning I cut grass for the rabbits, but the wind was very strong since last night, with rain and snow, so I gave up on doing outside work and just stayed indoors and read and prepared lunch, etc. My wife went over to the main house after breakfast and wove ropes for making *mushiro* mats from rice plant straws.

Today's expenditures:

2 chicken eggs	5 sen

At the main house: Kanezō cut hay for the horse early, and after breakfast he made ropes with *migo* (*migonawa*). In the afternoon he made a *ketsu mosoro* for the horse. This is basically a *mushiro* mat to put on the horse, for the horse to wear. In the evening, Kanezō rode the horse over to his natal house on an errand. My brother's wife went to a hospital in Akita City to be examined by a doctor. My mother cooked.

I'll explain how to make a horse *ketsu mosoro*. Sometimes people simply use a *mushiro* mat made from rice plant straws, depending on the size of the horse. But sometimes the *mushiro* must be sized, to make it smaller. You lay the mat down lengthwise and tie ropes to it in four places on the longer sides, first at about 5 *sun* (15.2 cm) from the edge and next about 3 *sun* (9.1 cm) farther along. You need to use two ropes to tie it at each spot and attach each to the mat at 5 *sun* (15.2 cm) increments, weaving the end of the rope into the mat, going in about 2 *sun* (6.1 cm) on both sides. After doing this four times, you put the part you wove first over the horse's head and tie those ropes. So, you also have the *muna obi* [for the chest], the *hara obi* [for the belly] and the *ketsuo* [for the rump]. There's one rope for each of the first two and there are two for the last. The part of the *hara-obi* rope that touches the horse's belly must be triple-braided (*mitsu ami*). You put this on the horse by attaching it to the *uma mosoro* on the left side as you face the horse's head. Then you tie it up tightly on the right side. The *muna obi* goes around the horse's chest—attached on the left side and tied on the right. You've got to attach the *ketsuo* part on both sides, so you go around both hind legs (under the rump) and bring the ropes back around and out again at the belly area and tie them.

November 12, Tuesday

Rain
Strong west wind
6°C in the morning, 8°C in the afternoon

Breakfast:	daikon miso soup, *takuan*, pickled napa cabbage, pickled eggplants

Lunch: *gunji* [goby] from Lake Hachirō cooked in miso soup, *takuan*, pickled eggplants

Supper: miso soup with taro pieces in it, pickled napa cabbage, pickled eggplants, etc.

In the early morning I cut grass for the rabbits. Since the weather today was just like yesterday's, my wife and the kids and I all went over to the main house with our lunches and did some work there. I wove *noma* mats to put over the small storehouse [at my place] all day. In the evening, I helped with persimmons at the main house—peeling and such [for hanging them to dry]. My wife wove *mushiro* mats all day.

At the main house: Kanezō cut hay for the horse early. He made ropes for tying up the center of *tawara*.[7] My brother's wife will be hospitalized for a while from tomorrow, so she had to prepare for that. My mother cooked.

November 13, Wednesday

Rain
Northwest wind
6°C in the morning, 8°C in the afternoon

Breakfast: miso soup with pumpkin pieces in it, boiled *seri* with soy sauce for flavor, pickled eggplants, etc.

Lunch: roasted salted trout, pickled napa cabbage, pickled eggplants (My wife and kids ate lunch at the main house.)

Supper: We poured hot water over cold rice [made *ochazuke*] and ate that. Side dishes: same as lunch.

Rain again today. During the morning I braved the rain and picked daikon. Then I separated them into four groups: (1) for making *takuan*, (2) for pickling whole, (3) for making *nadazuke*,[8] and (4) for eating in soup. Those for eating in soup must be stored, so I dug holes in a field and buried them. I picked the stems off the daikon that'll be used for pickling whole and for making *nadazuke*. I made lunch, and after eating I went over to the Nagane market to get some rubber boots. I bought a pair for myself and one pair of child's boots. My wife wove *mushiro* mats at the main house all day again. She and the kids ate lunch there.

Today's expenditures:
rubber boots (1 pair, for an adult) 2 yen 20 sen
rubber boots (1 pair, for a child) 50 sen

At the main house: One leg of the device for weaving *tawara* (*komozutsu*) broke so Kanezō fixed that early this morning. He wove *tawara* the rest of the day. My brother's wife will be hospitalized for a while from today, so she and my brother went to Akita City together. It turns out that she is four months pregnant, but the fetus seems to have died. It appears that the doctor will have to remove it. Ichisuke minded the shop, and my mother cooked and such. Between tasks, she went over to Kinshirō's house for a *nanagurayohi.*[9] The young wife (*yome*) at Kinshirō's house died in October last year due to some kind of uterine problem, at the age of thirty-four, so to hear what she had to say they called an *idako*[10] to the house. Most of the cost is supplied by relatives of the family [when an *itako* is called]. Women pair up and go to other villages and walk the streets, calling "*Nanagurayohi no hoguwandai shikoshi agete kunen hi?*" ["Please give a little for a *nanagurayohi,* won't you?"], and receive rice from people and then return to the house. It's common for village women to go over and listen to what the *idako* says, and shed a lot of tears, and then go back home.

November 14, Thursday

Cloudy, later sunny
1°C in the morning, 12°C in the afternoon

Breakfast:	miso soup with taro pieces in it, pickled napa cabbage, *takuan,* pickled eggplants, miso-preserved eggplants
Lunch:	roasted salted trout, pickled napa cabbage, pickled eggplants, *takuan*
Supper:	miso soup with taro pieces in it, pickled napa cabbage, pickled eggplants, *takuan, gunji* cooked with soy sauce, *takuan,* etc. (The *gunji* fish came from the waters of Lake Hachirō—I received them from an acquaintance in Momokawa Village.[11])

This morning I washed daikon, bundled them whole (four per bundle) for making *takuan* using rice straws, and hanged them on poles to dry—a total of 150 daikon. I also tied up daikon stems and hanged them to dry. These will be for feeding the rabbits in winter. In the afternoon, I cut grass for the rabbits, and then split the tree stumps and roots I dug up the other day, and I did *sarashi* to bitter persimmons I received from the main house. To do *sarashi* to persimmons means to remove their bitterness. I filled a big wooden bucket with water heated to about 30°C and put the persimmons in that. Then I put some wood ash in there and stirred it up and dissolved it as best I could, and packed straw (*wara*) on top to cover everything. Then I poured in water I'd heated to 35–40°C to fill the bucket, and wrapped the entire thing with straw and let it sit like that.

My wife also washed daikon in the morning, and she went over to the main house to weave *mushiro* mats in the afternoon.

Today's income:
pay for tanning two rabbit pelts 50 sen

At the main house: Kanezō cut hay for the horse early and harvested burdock roots the rest of the day. My mother cooked and washed daikon throughout the day.

November 15, Friday

Cloudy, sometimes sunny
Strong south wind
10°C in the morning, 17°C in the afternoon

Breakfast:	miso soup with potatoes and *egeshi*[12] in it, pickled napa cabbage, pickled eggplants, *takuan,* etc.
Lunch:	roasted salted *gunji* fish, pickled napa cabbage, *takuan,* pickled eggplants, etc.
Supper:	miso soup with taro pieces in it, *gunji* cooked in a pot with soy sauce, other dishes: same as morning

Today I massaged rabbit pelts, dug up some Chinese yams, and poured more hot water into the bucket of persimmons from yesterday. My wife wove *mushiro* mats during the morning, and she took the kids over to my old school for its festival (*gakugeikai*) in the afternoon.

At the main house: Kanezō cut hay for the horse early and harvested burdock roots after breakfast. Then, he also went to the school for the festival in the afternoon. My mother dealt with the daikon more, and she cooked and such.

November 16, Saturday

Cloudy
Northwest wind
6°C in the morning, 5°C in the afternoon

Breakfast:	miso soup with burdock root and daikon, pickled eggplants, pickled napa cabbage, *takuan*
Lunch:	roasted salted *gunji,* the rest: same as morning
Supper:	miso soup with taro root pieces inside, others: same as morning
Snack:	persimmons

This morning I harvested the last of my Chinese yams. I set some aside to use for planting next year, and buried them in holes for safekeeping. The rest are for selling. I pulled a persimmon out of the bucket I put them in yesterday and ate it to see how they're doing. All the bitterness was gone. Then I prepared to make charcoal for home use. I used a round hole I'd dug before—4 *shaku*

(1.2 m) in diameter and 2 *shaku* (60.1 cm) deep. I put rice bran in to cover the bottom, about 2 *sun* (6.1 cm). Then I put two pieces of split wood in the hole, side by side, about 5 *sun* (15.2 cm) apart, and some dried leaves between them, and then two more split logs about 3 *sun* (9.1 cm) apart, and then I lit the leaves on fire. After that, I put a bit more rice bran in the hole. I put a tin smokestack on top of the whole thing. It's about 3–4 *sun* (10.6 cm) wide and 2 *shaku* 5 *sun,* or about 3 *shaku* (approx. 83.3 cm) tall. I piled a bunch of the roots and tree stumps I dug up not too long ago around the smokestack, mixed with more rice bran, and stopped once the wood and stuff had reached the top of the smokestack. That made for a stack about 2 *shaku* (60.1 cm) high, with the smokestack in the center. Then I put rice bran into the hole, up to the top, and left it like that.

In the afternoon, I dug up burdock roots. Then I harvested napa cabbages and collected urine and such. My wife has been suffering from foot/leg pain, so she didn't go to the market to sell vegetables for the last few days, but she started going again from today. She was back by noon. In the afternoon, she went into the village to collect *temagome,* and she exchanged some rice for malted rice (*kōji*), etc. You can get 1 *shō* (1.8 L) of unpolished rice (*genmai*) for the same amount of *kōji. Temagome* is rice received in autumn for helping another household with their rice harvesting.

Today's income:

napa cabbages (5 *kan* (18.75 kg))	50 sen
taros (2 *shō* 5 *gō* (4.5 L))	25 sen

> At the main house: To dry daikon for making *takuan,* Kanezō tied long wooden poles together on the south side of the storehouse and hanged daikon on them early this morning. He also hanged some on persimmon tree branches. After breakfast he dug up burdock roots, and he harvested Chinese yams in the afternoon. My mother made *tsukemono*[13] and cooked and such.

November 17, Sunday

Cloudy
North wind
1°C in the morning, 4°C in the afternoon

Breakfast:	miso soup with daikon and burdock root, pickled napa cabbage, *takuan,* pickled eggplants, etc.
Lunch:	pumpkin cooked with salt, miso-preserved daikon, pickled napa cabbage, *takuan,* pickled eggplants, etc.
Supper:	basically, the same as yesterday

This morning I rubbed the rabbit pelts and pulled out the wooden board from before and scraped them with pumice. After breakfast, I cleaned up, repaired the handle of the shovel I use to dig up potatoes because it had broken, and harvested burdock roots. In the afternoon, I harvested more burdock roots and collected daikon leaves from the fields for feeding the rabbits and did other various field chores. I also picked napa cabbages. My wife returned by noon, and she made *nadazuke*. She also pickled whole daikon during the afternoon.

Today's income:

Chinese yams (1 *kan* 500 *momme* (5.6 kg))	50 sen
napa cabbages (5 *kan* (18.75 kg))	50 sen

Today's expenditures:

salt	30 sen
candies	5 sen

At the main house: Kanezō cut hay for the horse early, and after breakfast he harvested Chinese yams. He dug up burdock roots in the afternoon. Today he finished all the harvesting that requires digging. In other words, everything has been harvested in the main house's fields. My mother worked on making *tsukemono,* and she cooked and such.

November 18, Monday

Sunny
North wind, later out of the southwest
–3°C in the morning, 8°C in the afternoon

Breakfast:	miso soup with daikon and taro roots in it, side dishes: same as yesterday's lunch
Lunch:	daikon miso soup, pickled eggplants, *takuan*
Supper:	basically, the same as yesterday's supper

First this morning, I dug up more burdock roots. After breakfast, I cleaned up and then harvested more. I finished all that by ten o'clock, so I dug holes and buried burdock roots in them for storage, and harvested napa cabbages. In the afternoon I borrowed the main house's horse and used it to haul those napa cabbages over to Funagawa and sell them there. I got home at four-thirty, and then I harvested more napa cabbages and some cabbages as well. My wife returned by noon. In the afternoon she went into the village on errands and visited a relative who has been sick.

Today's income:

Chinese yams (1 *kan* 200 *momme* (4.5 kg))	40 sen
napa cabbages (20 *kan* (75 kg))	2 yen

Today's expenditures:

soy sauce	5 sen
sweets	5 sen

At the main house: Kanezō ate breakfast right after getting up and went out to work as a *dokata*.[14] The work was helping to repair the catchment pond of the temple, Enmeiji. The part of this catchment pond that the main house is responsible for is Sakurada 250-kari.[15] Money is collected from the landowners and from the smaller-scale farmers (*kosakunin*), and it seems that there's also financial support from the prefecture. Today my mother made *tsukemono* and cooked and such.

November 19, Tuesday

Cloudy with occasional sunshine
Southwest wind, growing stronger in the evening
0°C in the morning, 10°C in the afternoon

Breakfast:	potato miso soup, pickled napa cabbage, pickled eggplants, *takuan*, etc.
Lunch:	pickled napa cabbage with soy sauce for flavor, *takuan*, pickled eggplants
Supper:	miso soup with taro root pieces in it, side dishes: same as lunch

This morning I cleaned all the mud off the big *tōguwa* hoe and the shovel I use for harvesting taros and Chinese yams. After breakfast, I cleaned up and then went to the main house to collect daikon leaves, which I tied up with ropes and hanged in the sunshine on the little shed for drying before coming home. After that, I laid a *mushiro* mat down in front of the house and pounded buckwheat. I finished that by lunchtime. After lunch I cleaned up and threshed all that buckwheat in the thresher. My wife returned by noon, and during the afternoon she helped me put away the buckwheat I'd threshed. Then we pounded soybeans [see Fig. 0.44].

First, you knock all the dirt off the roots, and then you lay the plants out on *mushiro* mats and pound them with a stick. The pods will break open and the beans will fall out. After that you pound the beans again with a thin stick and more will come out. Then you just sift everything and separate the beans from the rest and put them away and you're done.

Buckwheat and soybean shells and such can be piled up outside (for composting). In the evening, I harvested more napa cabbages.

Today's income:

cabbages (1 *kan* 400 *momme* (5.25 kg))	25 sen
napa cabbages (7 *kan* (26.25 kg))	70 sen
1 young rabbit	50 sen

Today's expenditures:

loose-leaf paper	15 sen
candies and sweets	5 sen

At the main house: Kanezō did the same work as yesterday. My mother cooked. My brother went to Akita City to bring his wife home from the hospital—she was discharged today. Ichisuke minded the shop.

November 20, Wednesday

Cloudy, sunny later
North wind
4°C in the morning, 10°C in the afternoon

Breakfast:	napa cabbage miso soup, pickled napa cabbage, *takuan*, pickled eggplants
Lunch:	potatoes simmered in water with salt, side dishes: same as morning
Supper:	same soup as last night, burdock root *debu*, all else: same as morning

This morning I went to collect daikon leaves from someone else's fields. After breakfast, I cleaned up and then dug up some of the daikon I'd buried for storage because we didn't have enough to preserve for winter. I cut off their tops and washed them and such, and then I made lunch. In the afternoon I turned over the compost soil I made recently. And the protective encirclement around the house got damaged, so I had to repair that and clean up the mess. Then I harvested vegetables for selling. My wife came back about two o'clock and then she worked on preserving vegetables.

Today's income:

Chinese yams (1 *kan* 400 *momme* (5.25 kg))	40 sen
napa cabbages (6 *kan* (22.5 kg))	60 sen

At the main house: Kanezō cut hay for the horse early, and then after breakfast he went out to work as a *dokata* just like yesterday. My brother's wife tidied up her things from her hospitalization, and she put more rice plant pieces and scraps (*warashibe*) in the *shibebuton* [*futon*] and took care of the rice flour. My mother cooked and such. My brother made sweets, etc.

November 21, Thursday

Cloudy with occasional rain
Southeast wind
1°C in the morning, 10°C in the afternoon

Breakfast:	miso soup with taro pieces in it, pickled napa cabbage, pickled daikon, *nattō*[16] with salt
Lunch:	side dishes: same as morning, *nattō* we received from the main house (they bought it in Akita City)

Supper: cold rice in hot water, pickled vegetables: same as morning, burdock *debu,* chrysanthemum flowers (boiled and eaten with soy sauce) we received from my wife's brother's family

I caught a train to Akita City early this morning to see the seed and seedling trade fair. Now they call it the Akita Farmers' Festival (Akita no Nōminsai). It all began, they say, with the seed and plant exchanges started by great farmers like Ishikawa Rikinosuke.[17] This year's event was the 56th one. They do it once a year, rotating between the one city and nine counties in Akita. This time it was held jointly between Akita City and Kawabe County.[18] Each county basically hosts once in nine years. This year, about 20,000 individual items were entered and displayed, and about 30,000 people went to the festival each day, on average. It lasts for one week. And they say that each visitor has spent an average of about 3 yen at the festival or in the city.

My wife got home by noon. In the afternoon she had to go into the village on an errand.

Today's income:

Chinese yams (1 *kan* (3.75 kg))	35 sen
napa cabbages (6 *kan* (22.5 kg))	55 sen

At the main house: Kanezō ate breakfast early and then went out to do *dokata* work again. My brother's wife and my brother made sweets for another household. My mother cooked.

November 22, Friday

Rainy, sometimes stopping
Southwest wind
10°C in the morning, 14°C in the afternoon

Breakfast: miso soup with napa cabbage in it, all else: same as yesterday's supper
Lunch: miso soup with napa cabbage and *gunji* in it, and *shijigo* we received from the main house
Supper: miso soup with taro pieces and *gunji* in it, burdock root *debu, shijigo,* pickled napa cabbage, *takuan.* We received the *shijigo* and *gunji* from the main house.
Snack: apples

I was delayed yesterday because I had many things to do, so I stayed overnight at a friend's house. We talked all morning, and I had more errands in the afternoon. I finally got home at four-thirty. My wife was weaving *mushiro* mats for our use at the main house all day.

Today's expenditures:

participation fee for the seed and seedling exchange event[19]	4 yen
train fare	1 yen
a new diary for 1936	35 sen
loose-leaf writing paper	35 sen
folding umbrella (*kōmori gasa*)[20]	80 sen
meals	60 sen
a small gift for my friend (five packs of Golden Bat cigarettes)	35 sen
magazine	10 sen

At the main house: Kanezō went out early and cut hay for the horse. Later, he took his lunch with him and went up to Kayayama [miscanthus mountain] to participate in the land allotment proceedings. My brother and his wife were in the shop. My mother cooked. So, Kanezō went up there for the allotment proceedings, and the spot he got assigned to (in the drawing) is called Agumizaka. There are five big areas of Kayayama: Ōhira, Chizumi no Ue, Agumizaka, Yoshiyara, and Shimo no Ōhira. These will be further divided up among the people (households) assigned to each. Kayayama is considered communal village property; each household in the village has a right to receive a plot each year. And those who need no miscanthus in any given year don't need to enter the drawing.

November 23, Saturday

Partly cloudy
Strong southwest wind, later out of the northwest
4°C in the morning, 7°C in the afternoon

Breakfast:	miso soup with taro pieces in it, pickled napa cabbage, burdock *debu*, etc.
Lunch:	cold rice in hot water, side dishes: same as morning
Supper:	rabbit meat stewed with tofu and green onions in soy sauce, pickled napa cabbage

Early this morning I harvested napa cabbages. After breakfast, I cleaned up and collected urine from the latrine. Then I dug out the ashes and burned rice bran from the other day—from the pit—and pulled out the charcoal. Well, since I went into Akita City and lost time, it seems to have pretty much been a failure. I couldn't even get the usual one-third of the wood back as charcoal. I worked again on repairing the winter enclosure for the house, and then I made lunch. After lunch I cleaned up and worked once more on the enclosure, and was able to finish that. One rabbit died of some illness, so I skinned it and took the meat off the bones and pounded the bones and saved it for supper. Then I harvested more napa cabbages. My wife came home at two-thirty. She didn't do any farmwork after that; she cooked, instead.

Today's income:

Chinese yams (1 *kan* (3.75 kg))	35 sen
napa cabbages (6 *kan* (22.5 kg))	60 sen

Today's expenditures:

soy sauce (1 *shō* (1.8 L))	15 sen
tofu	6 sen
toothbrushes	15 sen
loose-leaf paper	10 sen

At the main house: Kanezō cut hay for the horse early—one horse load of miscanthus in the morning and one in the afternoon. One horse load of miscanthus usually amounts to 12–20 bundles. My brother and his wife were in the shop. My mother cooked and such, and she also went out on errands.

November 24, Sunday

Cloudy
Northwest wind
1°C in the morning, 5°C in the afternoon

Breakfast:	taro miso soup, pickled napa cabbage, pickled eggplants, etc.
Lunch:	pickled cabbage
Supper:	*mochi* with sweet red bean paste (*an*) on them and in hot broth, boiled napa cabbage with soy sauce
Snack:	persimmons

Early this morning I nailed a rabbit pelt to the board (for tanning) and then we ate breakfast. After cleaning up, I harvested napa cabbages and took them to my younger sister's family's house near the station, to sell them. After coming back, I put *noma* mats on the roof of my house. Then we had lunch. In the afternoon, I put more *noma* mats on the roof, and over the washing area (*minja*). Then I harvested more napa cabbages. My wife came home at two o'clock. After having some lunch, she went into the village to sell napa cabbages, and she put more rice plant scraps and bits (*shibe*) into Fusae's *shibebuton*, and then came back.[21]

Today is a special god festival day (*kami no ochitsuki*), so we pounded *mochi* and ate it. I took 1 *shō* (1.8 L) of *mochi* rice to the main house; they made *mochi* for us with that.

Today's income:

napa cabbages (17 *kan* (63.75 kg))	1 yen 56 sen

Today's expenditures:

refined sugar	10 sen

At the main house: Kanezō ate breakfast early and then headed out to do *dokata* work again. My brother and his wife spent the day in the shop—they really didn't have much to do. My mother did *mochi* preparation work and such.

November 25, Monday

Light rain, sometimes stopping
South wind
4°C in the morning, 7°C in the afternoon

Breakfast:	leftover *mochi* from yesterday afternoon, miso soup with napa cabbage in it, pickled napa cabbage, pickled eggplants, etc.
Lunch:	tofu and green onions cooked with soy sauce, pickled napa cabbage
Supper:	taro miso soup, burdock root *debu,* pickled eggplants, etc.

Today I wove more *noma* mats in the morning, and then I put them on the roof of our little house. In the afternoon I cleaned out the rabbit hutch and replaced the straw, and then I had to go into the village on some errands. My wife didn't go out to sell vegetables today; she went to the main house from the morning to weave *noma* mats instead, due to the rain. From the afternoon she stayed at our house and cooked and such—washed daikon, and made *debu* with burdock roots, etc.

At the main house: Kanezō went out again to do *dokata* work, but it was canceled due to the rain, so from that point he stayed at the house and made *tawara* with straw. It was the same for him in the afternoon. In the early evening, he took the horse up to Kayayama and brought back a horse load of miscanthus—about 60 bunches in total. My brother's wife caught a cold, so she was in bed. My brother stayed in the shop. My mother cooked and such.

November 26, Tuesday

Occasional rain
Northwest wind
8°C in the morning, 10°C in the afternoon

Breakfast:	daikon miso soup with *arame* seaweed in it, burdock root *debu,* pickled eggplants. The *arame* was collected at the shore.
Lunch:	pumpkin cooked with salt, pickled napa cabbage with soy sauce, pickled eggplants, and burdock root *debu*
Supper:	taro miso soup, burdock root *debu,* and pickled eggplants

After breakfast, I cleaned up, and then I fixed the enclosure around the washhouse because it had broken. That continued in the afternoon. After that, I put poles in the field for making a winter enclosure for the summer cabbages. Then I picked napa cabbages, etc. My wife came back by noon, and she made *moppe*[22] for the kids in the afternoon.

Today's income:
napa cabbages (7 *kan* 500 *momme* (28.1 kg)) 73 sen

Today's expenditures:

red flannel cloth	23 sen

At the main house: Kanezō cut hay for the horse early and then he went to do *dokata* work again. Even when he spends the day doing that, he takes the horse to the washing area every evening. My brother's wife was sick in bed today, too. My mother cooked.

November 27, Wednesday

Sunny
Strong northwest wind
5°C in the morning, 6°C in the afternoon

Breakfast:	potato and daikon miso soup with *arame* seaweed in it, pickled napa cabbage with soy sauce, pickled eggplants
Lunch:	side dishes: same as morning
Supper:	miso soup with taro pieces in it, pickled eggplants, *takuan*

After breakfast I cleaned up. Then I bundled eggplant plants with rope. I used those to build an enclosure to protect the summer cabbages during the winter. I also collected and bundled the poles I used for raising eggplants and tomato plants during the year. Then I made lunch. In the afternoon I worked on the winter enclosure for the house. I also harvested napa cabbages. My wife got back around two o'clock, and after that she collected cedar and cooked and such.

Today's income:
napa cabbages (6 *kan* 500 *momme* (24.4 kg)) 65 sen

At the main house: Kanezō did the same work as yesterday, and so did everyone else.

November 28, Thursday

Sunny
0°C in the morning, 7°C in the afternoon

Breakfast:	daikon miso soup, side dishes: same as yesterday morning
Lunch:	side dishes: same as morning
Supper:	soup stock made from rabbit bones and meat, with grated Chinese yam on rice (*tororo meshi*), side dishes: same as morning

After breakfast, I cleaned up, and then I got back to what I was doing yesterday. After finishing, I cleaned up the house. Then I made lunch. In the afternoon I split two loads of firewood and put half of that in the house. I left the rest out because it was wet—from the rain. That happened because I hadn't stacked them well enough. After doing that, I scraped a rabbit pelt and kept the bones and flesh for eating. My wife returned at two o'clock, and she helped me with the firewood. Then she prepared rice with (grated) Chinese yam on top (*tororo meshi*).

Today's income:

napa cabbages (4 *kan* 500 *momme* (16.9 kg))	42 sen
cabbages (1 *kan* 600 *momme* (6 kg))	30 sen

Today's expenditures:

postage stamps	5 sen
candies	4 sen
a small pot (urn)	10 sen

At the main house: Kanezō did the same things as yesterday. There was a *nanakurayohi*[23] at Yoshida Kuraji's house, so my brother and mother went for that. My brother's wife took care of the cooking instead.

November 29, Friday

Rain, changing to snow
Northwest wind
1°C in the morning, 4°C in the afternoon

Breakfast:	miso soup with taro and daikon in it, burdock root *debu,* pickled napa cabbage, miso-preserved eggplants
Lunch:	side dishes: same as morning
Supper:	same soup as morning, pickled napa cabbage, pickled eggplants, etc. We put *arame* in the soup. I received it from an acquaintance who collected it at the shore of Wakimoto.

I woke up before sunrise and made breakfast. After eating, I went to help rebuild an embankment at the temple (Enmeiji). Even though the weather was rainy and snowy, we worked. I hauled loads of dirt. My wife went to the main house and wove *mushiro* mats.

Today's expenditures:

kerosene (4 *gō* (720 ml))	15 sen
brown sugar	5 sen

At the main house: Kanezō did the same things as yesterday. He did *dokata* work at Enmeiji, like me. My brother was in the shop. His wife is still recuperating after her hospitalization. My mother cooked.

November 30, Saturday

Cloudy
Northwest wind
2°C in the morning, 7°C in the afternoon

Breakfast:	potato miso soup with *arame* in it, pickled napa cabbage, pickled eggplants, *takuan*
Lunch:	burdock root *debu,* miso-preserved daikon, pickled napa cabbage, *nadazuke, takuan*
Supper:	taro miso soup (we put *arame* in it), *nadazuke, takuan*

I went out early to do *dokata* work again. This time I didn't haul dirt—I mixed soil and packed it into the wheelbarrows. My wife stayed home and did the laundry, collected urine from the latrine, wove *noma* mats, washed green onions and bundled them up, harvested cabbages, and went to the main house to weave *mushiro* mats.

At the main house: Kanezō went out to do *dokata* work again. He hauled dirt today. He finished early, so he went back to the house around two-thirty. After that he cut hay for the horse and such. The others did the same as usual.

Figure 0.51. Making straw boots (*wara-gutsu*) (February 4).

Figure 0.52. The author's elder brother (May 1936, in front of the main house). From Yoshida's personal collection: not included in 1938 book.

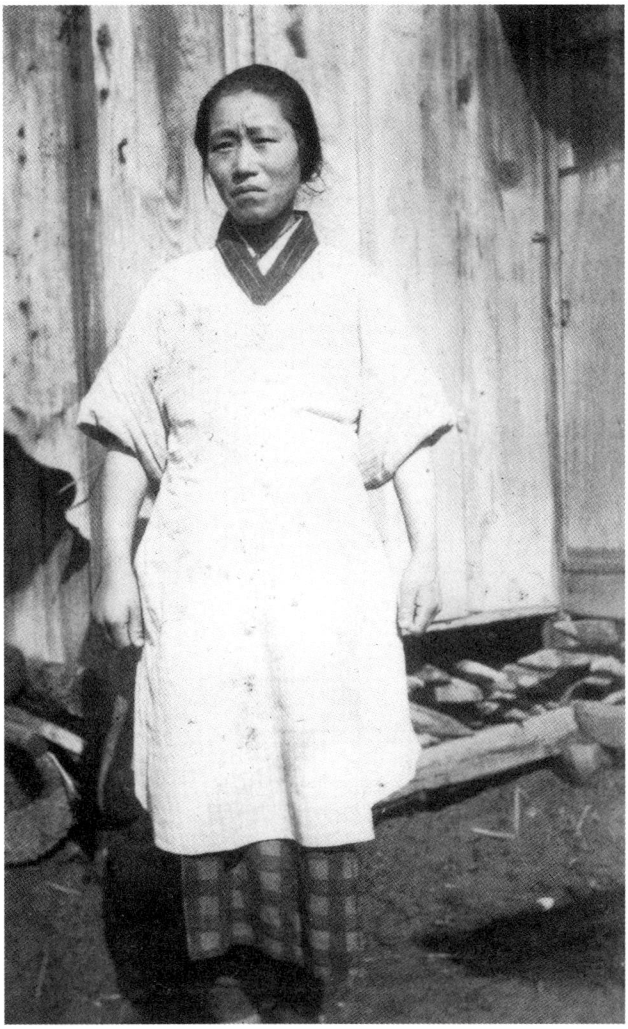

Figure 0.53. The author's elder brother's wife (May 1936, in front of the main house). From Yoshida's personal collection: not included in 1938 book.

Figure 0.54. The author's nephew, Ichisuke (May 1936, in front of the main house). From Yoshida's personal collection: not included in 1938 book.

Figure 0.55. The author's eldest daughter, Fusae, with his mother (June 1936, the year Fusae started elementary school). From Yoshida's personal collection: not included in 1938 book.

Figure 0.56. The author's wife and daughters enjoying potatoes for lunch (February 1937). From Yoshida's personal collection: not included in 1938 book.

Figure 0.57. The Wakimoto train station area (c. 1935), with part of the hamlet of Ōkura, and Mt. Kanpū, beyond. From Yoshida's personal collection: not included in 1938 book.

Figure 0.58. The Wakimoto shoreline (c. 1935), viewed from the area where the castle once stood. Ōkura is at the far left. Lake Hachirō lies in the distance. From Yoshida's personal collection: not included in 1938 book.

Figure 0.59. Yoshida striking a studious pose at the age of nineteen. From Yoshida's personal collection.

Figure 0.60. A group photo taken at the party to commemorate the publication of Yoshida's first book (March 1935). From Yoshida's personal collection. Shibusawa Keizō: middle row, far left. Ishiguro Tadaatsu: middle row, far right. Saburō and brother Shirō: middle row, third and fourth from right, respectively. Takahashi Buntarō: back row, second from right. Ōnishi Goichi: back row, third from right.

Figure 0.61. Yoshida's camera. Photo credit: D. C. Wood.

Figure 0.62. The Japanese Society of Ethnology facility in the village of Hōya. From Yoshida's personal collection.

Chapter 10: December 1935

December 1, Sunday

Cloudy, later rainy
–2°C in the morning, 4°C in the afternoon

Breakfast:	miso soup with taro root pieces in it, pickled napa cabbage, *nadazuke, takuan,* pickled eggplants
Lunch:	burdock root *debu,* pickled napa cabbage, miso-preserved daikon, *nadazuke,* etc.
Supper:	same soup as morning with *arame* seaweed in it, pickled napa cabbage, *takuan,* pickled eggplants, *nadazuke*

I did exactly the same work today as yesterday. My wife went out to sell vegetables, and she got back at one-thirty. After that, she didn't do anything out of the ordinary—she just cooked and such.

Today's income:

cabbages (3 *kan* (11.25 kg))	40 sen
green onions (6 bunches)	10 sen

Today's expenditures:

salted salmon	50 sen
sweets	5 sen

At the main house: Same as yesterday. Kanezō went to do *dokata* work—he hauled dirt.

December 2, Monday

Light rain
Strong northwest wind
7°C in the morning, 6°C in the afternoon

Breakfast:	potato miso soup with *arame* in it, pickled cabbage, pickled eggplants, *takuan*

Lunch:	roasted salted salmon, burdock root *debu,* pickled eggplants, *takuan*
Supper:	rabbit meat stewed with tofu in soy sauce for flavor, *nadazuke*

Today I did the same work as yesterday, but rain fell in the afternoon, so I rested. My wife picked cabbages and washed them and such, but she didn't do much else.

Today's expenditures:

tofu	6 sen
refined sugar	5 sen
sweets	15 sen

At the main house: Everyone did the same things as yesterday—my brother and Kanezō, too. They did *dokata* work and rested in the afternoon. Later they cleared the dung and straw out of the horse stall and piled it in the compost heap.

December 3, Tuesday

Snow, stopping later
Northwest wind
0°C in the morning, 4°C in the afternoon
Snow accumulation: 2 *sun* (6.1 cm)

Breakfast:	miso soup with taro pieces in it, miso-preserved daikon, chopped cabbage and Chinese yam with soy sauce for flavor
Lunch:	rabbit and green onions cooked in soy sauce with tofu, *nadazuke,* pickled cabbage
Supper:	miso soup (stock made with rabbit bones and parts), *nadazuke*

Yoshida Kenji's funeral was held today. He was an *Eitoku Maru* crewmember.[1] I went out early this morning to do *dokata* work, but returned soon because of all the snow. After that, I collected urine and night soil and piled the stuff up to compost it, and repaired the rabbit hutch. One rabbit died last night, so I skinned it and cleaned the pelt and nailed it to a board and all that. In the afternoon I went to listen to a talk at my old school. The speaker was Tomita Fumio, of the Ministry of Home Affairs, and the title of his talk was "The Essence of the Economic Rehabilitation Movement."[2] My wife went down into the main part of the village to get some cotton during the morning. She stayed at home and sewed and such in the afternoon.

Today's income:

payment for *dokata* work	1 yen 40 sen (for two people)[3]

Today's expenditures:

having old cotton redone (1 *kan* 500 *momme* (5.6 kg))	60 sen
sweets	5 sen

At the main house: Kanezō also went to do *dokata* work but gave up and returned to the house due to the snow. Then he wove *noma* mats. He also went to the school for the lecture in the afternoon, and then he cut hay for the horse and such. My brother went to Akita City on an errand. His wife minded the shop. My mother cooked, etc.

December 4, Wednesday

Snow, later stopping but remaining cloudy
West wind
1°C in the morning, 4°C in the afternoon
Snow accumulation: 2 *sun* (6.1 cm)

Breakfast:	daikon miso soup, chopped cabbage and Chinese yam with soy sauce for flavor, *nadazuke,* pickled eggplants
Lunch:	roasted salted salmon, *nadazuke, takuan,* pickled eggplants
Supper:	miso soup with taro pieces, pickled napa cabbage, *nadazuke,* pickled eggplants

I had to give up again on *dokata* work today because of the snow. Instead, I stayed home and worked on storing cabbages for winter. First, I dug holes, 1 *shaku* 5 *sun* (45.5 cm) deep and 2 *shaku* (60.1 cm) wide, in a field with very good drainage and lined the holes with straw—laid flat on the bottom and standing all around the sides. Then I picked cabbages and pulled off all the roots and some of the outer leaves and stacked the cabbages upside down in the holes. Then I put more straw on top of the cabbages and covered them with soil, and that was it for that. After finishing, I harvested the last of the napa cabbages—they'll be preserved for family consumption. Once all that was done, I gathered the cabbage and napa cabbage leaves and covered them with *noma* mats. This will be for feeding the rabbits. After that, I picked some green onions and delivered them to someone in the village who had ordered them. My wife stayed at home and preserved napa cabbages and sewed and such.

Today's income:
green onions (with dirt on the roots) (2 *kan* (7.5 kg)) 30 sen

At the main house: Kanezō wove *noma* mats today—they'll be used for covering the fertilizer on the paddies and dry fields after he hauls it all out there. My brother and his wife were in the shop, but they didn't really do much. My mother cooked until dark.

December 5, Thursday

Occasional rain
Southwest wind
3°C in the morning, 8°C in the afternoon

Breakfast:	taro miso soup, *nadazuke,* pickled napa cabbage, pickled eggplants
Lunch:	roasted salted salmon, *nadazuke,* pickled napa cabbage, pickled eggplants
Supper:	rabbit cooked with green onions in soy sauce, *nadazuke,* pickled eggplants, *takuan*

Today I did *dokata* work again—hauling loads of dirt. I got home at three o'clock and then I went into the village to get paid for doing autumn day work, in rice. My wife went and got a chest of drawers from her natal household and carried it over to mine. She'd left it behind because our little house is just too small, but circumstances made it best to move it over to the main house. In the afternoon I wove *mushiro* mats.

Today's expenditures:

brown sugar	5 sen

At the main house: Kanezō went out to do *dokata* work, as I'd expected. He hauled loads of dirt, too—occasionally returning to cut some hay for the horse. My brother and his wife were in the shop. My mother cooked, as usual.

December 6, Friday

Cloudy
4°C in the morning, 10°C in the afternoon

Breakfast:	same as yesterday
Lunch:	a dish of rabbit cooked with soy sauce, *nadazuke,* pickled napa cabbage, burdock root *debu*
Supper:	miso soup with taro pieces in it, *nadazuke,* pickled napa cabbage, *takuan,* pickled eggplants, etc.

Again, today I did *dokata* work—the same as yesterday. Today an official from the prefectural civil engineering department came out to inspect the construction project. My wife separated bad potatoes from our harvest and made *mushiro* and *noma* mats.

At the main house: Kanezō did exactly the same *dokata* work as yesterday, but he didn't go back and cut hay for the horse because my brother got up late this morning.[4] But he did bathe the horse—that's a necessity. My mother had to go to another village on an errand, so my brother's wife did the cooking instead.

December 7, Saturday

Sunny
Strong southwest wind
2°C in the morning, 11°C in the afternoon

Breakfast:	potato and daikon miso soup, pickled napa cabbage, *nadazuke,* pickled eggplants, *takuan*
Lunch:	roasted salted salmon, *nadazuke,* pickled napa cabbage, pickled eggplants, *takuan*
Supper:	same soup as last night, all else: same as this morning

I did *dokata* work again today, like yesterday, and the same work. My wife went down to the mill to gather sawdust. She made two round trips. She also did laundry. In the afternoon, she went to another village to collect payment for vegetables sold before.

Today's expenditures:

salmon	10 sen
sweets	5 sen

At the main house: Kanezō did *dokata* work again. In the morning he removed dirt from holes, and in the afternoon, he hauled loads of dirt again. Today he had to work until four o'clock without a break. The other family members did the same as usual.

December 8, Sunday

Rain
Strong southwest wind, later out of the west
6°C in the morning, 11°C in the afternoon

Breakfast:	daikon and potato miso soup, all else: same as yesterday morning
Lunch:	roasted salted salmon, burdock root *debu, nadazuke,* pickled napa cabbage
Supper:	miso soup (stock made from rabbit bones) with green onions in it, the rest: same as lunch

Today I did *dokata* work again, but I quit and came back home before noon because of the rain. My wife cleaned green onions, and she rested in the afternoon. I did some rabbit pelt tanning work in the evening. My wife sewed.

Today's expenditures:

soy sauce	20 sen
konnyaku	5 sen

At the main house: Kanezō cut hay early, and then he went to his natal household to participate in a Buddhist *hōji* ceremony for the rest of the day. My brother went to another village to deal with a rotating credit association[5] matter. His wife was in the shop all day, and my mother cooked.

When I go out to do the construction work on the water tank, I always used to head out at seven o'clock in the morning. But over the past four to five days, I've come to get up and head out about an hour earlier. The thing is, a whole bunch of people have come back from doing *dekasegi* work in Hokkaidō, and the autumn farmwork is finished here, so there are a lot of people with plenty of time. But the number of workers who can be employed each day is set, so many end up having to go back home in the morning with no work. It's basically first come first hired, so those who want to work—women and men alike—head out early, as I do. The laborers get their daily work assignments at the registration place, and it's a real scramble—pushing and shoving—it really is quite a sight. Those who don't get a work assignment for the day go home saying, "Ah, damn it! How stupid!" Those who do get hired say, "Great! I can live another day! But, what about tomorrow?"—expressing their joy and their worry at the same time [in front of the employers].

And one day there was a woman who said, "If I just stay at home and weave *mushiro* mats all day I can get maybe 15 or 20 sen, but I have to use my family's straw. But if I come out here and work all day, I can get 40 sen (and not spend anything). So, even if I work all night and all day, it's still a better deal."

And one time a man said, "I can stay home and weave rope all day and get maybe 20 or 30 sen, but working here I can get 70 sen. I can't just stay home. How stupid that would be!"

And another day a woman who was turned away said to the foreman, "Oh, please hire me today! Sometimes two can do better than one, and if you eat one of my cabbages, you'll live a thousand years. Please, hire me today!" In this way, the construction project rolls along.

December 9, Monday

Snow
Northwest wind
–5°C in the morning, 3°C in the afternoon
Snow accumulation: 3 *sun* (9.1 cm)

Breakfast:	miso soup with pieces of taro in it, *nadazuke,* pickled napa cabbage, pickled eggplants, *takuan*
Lunch:	roasted salted salmon, *nadazuke,* pickled napa cabbage, burdock *debu*
Supper:	I'll explain below.

Today I did the same work as yesterday. My wife was at her natal household all day, helping to prepare for a wedding.[6] I went in the evening to observe the ceremony.

Today's expenditures:
hide[7] (a gift for the wedding) 50 sen

At the main house: Kanezō went out to do *dokata* work again—he hauled loads of soil. As for the others, it was the same as usual.

I'll write about how this evening's wedding ceremony went.

First, the guests sat around the hearth (*irori*) in the main room, and the bride came in and the woman accompanying her (called the *gejobaba* in local speech) guided her to the *butsudan*, where they prayed for a moment. Then they greeted the household head, and sat in the correct positions for the ceremony. Next, the groom came in and sat in his place. Next, the two *shakudori*[8]—one male and one female—came in. These two were about twelve or thirteen years old. Next, the go-betweens (*nakōdo*) came in, and then the father and mother of the groom, and the *jahenin*[9] and all the others came in and sat down, and the ceremony started.

As for the necessary utensils and materials, there was a small stand with a mound of polished rice grains on it, and they had stuck a pine branch, a piece of bamboo, and a plum (*ume*) branch into that. And they had crane (*tsuru*) and turtle (*kame*) figurines on the stand, and a big candle, which was burning. On a tray with legs there were three sets each of matching wooden lacquerware bowls for drinking saké, and there were two iron pots full of saké, and another tray with daikon, kelp (*konbu*), carrots, and shrimp and such on it as well.

First, at the *jahenin*'s command, the young boy and girl (the two young *shakudori*) each took a saké pot in hand and then gently touched the spouts together. Then the girl poured saké from one pot into a wooden bowl for the groom to drink—a bowl that he had already picked up from the tray in front of him. The bride received saké in the same manner, from the boy. Once the saké had been poured, the two young *shakudori* boy and girl used chopsticks to pick up pieces of daikon, kelp, carrots, shrimp, and fish and such from the tray, and offered them to the groom, and to the bride. But the groom and bride didn't eat these—they folded them up in paper and set them aside. After that, the *jahenin* sang something from Takasago.[10] Then some of the guests—actually, young fellows of the village—sang some more of the Takasago, in a kind of nasal voice, sounding just like a bunch of cows mooing. Once the Takasago singing had finished, the groom and bride were served saké for the second time, and then a third. After that, the *jahenin* sang some of Takasago again, and then the young fellows did, too, with their nasal sound. Then, the tray with the ceremonial wooden lacquerware bowls was passed to the parents of the groom. Then, the groom and bride each took a pot in hand and touched the spouts together and poured saké for the groom's parents—the bride serving his father and the groom serving his mother. And when the groom and bride touched the spouts together, the young men who were there guffawed nasally. Then, the *jahenin* sang something from Takasago a third time, and that was the end of the ceremony. Everyone returned to their seats, and the food trays were brought in. I'll draw a diagram to show how everything was arranged (see Fig. 10.1).

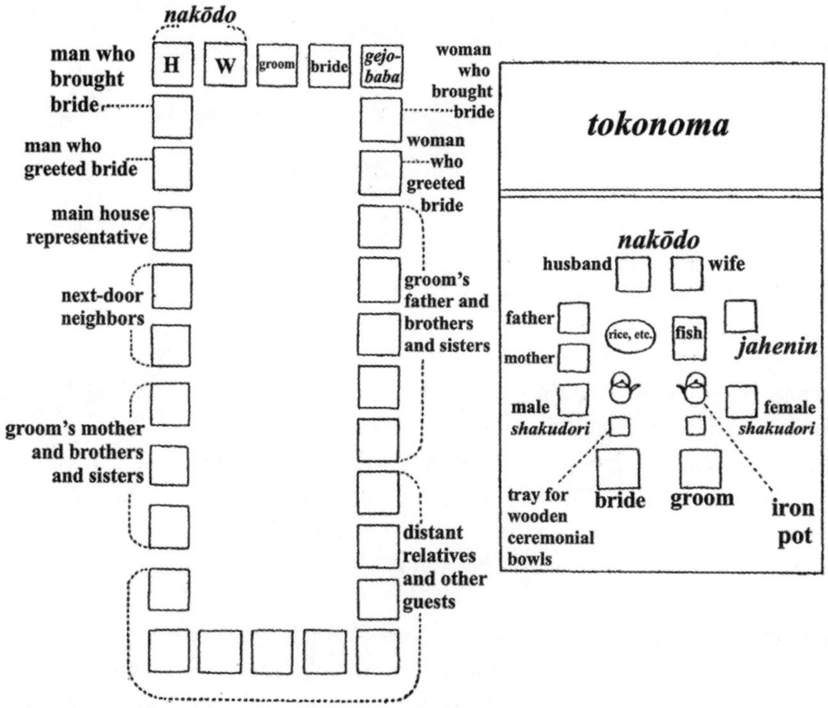

Figure 10.1. Village wedding seating arrangement and ceremony layout.

For each guest, there was a main tray (*honzen*), and a second tray, for the banquet. On each main tray there was a big plate with a raw candy sea bream, some bamboo shoots, a big sushi roll (*uji-maki*) and clams (*hamaguri*) on it. There also was a plate with wheel-shaped bread-like wheat bran (*fu*) cooked in soy sauce with sandfish (*hatahata*), ripe persimmons, pickled octopus tentacle slices, and a *sabeji* dish.[11] And there was a plate of flounder sashimi. On each second tray, there was a plate with a raw *ujimaki* candy roll, some cooked salmon, spiral shellfish (*tsubu*), and *yamashina*[12] and such on it. There also was a plate with a roasted fish[13] on top of boiled cabbage, with soy sauce poured over it for flavor. And there was a plate of cowpeas (*sasage*) that had been cooked with sugar, burdock root *debu,* a dish of ray (*kasube*) cooked with daikon and green onions, *namasu* made from daikon, carrot, and *hatahata* eggs, a *nishime* dish of carrot, sweet potatoes, *konnyaku,* and *kagori,*[14] a dish of crucian carp cooked in miso, *kagori* cooked in soy sauce with *konnyaku,* and of course saké.

After drinking saké, tofu miso soup and cooked white rice were served. Following that there were 7 *sun* (21.2 cm) in diameter white and red *manju* (one of each), and then each guest received red bean rice (*sekihan*) and such to take home. Also, at the end of the ceremony, one or two big 1 *shō* (1.8 L) bottles of saké and red rice wine were given to the young village fellows who sang.

In the old days, the bride and groom would walk wearing straw sandals (*waraji*), and young village men would carry her dresser (to her new home), but over time people gradually stopped wearing *waraji* and switched to *geta,* and folks started using a horse to carry the (bride's) dresser. The village kids [boys] would always get really excited at the sight of a groom and bride walking down a snowy road wearing *geta.* They'd shout, "Here comes the bride! Here comes the groom! She slipped in her *sakura-geta* and her big *bebe* split right open!" (*Bebe* means "genitals."[15]) But times have changed. Now they ride in horse-drawn carts, and if they walk, they wear rubber boots, and the dressers are carried on horse-drawn carts. And wealthier families use automobiles if they can. That was the case today—two automobiles, in fact.

December 10, Tuesday

Snow, with a real snowstorm later
Northwest wind
–3°C in the morning, –3°C in the afternoon

Breakfast:	taro and potato miso soup, *nadazuke,* pickled eggplants
Lunch:	same as yesterday
Supper:	green onion miso soup, *nadazuke,* pickled napa cabbage, pickled eggplants
Snack:	yesterday's sweets, *manju*

Despite the snow, I went out to do *dokata* work this morning as usual. And despite the snow, many laborers were out and had already been assigned work duties by the time I got there. I had to do different work today from yesterday because I was five minutes late. I had to scoop muck out of holes in snow falling so heavily that I couldn't see more than 1 *sun* (3 cm) in front of me, and I ate lunch standing under a roof with no walls while snow swirled all around me. The snow was so bad that I went home at midday. What I got for working from six o'clock in the morning to about noon—six hours of hard labor in a snowstorm—was a mere 35 sen.[16] It's not even enough to buy food for our family for a day. No work in the afternoon—I went into the village for some R & R. My wife cooked and such.

Today's expenditures:

kerosene (*sekiyu*)	10 sen

At the main house: As I'd expected, Kanezō also did *dokata* work. He hauled loads of dirt. In the afternoon, he wove ropes at the house. My brother was in the shop and his wife cooked. My mother went to a sutra-chanting association (*nenbutsukō*) meeting at the temple.

December 11, Wednesday

Snowstorm
Strong southwest wind
–6°C in the morning, –5°C in the afternoon

Breakfast:	daikon miso soup, *nadazuke,* pickled napa cabbage, pickled eggplants
Lunch and supper:	same as yesterday

Because of the very heavy snowfall, I gave up on *dokata* work. And I felt like I'd caught a cold or something, so I stayed in bed pretty much all day. When I woke up in the morning and looked outside, I just saw powdery snow blowing around everywhere—so much that the house was covered in it. It was just completely white outside. The rabbit hutch was under the snow, so I cleared all the snow around it and dug it out and fed the rabbits. Then I dug a path to the latrine and other important spots outside. My wife sewed.

Today's expenditures:
peanuts [*nankin* or *kanto* beans] 5 sen

At the main house: Kanezō cut hay in the early morning and made straw boots (*waragutsu*) after breakfast. In the afternoon he pulled *migo* out of rice straws. My brother went to another village on an errand. His wife minded the shop and sewed. My mother cooked.

December 12, Thursday

Rain
Strong west wind
–5°C in the morning, –3°C in the afternoon

Breakfast:	daikon miso soup, *nadazuke,* pickled napa cabbage, pickled eggplants
Lunch:	rabbit meat cooked with *konnyaku* and green onions in soy sauce, side dishes: same as morning
Supper:	miso soup with taro root pieces in it, side dishes: same as morning

I cleared snow and fed the rabbits early. A friend came to see me about something after breakfast, so I spent some time with him. In the afternoon, I went to polish rice. I also had to buy some charcoal. My wife was sewing all day. She also cooked and such. The reason why my friend Shūzō came to see me was this: He's a stoneworker, and when he was younger a big stone fell on him and his leg got broken; he was already poor from the start but due to the injury he wound up deeply in debt, and he finally recovered and is ready to work hard but no matter how hard he works he can't support his aging parents and his sister and his wife. And he hasn't been able to pay off his debts, so finally he lost ownership of his house. So now, his family of five is really in trouble because they are facing extreme poverty. He came to see me in search of some way out of the situation. Of course, I'm also poor—just living day to day—so it goes without saying that I can't do anything to save him. The two of us just commiserated and parted.

Today's expenditures:
1 *tawara* of charcoal from Hokkaidō (10 *kan* (37.5 kg)) 1 yen 90 sen

Figure 10.2. The Ōkura hamlet granary. From Yoshida's personal collection: not included in 1938 book.

At the main house: Kanezō braved the cold to go out and do *dokata* work this morning, but it was so cold that he gave up and went back before noon. In the afternoon he hauled firewood using the horse. The wind was so strong that two bundles blew right off. He carried those back, too. My brother and his wife were in the shop, and my mother cooked and cut hay for the horse, etc.

Today I heard that the granary (*gōgura*) has been completed (see Fig. 10.2). Now villagers can start storing their unpolished rice in it. The way this granary is operated differs from the old way—it's a cooperative organization. The work shed (nearby) is for communal use, so it follows naturally that the granary would be as well. One share of the granary costs 5 *kan* (18.75 kg) of unpolished rice, they say. But I heard that nobody has put any substantial amount of rice in yet. I can believe that—there aren't many farmers with more than enough rice to feed their own family to the end of the year, once autumn has passed and they've paid the landlords their shares and returned what they borrowed from the agricultural warehouse. "I don't even have enough rice to eat! How can I put any into the granary?" ask the poor farmers with facetious smirks on their faces. There are about eighty households in my village,[17] and about 80 percent are peasants.[18] Only about seven or eight of these are living all year on what they grow in their own rice paddies. Even so, people in other Wakimoto hamlets say that mine is doing better economically than the others.

December 13, Friday

Occasional sunshine
Northwest wind
–3°C in the morning, –2°C in the afternoon

Breakfast:	daikon miso soup, pickled napa cabbage, *nadazuke,* miso-preserved eggplants
Lunch:	roasted salted salmon, side dishes: same as morning
Supper:	potato and taro stem miso soup, side dishes: same as morning

Early this morning I went out to do *dokata* work, but it didn't go well because of all the snow, so I gave up and came back home. On the way, I stopped at a cousin's house for a visit. In the afternoon I made a pair of straw boots, and then I went out to buy some salt. I also brought down about one-half of the daikon I'd hung up to dry and took them back to the house and cut off the green stems and the little tails on the tips. It's part of the preparation for making *takuan.* My wife collected urine from the latrine in the morning, and then she wove rope and such. In the afternoon she moved the contents of one barrel of *nadazuke* to a different barrel, and thoroughly washed the first barrel, and she also made a pair of straw boots.

Today's expenditures:

salt (4 *shō* (5.8 L))	40 sen

At the main house: Kanezō cut hay for the horse early, and pounded *migo* and used them to weave *migonawa* ropes after breakfast. In the afternoon he pulled potatoes out of the shed and sorted them, setting aside the ones that were going bad, and then took all the good ones and packed them in *tawara.* Then he strung ropes up over the stove in the earth-floored area of the house (*doma*) and hoisted those potato-filled *tawara* up there for winter storage. After that, he wove *migo* ropes again. My brother was in the shop and his wife cooked and such. She also helped Kanezō sort the potatoes. My mother went to another village on an errand, etc.

December 14, Saturday

Snow
–2°C in the morning, 2°C in the afternoon
Snow accumulation: 5 *sun* (15.2 cm)

Breakfast:	same as yesterday
Lunch:	rabbit meat cooked with green onions and soy sauce, pickled napa cabbage
Supper:	*hatahata* we received from mine and my wife's natal households cooked with miso, *nadazuke*

I gave up on *dokata* work today. This morning I used a winnower (*chiritori*) to separate small bits of rice from rice bran. Then I boiled water, and washed out a wooden barrel carefully and spread some salt in the bottom. Then I mixed rice bran with salt and put a layer of that in the barrel, and then separated dried daikon by size and laid large daikon in the barrel. Once I'd put down a good layer, I covered the daikon with a mixture of rice bran and salt—so the daikon couldn't be seen. Then I put some dried persimmon peels in, and then another layer of daikon—in the same way but pointed in the opposite direction. Then I pressed it all down with my feet and put in more rice bran mixed with salt, and more persimmon peels. Once the large daikon were all in there, I put the smaller daikon in. This is how *takuan* is made. And it's important to increase the amount of salt-mixed rice bran you use as you go upward.

Once I'd packed everything in the barrel, I put a bunch of dried daikon stems and leaves on top, then some salt, and a lid and a big rock on top of the lid. That's basically it. The amount of salt and rice bran you use depends on the weight of the daikon and the time of year you plan to eat them. To preserve 2 *kan* 800 *momme* (10.5 kg) of daikon today I used 1 *shō* (1.8 L) of salt and a little over 2 *shō* (3.6 L) of rice bran. We plan to eat the daikon from March to around June of next year. I'll have to put some water in later, depending on the dryness of the daikon. They will be fine for about two weeks as they are, but after that the water content of the barrel will need to be adjusted. The daikon I used had been dried for about 30 days, so I added 2 *shō* (3.6 L) of water today. I estimate I'll need to add about another *shō* (1.8 L) later. The stone I used as a weight for the roughly 40 × 40 cm 1 *to* (18 L) barrel (*itto taru*) weighs about 6 *kan* (22.5 kg). In the afternoon, I shoveled snow and pulled out food for the rabbits and fed them and gave them water and cut fresh straw for their hutch and put that in there for them, and then I rested a bit. My wife went to another village on an errand in the morning and didn't return until evening.

At the main house: Kanezō ate breakfast early. Then he went to Funagawa with the horse to buy *hatahata*. He said he paid 3 yen 80 sen for the load he brought back. He was back at the house by noon. In the afternoon he wove *migonawa* ropes. My brother stayed in the shop, and his wife prepared the *hatahata*. She cut off the heads and gutted them and put them in buckets and filled them with water. She also cooked and such. My mother went to my younger sister's house to help because my sister is sick.

Hatahata are the fish that the villagers crave the most. That goes for all people of Akita Prefecture, in fact. Come every December, everyone in Akita—especially the fishermen—start talking about how many *hatahata* there will be, and how cheaply they might sell, hoping for low prices. But I can say that in recent years the catches off the Oga coast don't seem to be as big as in the past. Before, as soon as the word of good *hatahata* hauls went out, the farmers would head to the ports with their horses and buy what they could carry and then take the loads back to the station area and sell them to the big brokers for a very good profit. After two or three weeks, a farmer might have anywhere from 50 to 300 yen to show for his efforts. In this way, the *hatahata* brought great benefit to all the people of Oga—to the fishermen, the farmers, and to all consumers.

The farmers would normally pay about 2 yen per horse load at the docks and then sell those for 4 or 5 yen to the brokers. And they could fill their wooden boxes with *hatahata* at the docks and bring them back and siphon some fish off to make the boxes only about 80 to 90 percent full and sell those loads to the brokers, and if they could manage to carry two or three loads each time, they'd take a little extra off each load. And those farmers in good locations could possibly make two or three trips a day. Three loads bought for 2 yen each and sold for 4 yen each would yield a profit of 6 yen. And, if you can get an extra load out of all the fish you've siphoned off each time, you'll have 10 yen. Make three trips, and you've got 30 yen. That's why the villagers were always waiting for the *hatahata* to come in. And the farmers would always buy roasted *mochi* from the shops (near the station) on days when they did really well and bring those back for the kids. When I was little, we'd all be waiting, and we'd jump for joy to get those *mochi.*

But times have changed. Now the fishermen have their own cooperative, and the brokers drive their own tractors straight to the docks to negotiate directly with the cooperative and buy their fish. So now, we—local traders and Oga residents alike—can no longer get *hatahata* so cheap anymore. We have to pay very high prices for the *hatahata* caught right off our own shores. But *hatahata* are really famous fish of Akita. They're even mentioned in the Akita Ondo: "Hachimori has *hatahata,* Oga has *buriko.*" *Buriko* is *hatahata* roe.[19] As for why it's called *buriko,* I have no idea.

December 15, Sunday

Snowstorm
West wind
–3°C in the morning, –2°C in the afternoon
Snow accumulation: 2 *sun* (6.1 cm)

Breakfast:	cabbage miso soup, miso-preserved eggplants, *nadazuke,* pickled napa cabbage, etc.
Lunch:	nothing
Supper:	miso soup with *hatahata* in it, *nadazuke,* pickled napa cabbage

My wife went out early to buy *hatahata,* but there weren't any. She got back around three-thirty. Other than cook, I didn't do anything in particular—just read books and such. But I did clear paths in the snow and dig up the rabbit food and fed the rabbits and such. In the evening, I went to the village study hall (*gakudō*) to listen to a lecture by Kon Sakunosuke of the Society for Cultivation of Constitutional Justice (Rikken Yōseikai).[20]

> At the main house: Kanezō went out to do *dokata* work despite the snowstorm. My brother's wife cooked and such. She also put *hatahata* in buckets and filled them with water and such.

December 16, Monday

Snow, with occasional sunshine
West wind
–3°C in the morning, 2°C in the afternoon

Breakfast:	miso soup with potatoes and grated daikon in it, *nadazuke,* miso-preserved eggplants
Lunch:	roasted *hatahata, nadazuke,* pickled napa cabbage
Supper:	miso soup with taro root pieces in it, roasted *hatahata* with soy sauce on it, *buriko* (*hatahata* roe) with boiled cabbage—soy sauce poured on for flavor

In the morning I fed the rabbits and went to my younger sister's house near the station to get some *hatahata.* I received about 100 fish. Her family got them from relatives in Funagawa, it seems. I also had to go to my wife's brother's place to get some *hatahata* that we'd asked them to buy for us. In the afternoon I dug through the snow on part of my fields and pulled out some green onions and washed and bundled them. I also dug up some cabbages and burdock roots and bundled them. All this was preparation for my wife to take them out and sell tomorrow. She did laundry and cooked and such, and cleaned *hatahata.* We bought two full barrels of *hatahata* today—one had 272 fish in it. There will usually be anywhere from 250–300 in a barrel, depending on their size and how they're packed.

Today's expenditures:

2 barrels of *hatahata*	1 yen 75 sen

At the main house: Kanezō went out to do *dokata* work as usual. He hauled loads of dirt. My brother's wife cleaned *hatahata*—they'd received two barrels from my younger sister's family. My brother spent the day in the shop. My mother didn't return today either—she was at my younger sister's house, helping to clean *hatahata* and such.

December 17, Tuesday

Cloudy
-2°C in the morning, 4°C in the afternoon
Snow accumulation: 3 *sun* (9.1 cm)

Breakfast:	same soup as last night, miso-preserved eggplants, *nadazuke,* pickled napa cabbage
Lunch:	rabbit meat cooked with green onions in soy sauce, pickled napa cabbage
Supper:	miso soup with *hatahata* in it, side dishes: same as morning

After breakfast I cleaned up, fed the rabbits, saw my wife off down at the road, dug paths in the snow, collected water, dug through the snow to check on the green onions, and more. Then I made lunch and cleaned up, collected urine, and cut the heads off many *hatahata* and gutted them and dumped the heads and guts into the night-soil cistern. Then I read for the rest of the day. My wife went out to sell vegetables again today and she returned at about two o'clock. She cleaned *hatahata* after lunch.

Today's income:

green onions (22 bundles)	40 sen
burdock roots (1 *kan* 500 *momme* (5.6 kg))	30 sen
cabbages (1 *kan* 400 *momme* (5.25 kg))	28 sen

Today's expenditures:

dried alum (1 bag)	30 sen
laundry detergent	20 sen
headache medicine	20 sen
chinomichi gusuri[21]	30 sen

At the main house: Kanezō went out to do *dokata* work again—hauling loads of dirt. He returned at three o'clock and then cut hay for the horse. My brother went somewhere at his own pleasure and his wife minded the shop. She also cooked and such.

December 18, Wednesday

Cloudy, sunny, cloudy again
Strong northwest wind
1°C in the morning, 4°C in the afternoon
Snow accumulation: 3 *sun* (9.1 cm)

Breakfast:	daikon miso soup, *nadazuke,* pickled napa cabbage
Lunch:	roasted salted *hatahata, nadazuke,* pickled napa cabbages
Supper:	miso soup with taro root pieces in it, side dishes: same as lunch, *mochi* we received from the main house

In the morning I went out to take some photos, and then I went to the post office down in the main part of the village to send them off because that has to be taken care of right away. I ate lunch at the main house and then I wove ropes. In the evening I received a box of *mochi* and took that home with me. My wife washed *hatahata* and packed them in salt for preservation. In the afternoon she did some sewing. From the time of the autumn harvest up to around when I started doing *dokata* work, we were eating 1 *shō* (1.8 L) of rice mixed with about 5 *gō* (900 ml) of rolled oats daily. But the oats are all gone, so we're just eating 1 *shō* 2 *gō* (2.2 L) of rice every day.

Today's expenditures:

postage on small package	10 sen
scarves (*erimaki*) for the kids' kimonos	15 sen

At the main house: Kanezō went to do *dokata* work again but returned early due to the rain. Then he cut hay for the horse, wove *migo* ropes, and pounded *mochi*. My brother's wife did *mochi* preparation. My mother returned in the afternoon and helped. Today—the old November twenty-third—is *deshiko mochi*[22] day in the village, so we made and ate them.

December 19, Thursday

Snowstorm
Strong northwest wind
–5°C in the morning, –2°C in the afternoon
Snow accumulation: 2 *sun* (6.1 cm)

Breakfast:	potato and daikon miso soup, *nadazuke*, pickled napa cabbage, miso-preserved eggplants
Lunch:	roasted salted salmon, *nadazuke*, pickled napa cabbage, etc.
Supper:	daikon miso soup, roasted salted *hatahata*, *nadazuke*, pickled napa cabbage

This morning I dug through the snow and pulled out cabbage leaves for feeding the rabbits, and I used *fumidawara*[23] to trudge through the snow and stomp paths for walking. Then I went into the village on errands and stopped at a relative's house to help tan some rabbit pelts. I also made a new lid for the *hatahata* bucket—for pressing down with a stone. My wife spent the day sewing and cooking and such—the usual.

Today's expenditures:

postage stamps	10 sen
brown sugar	10 sen

At the main house: Kanezō went out to do *dokata* work again—hauling loads of dirt. He returned around two o'clock and then he cut hay for the horse. My brother's wife minded the shop and my mother cooked and such.

December 20, Friday

I could see the sun sometimes today.
Northwest wind
–1°C in the morning, 0°C in the afternoon

Breakfast:	daikon miso soup, *nadazuke*, pickled napa cabbage
Lunch:	*soba nereke* with soy sauce for flavor, roasted salted *hatahata*, *nadazuke*, pickled napa cabbage

Supper: miso soup with daikon and salted *hatahata* in it, side dishes: same as lunch, and miso-preserved eggplants

This morning I did the same as always for the rabbits. And I went and bought a 2 *to* (36.1 L) barrel and took the lid off and put it in water and such. This is going to be used for making *takuan*. After that, I brought down a bunch of dried daikon and took them home and cut off the stems and the little tails on the tips, and I put a new handle on the cleaver. Then I went over to the water tank construction project office to get what they owed me for my labor. In the afternoon, I went to a relative's house in Tayazawa, hoping to receive some rice that I figured they owed me for helping them harvest their paddies, but there was no more rice, so I came back home. After that, I went and bought two bags of rolled oats at the store. My wife did some work in the kitchen, and brought in daikon and burdock roots and cleaned them. She also sewed and such.

Today's income:
dokata labor wages 4 yen 62 sen

Today's expenditures:
wooden barrel 1 yen
loose-leaf paper 45 sen
candies 5 sen
rolled oats 4 yen 60 sen

At the main house: Kanezō went out to do *dokata* work again—hauling loads of dirt. My brother and his wife were in the shop. My mother cooked.

December 21, Saturday

Snowstorm
Northwest wind
–2°C in the morning, 0°C in the afternoon
Snow accumulation: 3 *sun* (9.1 cm)

Breakfast: daikon and burdock root miso soup, miso-preserved eggplants, *nadazuke,* pickled napa cabbage
Lunch: roasted salted *hatahata,* side dishes: same as morning
Supper: dog meat[24] cooked with miso, *hatahata* innards cooked with miso, *nadazuke,* pickled napa cabbage (We received the dog meat from a young fellow in the village.)

Every morning I check the rabbits and feed them, before breakfast. This morning, I took care of some indoor tools and put them away, and I shoveled snow off the roof of the shack. In the afternoon I dug up one row of green onions and dug small holes and buried potatoes in them. Then I made a pair of straw boots and pounded two bundles of straw for later use. My wife sewed all day.

The dog meat we ate we received from a relative of mine I call "Boy."[25] He said he got it from an acquaintance in Urata, and that in return he promised to tan the pelt of that dog for the guy who gave the meat. He asked me to teach him how to do it.

At the main house: Kanezō went out to do *dokata* work again but the snowstorm was too much for him, so he gave up and went back to the house. After that, he wove *migo* ropes and cut hay for the horse, and went out to buy rice bran. My brother's wife minded the shop. My mother cooked. My brother's wife also helped with that.

December 22, Sunday

Snowstorm
Northwest wind
–3°C in the morning, –1°C in the afternoon
Snow accumulation: 5 *sun* (15.2 cm)

Breakfast:	daikon miso soup with *anpura*[26] in it, *nadazuke,* pickled napa cabbage
Lunch:	*anpura* cooked in water, roasted *hatahata, nadazuke,* pickled napa cabbage
Supper:	burdock root and daikon miso soup, roasted salted salmon, *nadazuke,* pickled napa cabbage, *takuan*

Early this morning I did the same as usual. After breakfast, I dug up a bunch of green onions that were buried under the snow and took them down to the pond, where my wife and I washed them. Once that was done, I bundled them up. I got home at about three o'clock. There were 51 bundles. Then I put a rabbit pelt and the dog pelt in water for tanning—as I was asked to do. After helping to wash the onions, my wife sewed and relaxed a bit. And she cooked, too.

Today's expenditures:

brown sugar	10 sen

At the main house: Kanezō went out again to do *dokata* work but he gave up and went back to the house because of the snow. After that, he wove *tawara.* They'll be for rice. Normally, a person can only manage to make ten of these in a day, working from morning to evening. My brother's wife went to her natal home on an errand, and my brother went out to deal with a rotating credit association matter again. The kids minded the shop. My mother cooked and such.

It's very unusual, but this year two members of my village have died at sea. First, Yoshida Kenji drowned after his ship, the *Eitoku Maru,* went down, but his body was never found. And now, Yoshida Kakichi—nicknamed "Tarō"—who went up to Hokkaidō for *dekasegi,* has drowned after his ship sank. His body wasn't found, either.

December 23, Monday

Snowstorm, but sun sometimes visible
West wind
–4°C in the morning, 2°C in the afternoon
Snow accumulation: 2 *sun* (6.1 cm)

Breakfast:	pumpkin cooked with miso, *nadazuke,* pickled napa cabbage, *takuan*
Lunch:	roasted salted *hatahata, nadazuke,* pickled napa cabbage, *takuan*
Supper:	daikon and burdock root miso soup, side dishes: same as lunch

Same as always in the early morning. After breakfast, I cleaned up and then I wrote New Year greeting cards. Then I made lunch. After cleaning up, I washed out the barrel (I bought) and sterilized it with boiling water and used it to make *takuan.* To make *takuan* to be eaten from around August to around October next year, I used a mixture of 2 *shō* (3.6 L) each of rice bran and salt, and 3 *kan* 500 *momme* (13.1 kg) of daikon. After finishing, I poured 3 *shō* (5.4 L) of water into the barrel. The way I did it was the same as before. And I removed the stone from atop the daikon I packed in a barrel to make *takuan* some days ago. I did that to smooth out the surfaces of the daikon. After that I collected water, and cut bamboo for making strips to wrap around barrels. My wife went out again today to sell green onions. She came home around three o'clock, and did some cooking.

Today's income:

green onions (51 bundles + 500 *momme* (1.9 kg))	1 yen 10 sen

Today's expenditures:

kerosene (1 *shō* (1.8 L))	30 sen

At the main house: Kanezō went out and did *dokata* work, hauling loads of dirt. He got back at three o'clock, and then he cut hay for the horse. My brother's wife went to her natal household again. Her younger sister will marry soon, so she accompanied her sister to the Nagane market to buy things. My brother went out on some sort of errand. Ichisuke minded the shop. My mother went to another village [or hamlet] on an errand, and she also went to my brother's wife's natal home and to the household into which my brother's wife's sister will marry, at their invitation. My mother drank some saké there before going home. Then she cooked.

December 24, Tuesday

Intermittent snow and sunshine
West wind
–4°C in the morning, 2°C in the afternoon
Snow accumulation: 1 *sun* (3 cm)

Breakfast:	daikon and burdock root miso soup, *nadazuke,* pickled napa cabbage, pickled eggplants, *takuan*
Lunch:	soba *nereke,* roasted salted *hatahata,* pickled napa cabbage, pickled eggplants, *takuan*
Supper:	*okayu* [gruel] made with cold rice, side dishes: same as lunch

Same early morning work as always. After breakfast, I fashioned the bamboo (I cut yesterday) into strips (*taga*) and fit them on the water barrel. Then I made two more for the night-soil barrel and put them around it. Then I dug up the green onions I left under the snow two days ago. This went on into the afternoon. I cleared the snow and chopped the soil with a hoe, and then pulled the onions out with my hands, which were completely numbed by the snow. But the ground was totally frozen in places, so I couldn't get all the onions. In most years December isn't so very snowy, but this year we had a terrific snowstorm in the early part of the month, so a layer of ice formed on the ground, and because of that most of the farmers haven't been able to dig up all their onions and burdock roots. My wife did laundry, and pulled out some daikon to use for making soup and cleaned them.

At the main house: Kanezō went out to do *dokata* work again—hauling loads of dirt. My brother's wife went to her natal home again, this time to help with the kimono for her younger sister's wedding. My brother minded the shop and my mother cooked.

The other day, an inspector of the Mitsui Hōonkai,[27] Mr. Nishida, the Akita Prefecture Kosakukan,[28] and a Hiraka County[29] Nōkai technician, came to the village to visit Yoshida Kōtarō. But I don't know exactly why they came to see him.[30] Yoshida Kōtarō is the son of one of the wealthiest families in the village, and he religiously professes the teachings of Ishikawa Rikinosuke. To what extent he follows them in practice, I'm not really sure, but he is a sort of missionary for Ishikawa's philosophy. He owns about 2 *chōbu* (approx. 2 ha) of fields in the foothills of Mt. Kampū, and over one-half of that is a fruit orchard. This orchard has been there for a while. He named the entire farm Kinnōsha Kōshin-en, built a magnificent house that cost 1,000 yen, and unreservedly poured money into the project, and moved his younger sister and her husband and his younger brother and his wife into the house to co-reside and to run the farm as a kind of self-sufficient agricultural project in line with Ishikawa Rikinosuke's philosophy. But everyone says he isn't really working on the farm—he's just the manager.

December 25, Wednesday

Cloudy
Southwest wind, later out of the northwest
–1°C in the morning, 2°C in the afternoon

Breakfast:	exactly the same as yesterday morning
Lunch:	roasted salted *hatahata, nadazuke,* pickled eggplants, *takuan*
Supper:	miso soup with potatoes and taro stem pieces in it, side dishes: same as lunch

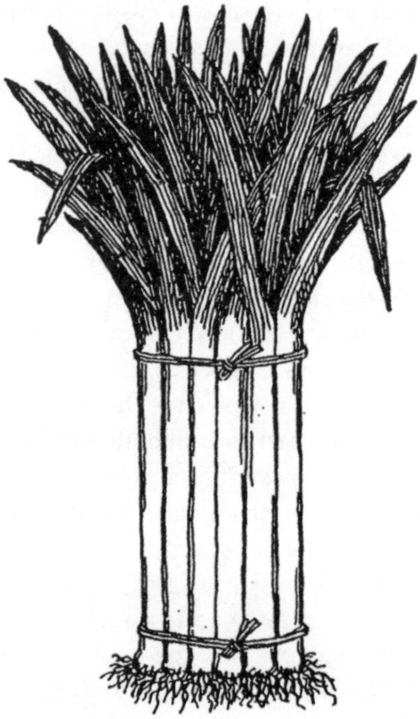

Figure 10.3. A green onion bundle.

Same as always early this morning. I dug more green onions up from beneath the snow and my wife and I washed them together, and after that I bundled them. My wife sewed. I bundle green onions like this: I take four relatively thin ones—these will go on the inside—and then six thick ones to go around the outside, to make for ten in total, and I tie them with straw in two places to make a bundle (see Fig. 10.3).

Once that was done, I dug a hole in the snow and buried the onions I dug up yesterday in there, under the snow. After that I dug paths is the snow—one leading down to the road. The way I stored green onions this year differs from my usual way. Snow fell while they were still in the field, so I had to just lay them down and cover them with snow. Normally every year I make use of the hotbox pits. I stand the green onions in the pits, working from one side to the other, packing them in pretty tightly, with 2–3 *sun* (approx. 7.6 cm) of space between them, with the green parts pointing up and the roots pointing down, just as they'd be in the field. Once I've packed in one row, I put about a 5 *sun* (15.2 cm) deep layer of sawdust down beneath them, thus raising them up, and then some straw on top, and then keep going. Once a hotbox pit is full, I stop and put a cover over it, and put *noma* mats on top.

At the main house: Kanezō didn't go out to do *dokata* work today; the water tank repairing project has been put on hold. I hear they'll start again after the snow melts. Instead, Kanezō cut hay for the horse early, and after breakfast made (straw) horseshoes. In the afternoon he got ready to do *koifugi*. He prepared a horse-drawn sled, and he repaired the saddle. To do *koifugi* means to haul compost (manure). Today my brother's wife went to her natal home again to help with preparations for her sister's wedding, and my brother was in the shop. My mother cooked. She also washed barrels for making *takuan,* etc.

December 26, Thursday

Sunny
Northwest wind
–11°C in the morning, 6°C in the afternoon

Breakfast:	same as yesterday
Lunch:	roasted salted salmon, *nadazuke,* pickled napa cabbage, pickled eggplants
Supper:	miso soup with daikon and burdock root and such in it, roasted salted *hatahata,* side dishes: same as lunch

This morning, I got up a little earlier than usual and made a ladle (*hishaku*)[31] for collecting urine and night soil (see Fig. 10.4). In local speech we call it a *geshi bishaku.* It's a wooden box and it's kind of angled, and it measures 5 *sun* (15.2 cm) on each side and is about 3 *sun* (9.1 cm) tall. I attached a stick about 4 *shaku* (1.2 m) long to the box at one corner. As the diagram shows, you cut a notch in one upper corner, about one-fifth of the height, and then fit the stick into that, going all the way to the bottom corner, and attach it with a nail. People make these differently; some make them deeper, or larger in diameter, etc. After that, I fed the rabbits, and then we ate breakfast.

After breakfast, I cleaned up and went to the main house to borrow a large sled. On this I loaded four barrels for collecting night soil (called *geshi daro*) and used ropes to tie some wooden poles and boards to the sled (to hold the barrels in place) and headed to my sister's house near the station to collect night soil, pulling the sled behind me. A bunch of snow had fallen off the roof of the house next door—right on top of the opening of the septic tank. It was like a mountain, so I had to use a shovel to clear it all and then I could scoop everything out of there. But the opening is very small, so it was extremely difficult to do that—a real struggle. I wound up getting pee and poop splattered all over my hands and face and clothes. Some flew into my mouth, so I just went ahead and savored the flavor. It was sort of salty and sort of sour—a flavor I really can't describe with words. As poop goes,

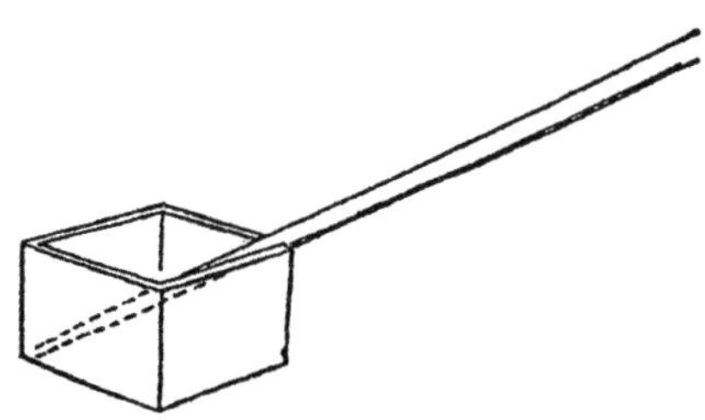

Figure 10.4. The ladle I made today.

older, well-decomposed stuff, and newer, fresh stuff, taste different. The latter is neither salty nor bitter—it's kind of sweet, with maybe just a hint of salt. I tasted such fresh poop about eight years ago, after making a bet with my younger brother [Shirō], who's now in Tokyo. My wife served as the witness. We were planting potatoes in a field [and made a bet], and I wound up having to lick poop and he ate ten raw unpeeled potatoes. He gobbled those spuds faster than I could, and he got all filled up. I remember the bitter expression on his face as he forced himself to eat all those potatoes—I think he suffered more than I did. Even now, as I write, I feel like I can see his face before me.

Today while I was scooping up all that stuff, the old lady at the Iwaya Inn (*ryokan*) next door, said, "Whoa! What a stink! You should do that kind of thing early in the morning. I can't even eat with that stench!"[32] It was all because she was having breakfast at about ten o'clock. So, I filled all four barrels and loaded them on the sled and pulled it back home and made lunch and ate. Then, I went back again and filled the barrels once more. Then I washed the barrels out with water and returned everything to the main house. After that, I dug up some cabbages I'd stored under the snow and prepared them and other vegetables for selling tomorrow. My wife went out to sell green onions and burdock roots this morning and returned at one-thirty, and after that she cooked and such.

Today's income:

green onions (32 bundles)	60 sen
burdock roots (1 *kan* 500 *momme* (5.6 kg))	30 sen

Today's expenditures:

mandarin oranges (*mikan*) for the kids	5 sen

At the main house: Kanezō cut hay early, and today he finally started hauling compost (doing *koifugi*). He began with the Sanbonmatsu paddies. My brother went out on some sort of errand, and his wife was in the shop, working on a bridal kimono. My mother cooked.

December 27, Friday

Sunny
Northwest wind
–10°C in the morning, 4°C in the afternoon

Breakfast:	daikon and burdock root miso soup, *nadazuke,* pickled napa cabbage, pickled eggplants, *takuan*
Lunch:	roasted salted salmon, side dishes: same as morning
Supper:	miso soup with taro stem pieces in it, roasted salted *hatahata,* side dishes: same as morning

The main thing I did today was tan the rabbit pelts I put in water the other day. The fur had come off three of the nine pelts, but the others were okay. I was able to finish them by lunchtime.

I made lunch, and after eating I did the laundry and preserved vegetables using dried alum (made *myōbanzuke*) and by the time I'd finished everything, it was dark. My wife came home at two o'clock and boiled water and helped me.

Today's income:

green onions (30 bundles)	55 sen
burdock roots (1 *kan* 500 *momme* (5.6 kg))	30 sen
cabbages (1 *kan* 100 *momme* (4.1 kg))	19 sen

Today's expenditures:

sweets and candies	5 sen

From today we started mixing rolled oats in with our rice and eating that. We mixed 7 *gō* (1.3 L) of rice with 6 *gō* (1.1 L) of oats. I estimate that we'll be able to do this for a good while.

At the main house: Kanezō cut hay early and prepared to haul compost. He spent the whole day on that. My brother's wife sewed while cooking. My mother went to my sister's family's house [near the station] to visit. My brother stayed in the shop.

December 28, Saturday

Cloudy
–4°C at 7:00 a.m., 4°C at 2:00 p.m.

Breakfast:	exactly the same as yesterday
Lunch:	roasted salted *hatahata*, *nadazuke*, pickled napa cabbage, pickled eggplants, *takuan*
Supper:	green onion miso soup, side dishes: same as lunch

Today I dug up more green onions and washed them together with my wife and then I bundled them. My wife pulled out a bunch of daikon and washed them and chopped them up for making soup. She had to go into the village in the afternoon on an errand. After I'd finished bundling the green onions, I dug up a bunch of burdock roots and tied them in 500 *momme* (1.9 kg) bundles. Then I pulled out some cabbages, and got everything ready for selling tomorrow.

At the main house: Kanezō did the same work as yesterday. He spent an hour right after lunch skinning and butchering a rabbit and pounding the bones. Ichisuke caught this rabbit up in the hills using a wire trap he'd made. My mother went on a (short) shrine pilgrimage. My brother's wife sewed while cooking. My brother was busy in the shop making sweets that someone had ordered.

December 29, Sunday

Cloudy, later rainy
Southeast wind
6°C in the morning, 9°C in the afternoon

Breakfast:	same as yesterday
Lunch:	cooked *anpura,* roasted salted *hatahata,* pickled napa cabbage, *nadazuke*
Supper:	rabbit meat, green onions and *konnyaku* cooked together in soy sauce, pickled napa cabbage, *nadazuke*

First, I fed the rabbits, and then after breakfast I cleaned up and changed the straw in the hutch and repaired it because it had broken again. Then I made lunch and killed one rabbit and skinned and butchered it and pounded the bones and prepared it for cooking. My wife came home at two o'clock, and cooked.

Today's income:

green onions (36 bundles)	65 sen
burdock roots (1 *kan* 500 *momme* (5.6 kg))	30 sen
cabbages (1 *kan* 500 *momme* (5.6 kg))	30 sen

At the main house: Kanezō cut hay for the horse early, and he spent the rest of the day weaving *noma* mats. These are for laying on top of the compost he hauled out to the paddies. My brother's wife was in the shop, where she sewed. My brother was in the shop, too. My mother made *takuan* while cooking. At the main house, they used a mixture of 5 *shō* (9 L) of rice bran and 3 *shō* (5.4 L) of salt to preserve 100 daikon as *takuan.* They'll add water later, they say. It's said that the day called *kanku* is the day to add water to *takuan.* Apparently, everyone in the village does so on that day. People say, *konogazuke ni kanku no mizu o hiru.*[33]

Today, a new bride came to the home of Yoshida Tokuji, a young fellow of the village. Also, a young woman of Yoshida Sankurō's household went as a new bride. She married into a family in Monzen Village.[34] The young woman who married Tokuji is from Funagawa.

December 30, Monday

(December 5 according to the old calendar)
Snow
Strong northwest wind
0°C in the morning, 4°C in the afternoon
Snow accumulation: 3 *sun* (9.1 cm)

Breakfast:	cabbage miso soup, side dishes: same as every morning
Lunch:	rabbit meat, *konnyaku* and green onions cooked in miso, boiled cabbage with soy sauce for flavor, roasted salted *hatahata*, pickled napa cabbage, *nadazuke*, etc.
Supper:	red bean rice (*sekihan*) we received from the main house and rice with grated Chinese yam on top (*tororo meshi*) we made at home, boiled cabbage with soy sauce for flavor, roasted salted *hatahata*, pickled napa cabbage, *nadazuke*

I fed the rabbits early, and after breakfast I cleaned up and then dug up the last of the green onions—there were just a few still under the snow. These I buried again under snow for storage. Next, I dug up some Chinese yams and washed them, and then made lunch, and ate and then cleaned up, and then I had to go into the village on an errand. My wife returned at two o'clock and got started right away on making supper. She made *tororo meshi* for us to eat. Today is December 5 according to the old calendar—the New Year festival day for Ebisu. Here in the village, they call it *Ebisu-ko no toshitori*.[35] We always celebrate it with red bean rice (*sekihan*). At my house we have no red beans (*azuki*), so we didn't make *sekihan*.

Today's income:

green onions (36 bundles)	70 sen
burdock roots (1 *kan* 500 *momme* (5.6 kg))	35 sen

Today's expenditures:

konnyaku	5 sen
brown sugar	15 sen
2 sheets of cotton	40 sen
cotton cloth	20 sen
hair oil	5 sen
candies	5 sen

At the main house: Kanezō went out early to cut hay, as usual, and then spent the day making *noma* mats, like yesterday. My brother went out on an errand and his wife stayed in the shop and boiled water.[36] My mother cooked.

December 31, Tuesday

(December 6 according to the old calendar)

Snow

–5°C in the morning, –1°C in the afternoon

Snow accumulation: 5 *sun* (15.2 cm)

Breakfast:	leftover *tororo meshi* from last night, all else: pretty much the same as last night
Lunch:	buckwheat *nereke* with soy sauce for flavor, roasted salted *hatahata, nadazuke,* pickled napa cabbage
Supper:	miso soup with burdock root pieces in it, side dishes: same as lunch
Snack:	*kohin* (*Kohin* is powder made by grinding roasted unpolished *mochi* rice in a mortar.[37])

Early this morning I fed the rabbits. Then I had to scoop water out of the toilet's septic tank because water from last night's rain had leaked into it. The location isn't good, so on rainy days water will get in there if I'm not always paying attention. It's a real pain. In the afternoon, I had to go into the village on an errand. My wife went to the main house in the morning to make *kohin*. It was for eating tonight. Here in the village people call today—December 6 according to the old lunar calendar—New Year festival day for the Hada god (*Hada no Kami no toshitori*).[38] They make *kohin* and celebrate. It's made by grinding roasted unpolished *mochi* rice grains (*genmai*) in a mortar.[39] In the afternoon, my wife made a *futon* for sleeping on. She also cooked and such.

At the main house: Kanezō cut hay early, and wove *noma* mats the rest of the day. My brother's wife stayed in the shop, but she made *kohin* with my mother and cooked and such in the evening. My brother went out again, to do something.

Chapter 11: January 1936

January 1, Wednesday

Some occasional sunshine
Northwest wind
–1°C in the morning, 2°C in the afternoon

Breakfast:	miso soup with daikon and burdock root in it, *nadazuke,* pickled napa cabbage
Lunch:	salted *hatahata* simmered in water, *nadazuke,* pickled napa cabbage
Supper:	same soup as morning, roasted salted *hatahata, nadazuke,* pickled napa cabbage
Snack:	*kohin*

I did the same as always early, and made some boots with straw (after breakfast). In the afternoon, I had to run some errands. Also, I had to feed the rabbits three times today, as always. My wife sewed a lot during the day, and she cooked and such.

> At the main house: Kanezō wove *tawara* with straw all day. My brother's wife went to her natal home again to help with her sister's wedding preparations. My brother minded the shop. My mother cooked and such.

January 2, Thursday

Cloudy, later rainy
Southeast wind
–4°C in the morning, 2°C in the afternoon

Breakfast:	*okayu* rice porridge made with cold leftover rice, fresh miso [to flavor the porridge], pickled napa cabbage, pickled eggplants
Lunch:	roasted salted *hatahata, nadazuke, takuan*
Supper:	miso soup with dried taro stem bits in it, side dishes: same as lunch
Snack:	*kohin*

Early this morning, I fed the rabbits. After breakfast, I wrote New Year greeting cards (*nengajō*) to people I hadn't sent one to, but who sent one to me. I also wrote captions/explanations for all the photos I needed to mail to the Attic Museum and ran them down to the post office in the village and mailed them. I ran like heck because it seemed like rain was about to fall. I got dizzy and felt anemic, so I rested about an hour and then headed back home. Once I got back, I just collapsed and rested in bed for a while. My wife was at the main house all day, grinding buckwheat and *kozagimai* in a stone mortar. *Kozagimai* are little bits of rice grains. My mother received 2 *shō* (3.6 L) of this from someone. My wife also ground 2 *shō* of unpolished rice (*genmai*). We'll grind this into powder later.

Today's expenditures:

postage stamps	10 sen
sweets	10 sen

At the main house: Kanezō cut hay early, and wove *tawara* the rest of the day. My brother's wife went to her natal home again to help with her sister's wedding preparations. And she accompanied her sister to her new home. In other words, she did *yometsure*. My brother went there too, to drink celebratory saké. Ichisuke minded the shop. My mother cooked and such.

January 3, Friday

(December 9 according to the old calendar)
Cloudy, snow later
Northeast wind
–1°C in the morning, 1°C in the evening

Breakfast:	daikon miso soup, *shōyu no mi, nadazuke,* pickled napa cabbage (We received the *shōyu no mi* from the main house.)
Lunch:	*soba nereke* with soy sauce for flavor, roasted salted *hatahata, nadazuke,* pickled napa cabbage
Supper:	same soup as yesterday evening, side dishes: same as lunch

I dug up some daikon and fed the rabbits early, and cleaned three rabbit pelts after breakfast. Then I did some laundry, and I managed to preserve some vegetables with dried alum. In the afternoon, I spread night soil on the fields that I'll use this year for growing green onions. It's kind of thin—watered down—because rainwater got into the septic tank. I spread a total of ten *ka* over the fields. My wife washed daikon and made soup and such, and she sewed and cooked other dishes as well.

Today's expenditures:

candies for the kids	3 sen

At the main house: Kanezō did *shinagogu* with straw. This means to remove the stamens and pistils. But he spent most of the day weaving *tawara*. My brother's wife didn't return last night, but she did late this evening. My brother went out on errands, and Ichisuke minded the shop. My mother went to the home of a sick acquaintance to pay her respects, and prepared for making red bean rice (*sekihan*) during the afternoon. Today is the day they call *Degogu-sama no toshitori*,[1] so here in the village people make *sekihan* and celebrate.

Back when I was little—before kids had things like (special) *geta* (for skating), or *gunkan gane*,[2] or skates or skis—we used to make *kura* sleds and go sledding on the snow. And at night we'd pour water on the shiny snow and ice and really slide fast while wearing straw boots. We called it *kuzutenka*. This has to be done on a slope, or it doesn't work. My friends and I usually slid down the Jundo slope.[3] It was a very long and steep slope.

When we slid down from the top of the slope on a sled, it gave us a gut-freezing thrill that was just terrifying. But with the passage of time, *geta* became popular, and then *gunkan gane*, and then skates, and skis, so kids stopped playing with *kura* sleds and making *kuzutenka* and sliding down wearing straw boots (see Fig. 11.1). You never see that anymore. Not only that, but the Jundo slope has been shrinking.

According to village elders, this slope used to be a wide path for horses loaded with compost. That's why we could enjoy playing on it so much. The reason why it has been shrinking is because the entire hill is made of sand, and one side of the big slope is a sandy cliff—called *mama* in local talk. And the land right under the cliff is the residential property of one Yoshida Chūzō. Every time rain falls, parts of the cliff will tumble down onto his property. Of course, if you just leave it alone, it will eventually smooth out and not fall anymore, but Chūzō says the sand is a nuisance, so he always gives it away. People happily receive it, and mix it with compost or add it to their rice fields to improve the soil, and even use it in their nursery paddies. So, since the sand that falls each time is taken away, several *tsubo*[4] of sand come crashing down with every big rainfall,

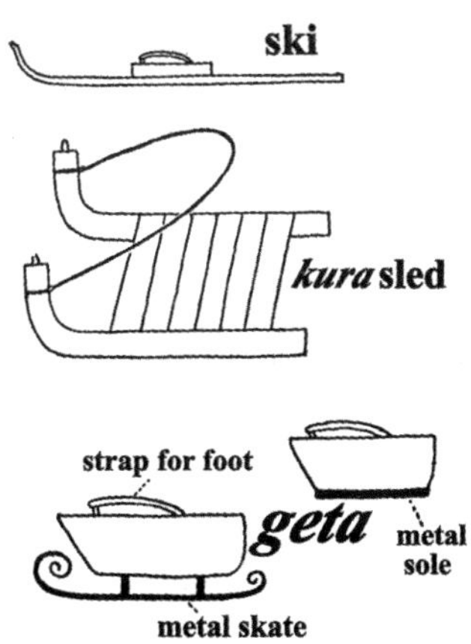

Figure 11.1. Sleds, skis, and *geta* for sliding on snow and ice.

making a huge noise. And every time, Chūzō and others take it away. This continues year after year, so even a huge slope like that one naturally vanishes. But all the same, it has been rebuilt again and again. The thing is, some people really need this sloped path to reach their paddies and fields, so each time a path is lost, they cut a new path on the other side for themselves to get up the slope. Then, sand starts falling onto another person's property, and then that path is lost and the sand is taken away, so they need a new path, so they make another one, and the cycle continues.

One side of the mountain, they say, belongs to the biggest landowner in the village—a man by the name of Katō. Because the sand that falls gets cleared away like that, the property of Yoshida Chūzō—which wasn't so huge when I was a kid—has become very big, and even sports a large work shed now. I don't know if the big Jundo slope was his family's property in the first place or not. But I've never heard anything about such sloped horse paths belonging to any specific person. Anyway, I'm going to just write down what I know to be true. There was another such path that disappeared. It was on a slope behind the Mishima Shrine. One side belonged to the shrine, and beneath the other side were dry fields owned by Takakuwa Yasuke. And the same thing that happened to the Jundo slope happened to this slope—it totally disappeared. So today, everyone just travels along a path they've made through the shrine's property, because there's no other way. Why is it that some villagers expand their own property, even by destroying public paths? And how is it that everyone else can just keep quiet about it, as if it doesn't affect them? I can't say for certain. The guy from the first example is referred to as a *shinkō jinushi* (a newly risen landlord) and the guy from the second example is generally thought of as a *yabatsu ningen*[5] (a totally unmanageable person), or as a tricky sophist,[6] by villagers. I don't know if everything they say about these guys is true, though.

January 4, Saturday

Cloudy
Strong west wind
–1°C in the morning, 1°C in the afternoon

Breakfast:	same as yesterday morning
Lunch:	roasted salted *hatahata, nadazuke,* pickled napa cabbage
Supper:	rabbit meat and green onions cooked in soy sauce, *nadazuke,* pickled napa cabbage

This morning, I pulled out some daikon and fed them to the rabbits. I spent the rest of the day weaving *tsukenawa* ropes at the main house. I used seven bundles of straw and made six bundles of 40 *hiro*[7] rope. My wife sewed today, and she helped me weave rope for a while in the evening. She also cooked and such.

Today's expenditures:

5 chicken eggs	15 sen
candies	2 sen

Figure 11.2. Yoshida Kintarō weaving *tawara* in front of his house (February 4). [Kintarō was a member of the "big main house" and one of Saburo's cousins.]

At the main house: Kanezō cleared out the stable early, and then wove more *tawara* (see Fig. 11.2). In the evening, he cut a bit of fodder[8] for the horse, and he also washed the animal. The horse has to be washed off every evening—it's basically a rule. My brother went out on an errand, and his wife minded the shop. My mother cooked and such.

January 5, Sunday

Snowstorm
Strong west wind
–5°C in the morning, –4°C in the afternoon

Breakfast:	potato and daikon miso soup, side dishes: same as every morning
Lunch:	roasted salted salmon, *nadazuke,* pickled napa cabbage
Supper:	daikon miso soup, roasted salted *hatahata, nadazuke,* pickled napa cabbage

Today was pretty much the same as yesterday for me. My wife went to her natal home and ground rice into powder and came back. She used polished rice that had been soaked in water once and then drained and dried, and pounded that in a wooden mortar with a pestle to make powder. This powder will be used to make *mochi* tomorrow. It takes all day.

At the main house: Kanezō wove *tawara* from the early morning. In the evening, he cut some more hay for the horse, and he took the horse down to Wakimoto beach to let it run around some. My brother's wife minded the shop while making rice powder. At the main house they don't use a wooden mortar—they grind rice in a stone mortar. My mother cooked and such.

January 6, Monday

(December 12 according to the old calendar)
Snowstorm
Strong northwest wind
–7°C in the morning, 9°C in the afternoon
Snow accumulation: 2 *sun* (6.1 cm)

Breakfast:	miso soup with salted daikon leaves, side dishes: same as every morning
Lunch:	roasted salted *hatahata, nadazuke,* pickled napa cabbage
Supper:	same soup as morning, side dishes: same as lunch (We received the salted daikon leaves from my wife's natal home.)
Snack:	grilled *mochi* with sweet bean paste in them

Today was just the same as yesterday for me. My wife cooks red beans with sugar [makes sweet *azuki*-bean paste] every day lately. Yesterday, she made *mochi* with rice powder and grilled them. She dissolved some salt in hot water and added that to the rice powder and kneaded it into *mochi,* put some sweet red bean paste inside each, and grilled them. Today is December 12 according to the old lunar calendar—the day we call *yama no kami no toshitori* [The New Year festival for the mountain god], and this is how we celebrate it. You need rapeseed (*nanohana*) oil[9] to grill these *mochi.*

Today's expenditures:

brown sugar	20 sen

At the main house: Kanezō started weaving *kagari nawa* (ropes) through the *tawara* he'd made in the early morning. (He finished making the *tawara* yesterday—about 50 in total.) You have to weave *kagari nawa* through the flat mat-like *tawara* [in order to tie them up and shape them later]. My brother went out on some errand, and his wife and my mother made and grilled *mochi* together all day.

Regarding these *kagari nawa,* there are two ways of weaving them through *tawara: nana kagari* and *hachi kagari.* They are the ropes that are visible on the surface of *tawara.* In other words, you go around the *tawara* with the rope once, weaving it through the straws, with 2 *sun* (6.1 cm) inside the *tawara* and 4 or 5 *sun* (12.1–15.2 cm) showing on the outside, and then 2 *sun* (6.1 cm) on the inside again, all the way around. If the rope is visible from the outside in seven places, it's called

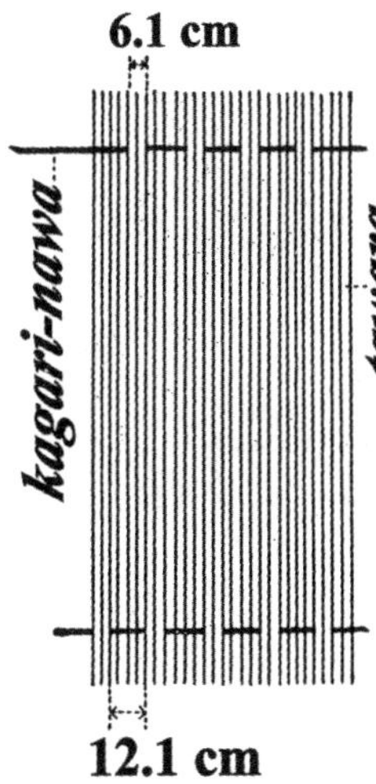

Figure 11.3. How to weave *tawara* (1).

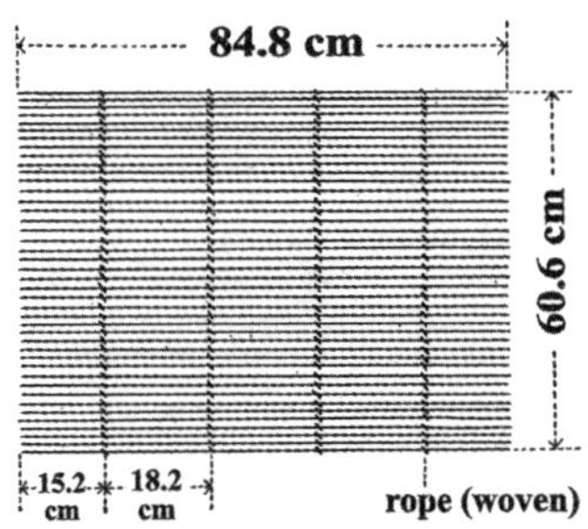

Figure 11.4. How to weave *tawara* (2).

nana kagari and if it's visible in eight places it's called *hachi kagari.* The diagrams make it clear (see Figs. 11.3 and 11.4).

Normally, measured with adult hands, you use a 9 *hiro*[10] *migo* rope to make one *tawara.* It's kind of a set rule. But if you use a 10 *hiro*[11] rope you'll have a little extra to work with. It's convenient. So, what you'll have in the end is a *tawara* about 2 *shaku* (60.6 cm) wide and about 2 *shaku* and 7 or 8 *sun* (approx. 83 cm) long. You weave the *migo* ropes through the *tawara* in four places, with 6 *sun* (18.2 cm) between them, and about 5 *sun* (15.2 cm) left over on each end.

When I was a kid, *yama-gami no mochi*[12] were always made on the twelfth day of December in my village by putting sweet bean paste on pounded *mochi.* We always ate those. But now, everyone makes grilled *mochi* instead, like I explained. In the main part of Wakimoto Village, they've been making grilled *mochi* since a long time ago. So, what happened is that in my hamlet, people stopped making *mochi* in the old way and adopted the other method.

Yama-gami no mochi, offered on December 12 every year in my village, were pounded *mochi* with *azuki* (sweet bean paste) on them when I was a kid. But nobody pounds *mochi* like that on this occasion today.

January 7, Tuesday

Occasional snowstorms
Northwest wind
–12°C in the morning, –4°C in the afternoon
Snow accumulation: 5 *sun* (15.2 cm)

Breakfast:	same as every morning
Lunch:	salted *hatahata* cooked in a pot, *nadazuke,* pickled napa cabbage
Supper:	miso soup with dried taro stem pieces in it, roasted salted *hatahata, nadazuke,* pickled napa cabbage
Snack:	grilled *mochi*

Again today, I dug paths through the snow and dug up food for the rabbits early. Then I spent the rest of the day weaving ropes. My wife sewed and played [with the kids] and cooked.

At the main house: Kanezō made straw horseshoes (*umagutsu*) from the early morning, and continued after breakfast. In the afternoon he dug paths to the rice paddies in the snow, for hauling compost out to them. Everyone with paddies in the Sakura area gathered to dig paths through the snow. In the evening, Kanezō cut horse fodder. My brother's wife minded the shop and cooked, etc., and my mother and brother went out on separate errands.

January 8, Wednesday

Occasional snow
North wind
–5°C in the morning, –1°C in the afternoon
Snow accumulation: 3 *sun* (9.1 cm)

Breakfast, lunch, and supper:	basically, the same as yesterday
Snack:	grilled *mochi*

My work for today was pretty much the same as yesterday. My wife worked in the kitchen, and she wove a bit of rope while working.

Today's expenditures:

magazine	25 sen

At the main house: Kanezō hauled compost from the early morning. My brother's wife stayed in the shop and ground powder with a stone mortar. My brother went out on an errand. My mother cooked, and she also cut some hay for the horse while working.

Today was the day for doing something called *kanzuguri* here in the village. This is where a veterinarian does acupuncture on horses and uses a hot iron on their hoofs. In the old days, a *hagurago* would travel from village to village and do acupuncture on the horses and treat their hoofs with hot irons.[13] Here in the village, *kanzuguri* means doing acupuncture on horses and using hot irons on their hoofs after entering the season called *kan*—the coldest time of the year. But times have changed. I'm not sure exactly when it happened, but the *hagurago* of older times have completely disappeared. Today, properly schooled veterinarians have replaced them. And whereas *kanzuguri* was done to all horses in the village in the past, it's done to very few horses today.

January 9, Thursday

Snowstorm
Strong northwest wind
–7°C in the morning, –6°C in the afternoon
Snow accumulation: 4 *sun* (12.1 cm)

Breakfast:	daikon miso soup, *shōyu no mi, nadazuke,* pickled napa cabbage, pickled eggplants
Lunch:	roasted salted *hatahata,* side dishes: same as morning
Supper:	miso soup with dried taro stems in it, side dishes: same as lunch
Snack:	grilled *mochi*

I had to go out on multiple errands today. My wife sewed and hanged rabbit pelts up to dry before the fire.[14] After supper, I sprayed them with water and pressed them down lightly and left them that way.

At the main house: Kanezō did exactly the same work as yesterday—same for my brother's wife and my mother.

January 10, Friday

Occasional snow
Northwest wind
–3°C in the morning, 2°C in the afternoon
Snow accumulation: 3 *sun* (9.1 cm)

Breakfast:	miso soup with cabbage in it, side dishes: same as every day
Lunch:	roasted salted *hatahata, nadazuke,* pickled napa cabbage
Supper:	daikon miso soup, roasted salted salmon, *nadazuke,* pickled napa cabbage

Yoshida Tomeji's eldest son (one year old) died today.

I rubbed the rabbit pelts, made straw boots, and spread a mixture of egg yolk and starch on the dog pelt, among other things. My wife sewed, and went out on errands, and cooked.

Today's expenditures:

1 *to* (18 L) of soybeans	1 yen 70 sen
candies	3 sen

At the main house: Kanezō cut hay for the horse in the early morning. He hauled compost after breakfast, and more in the afternoon. My brother and his wife were both in the shop, making sweets, and cooking. My mother went to her natal home [to help prepare] for a Buddhist *hōji* ceremony to mark the thirty-third anniversary of her grandfather's passing, and the thirteenth of her mother's passing.

Today, I read in a book by Mutō Tetsujō that along the Sea of Japan coast and in adjacent regions people call dragonflies *dabu*, and that in regions along the major mountain ranges people say *danburi*.[15] But even though my village is on the Sea of Japan coast, we say *danburi*. Kids will hold brooms up in the air, singing: "*Danburi, danburi,* land here! If you don't, I'll tell your parents!" while catching dragonflies.

January 11, Saturday

Cloudy
−4°C in the morning, 1°C in the afternoon
Snow accumulation: 1 *sun* (3 cm)

Breakfast:	potato and daikon miso soup, *nadazuke*, pickled napa cabbage, pickled eggplants
Lunch:	*soba nereke* with soy sauce and sugar for flavor, roasted salted *hatahata*, *nadazuke*
Supper:	daikon miso soup, burdock root *debu*, side dishes: same as lunch

I cleaned off the rabbit pelts that I rubbed yesterday with a pumice stone, and wove some ropes. I also cleared snow off the cover of the night-soil cistern and collected some urine [from the septic tank]. My wife sewed, dug up daikon, and prepared wild plants for putting in soup later.

Today's expenditures:

postage stamps	6 sen
brown sugar	5 sen

At the main house: Kanezō cut hay for the horse early, and hauled compost the rest of the day. He started carrying it out to the paddies at Bishamon. My brother went to participate in the *hōji* for my mother's natal family. His wife minded the shop while sewing and cooking.

January 12, Sunday

Cloudy, rain later
Southeast wind
2°C in the morning, 5°C in the evening

Breakfast:	miso soup with burdock root and potato in it, the usual side dishes plus burdock root *debu*, etc.
Lunch:	exactly the same as yesterday's lunch
Supper:	miso soup with dried taro stem pieces in it, side dishes: same as yesterday's supper

Today I wove rope again, but these are for making *sudare* screens, so I made them a little narrower than usual ropes. My wife did laundry and such in the morning, and sewed, etc., in the afternoon.

Today's income:
pay for cleaning rabbit pelts 50 sen

At the main house: Kanezō hauled compost from the early morning, but not in the afternoon, because the paddies are too soft and muddy for the horse to walk, so he cut hay for the horse and made straw horseshoes instead. My brother went out on errands. My mother didn't come back [from her natal home] yet. My brother's wife sewed and cooked and such while minding the shop. Ichisuke has been weaving rope for daily use a lot since his winter vacation started. Today he finally reached one *tsuyogu.* He took it all over to the agricultural cooperative warehouse [near the station] and sold it for 25 sen. He was overjoyed about that. One *tsuyogu* of rope is 10 bundles (*wa*), and each of these—as I explained before—is 40 *hiro* long. One *hiro* is 5 *shaku* (1.5 m). Village kids typically weave ropes like this instead of just playing all the time, and the money they get for their ropes is known as *homazu hinko.* Even though they're just kids, they're pretty good at it. An adult who can weave one *tsuyogu* in one day working from early morning is considered fast, but kids can weave five or six bundles. The eldest son of Yoshida Shūjirō—Ichirō (13)—is said to be able to weave one *tsuyogu* in a day, working from the early morning. It really is amazing. Moreover, their family isn't living hand to mouth—they're pretty wealthy—so the villagers are all very surprised at his ability.

January 13, Monday

Snowstorm
Strong northwest wind
–1°C in the morning and in the afternoon
Snow accumulation: 1 *sun* (3 cm)

Breakfast:	daikon and burdock root miso soup, *shōyu no mi, nadazuke,* pickled napa cabbage, burdock root *debu*
Lunch:	exactly the same as yesterday's lunch

Supper: miso soup with dried taro stem pieces in it, roasted salted *hatahata*, *nadazuke*, pickled napa cabbage, etc.

Today I wove ropes for making a *futarimokko* [a *mokko* carrying sling to be used by two people]. They had to be fine and tightly woven. My wife sewed and such all day.

At the main house: Kanezō cut hay for the horse early. After breakfast he made horseshoes. In the afternoon, he went out and dug paths through the snow at Bishamon. My brother was out. His wife minded the shop and ground buckwheat in a stone mortar. My mother cooked, and Ichisuke wove rope today, too.

In the old days, farming families with plenty of rice would polish enough table rice—after entering the coldest season (*kan*)—to last until June or July, and doing this was called *kantsuki*. According to old-timers, people would use a big tree trunk to make a mortar large enough to hold about 3 *to* (54.1 L) of rice, and mix in a little powder from dried white wall plaster, and then put a thing called a *dewa* in the center of the mortar, and aim inside the *dewa* while pounding the rice with a *tsugi* [a large pestle]. I've tried pounding rice like this before. If you pound it 1,200 to 1,300 times you can polish the rice to the level that farmers generally eat—about 70 percent polished white rice. They say that if you mix in some of that white wall plaster powder you can do it faster than that—you don't have to pound it so many times. What replaced those wooden mortars were stone mortars [standard: *usu*] called *kara-ushi*. These are stones with big hollowed-out depressions. They were made for pounding rice grains with the feet. To polish 3 *to* (54.1 L) of rice, you have to stamp with your feet 1,700 or 1,800 to 2,500 times. Even so, they say, this was much easier than pounding by hand.

In the old days, this is how they did *kantsuki*. Three *tawara* of rice is what a young fellow could manage to polish in a day using one of these 3 *to* rice mortars. Even at my age, I've polished three *tawara* worth of rice at once only three times. I guess I can be counted among those strong young fellows of old. But I used a *kara-ushi* stone mortar; there's no way I could do that by hand using a wooden mortar. It's amazing that those guys were able to polish three *tawara* of rice by hand in a wooden mortar. According to what people say—and I've done it, too—if you stomp 300 to 400 times, a lot of the rice bran will appear. But it's important not to stop there—you need to keep stomping until you've done it 600 or 700 times. If you're lazy[16] about it and just stop at 300 or 400, and then try again later, people say you'll never get it any whiter, even if you do it thousands of times. This is true. And if you don't concentrate, you may do something really foolish, like wrack your balls! It truly happens. You really have to pay attention when doing this.

After the period of *kara-ushi* stone mortars passed, along came the modern era. At first, everyone took their rice to the polishing centers, where their cash, their rice bran, and even spilled rice grains, would be taken from them. So, these rice polishing businesses really raked in the profits. More recently, the farmers came to realize that they were being ripped off, and they started establishing cooperative agricultural workspaces. So now, the operators of those rice polishing shops are gradually losing business. The process of producing unpolished *genmai* rice from harvested (unshelled) rice grains also followed this evolutionary course. There've been a lot of changes in the mortars over time up to now. And nobody polishes even one *tawara* worth of rice at home today—they take it to a rice polishing business or to a communal polishing center. And all the old mortars have faded away. Today, the farmers are wondering whether revolving hard clay mortars or roller polishers (*rōru-suri*) are better. For the farmers, this is a very important problem. Well, I have no idea what the outcome of this will be.

According to the chapter on plants in Mutō Tetsujō's book, *hakobegusa* [*Stellaria*] is potential food for humans.[17] In my village, people do eat it—mainly women. They say it's good for women who are having trouble producing milk.

January 14, Tuesday

Snowstorm
Strong west wind
−4°C in the morning, −1°C in the afternoon
Snow accumulation: 2 *sun* (6.1 cm)

Breakfast:	cabbage miso soup, side dishes: same as every morning
Lunch:	cold rice with hot water poured over it [*yuzuke*], roasted salted *hatahata, nadazuke,* pickled napa cabbage
Supper:	potato and daikon miso soup, side dishes: same as lunch

Today I kneaded the dog pelt and made boots from straw. I also had to go out on some errands. My wife did the same things she does every day.

Today's expenditures:

toothbrushes[18]	10 sen
Lion toothbrushing powder	10 sen
candies	5 sen

At the main house: Kanezō cut horse fodder from the early morning, and later prepared for hauling compost (*koe*) out to the paddies (doing *koefugi*). He spent the rest of the day hauling compost. My brother's wife and my brother were in the shop, and my mother cooked and such. Ichisuke wove ropes again today.

In the old days, people used to make a lot of straw horseshoes. Elders say that a person could make up to 24 in a single day, working from early morning. Once iron horseshoes became widely available, people only needed to make straw horseshoes for wintertime use, so now there isn't anyone who can make 24 in a day. Eventually, the techniques of making iron horseshoes improved, and people started adding anti-slip studs or cleats to the horseshoes for winter, and farmers came to feel that it was more economical to use such horseshoes instead of making them with straw, so they started buying iron horseshoes and putting them on their horses. Even the techniques for making straw boots for people are gradually disappearing from this village, and I suppose that those for making straw horseshoes will follow the same path. I never see anyone making straw *waraji* sandals in the village anymore. I do still see *kyahan* [leg gaiters] sometimes, though.

In Mutō's book, it says that *hanashōbu* (Japanese water irises) are called *shodome* in the Akita City area.[19] In my village, they're called *sodome.*

And just as *ayame* (Siberian iris) flowers are called *sodome,* women doing rice paddy planting in May are also called *sodome.* Women are the stars—the bosses—at rice planting time; they show off and act like they're real proud of themselves. Also, in the same book, it says that people make a roof using mugwort and such, and [beneath it] run a hot medicinal iris bath at May festival time [May 5]. In my village, we do have that same seasonal event, but we don't use the same flowers.

January 15, Wednesday

Snowstorm
Strong northwest wind
–5°C in the morning, –6°C in the afternoon
Snow accumulation: 2 *sun* (6.1 cm)

Breakfast:	daikon miso soup, side dishes: same as every morning
Lunch:	miso and pickled eggplants
Supper:	salted *hatahata* and dried taro root stems simmered together in water, *nadazuke,* etc.

Today I made a *futari mokko.* I made it 2 *shaku* 5 *sun* (75.8 cm) wide and 3 *shaku* 5 *sun* (1 m 6 cm) long. Until today, I always made *mokko* by sliding wooden sticks right through the rope loops along the sides (see Fig. 11.5). But today, I tried a different method. Instead, I slid bamboo sticks through the loops and attached those to long wooden sticks (see Fig. 11.6). If you slide the wooden sticks right through the rope loops, you might accidentally cut the ropes with a hoe, and you may be troubled by the *mokko* sliding around. But if you slide bamboo sticks through the loops and attach them to the wooden sticks, you won't cut the ropes, and the *mokko* won't move around. I was able to weave this one *mokko* with 60 *hiro* (90.9 m) of rope, measured by my own arms. I used 10 *hiro* (15.2 m) of the 60 *hiro* as *amizuna* [rope to be woven through the *mokko*]. I cut that into four equal parts and wove it through the *mokko* in four places. After that, I made straw boots, and cleaned off the dog pelt. My wife was sick in bed from the morning, so I cooked and such instead.

At the main house: Kanezō hauled compost (did *koefugi*) early. He quit a little before lunchtime, and then started pounding *mochi.* In the afternoon, he did some more *mochi* pounding, and then he went back to doing *koefugi.* My brother minded the shop, and his wife and my mother prepped for making *mochi,* etc.

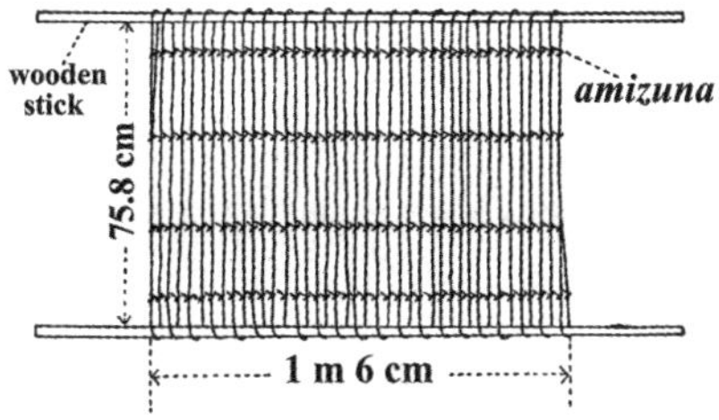

Figure 11.5. My way of making *mokko* up to now.

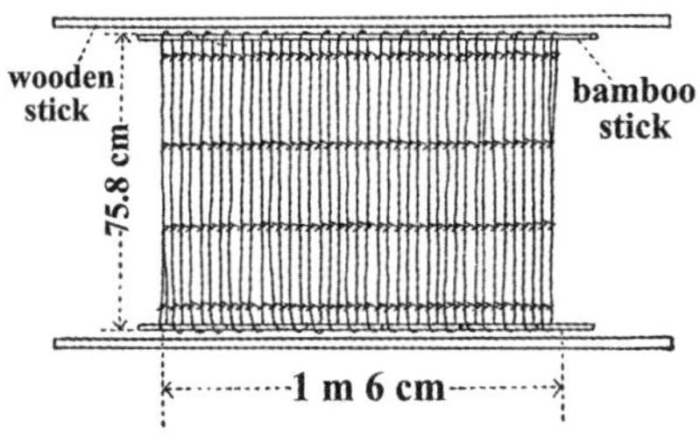

Figure 11.6. My new way of making *mokko.*

The *mochi* they made today is called *kanmochi.* It will be eaten on January 1, according to the old lunar calendar. But people also eat them around the time they start heading out to work in the fields and paddies—in the spring. *Mochi* made during the coldest season (*kan*) naturally freezes[20] very well. When rice is steamed for making *mochi,* sometimes there will be parts that we call *hanfuke,* or only partially cooked (*fukeru* means to steam).[21] If you make *mochi* with such rice, you get *janguri mochi.* And when you boil rice, it sometimes comes out like that, too. That's called *mekko meshi.*

Curiously, people with smallpox scars are disparagingly called *janguri* in the village.[22]

January 16, Thursday

Snowstorm
Strong northwest wind
-7°C in the morning, -5°C in the afternoon
Snow accumulation: 2 *sun* (6.1 cm)

Breakfast:	miso soup with dried taro stems in it, *nadazuke, shōyu no mi*
Lunch:	roasted salted *hatahata, nadazuke*
Supper:	same soup as morning, side dishes: same as lunch

I cooked breakfast, and everything else today, and stayed at home and wove ropes. My wife was sick in bed again.

Today's expenditures:

haircuts for the kids	20 sen
candies	3 sen

At the main house: Kanezō cut hay for the horse early, and after breakfast he did *koefugi*. In the afternoon, he dug up the soil in the compost pile spot. This will be used as compost soil. My brother's wife minded the shop and ground [rice] powder with a stone mortar while doing that. My brother went out on errands. My mother cooked, etc. She also cut the *mochi* they made yesterday into small pieces and tied them up with straw rope.

Today two villagers—Takakuwa Yasuke and Yoshida Shichisuke—moved. They switched houses. That is, Shichisuke sold his house and land to Takakuwa, and moved into Takakuwa's little house.

January 17, Friday

Snowstorm
Strong northwest wind
–7°C in the morning, –4°C in the afternoon
Snow accumulation: 2 *sun* (6.1 cm)

Breakfast:	cabbage miso soup, side dishes: same as every morning
Lunch:	nothing
Supper:	miso soup with dried taro stems in it, roasted salted *hatahata, nadazuke*

Again, I did all the cooking today—from morning. I also dug paths through the snow, dug up daikon and fed them to the rabbits, pulled out potatoes and set aside the frozen ones, and headed into the village on errands. My wife was sick in bed again. My mother came over to pay her respects to my wife (to do *mimai*).

At the main house: Kanezō returned to his natal home last night. I heard that he'll relax there for two or three days before coming back. My brother's wife went to the market to sell burdock roots and Chinese yams. My mother put rice into a wooden mortar and made *hadakiko* [rice powder] and cooked and such.

January 18, Saturday

Snowstorm
Northwest wind
–6°C in the morning, –2 in the afternoon
Snow accumulation: 2 *sun* (6.1 cm)

Breakfast:	cabbage miso soup, *nadazuke, shōyu no mi*
Lunch:	*soba nereke* with sugar, roasted salted *hatahata*
Supper:	green onion miso soup, roasted salted *hatahata, nadazuke*

I made breakfast this morning. After eating, I dug up some green onions and washed and bundled them. I bundled more in the afternoon. Then, I pulled out some burdock roots and bundled them, too. I carried all of these to the main house so they can be sold at the Wakimoto market tomorrow. My wife finally got out of bed today. She cooked and did some sewing, etc.

Today's expenditures:

matches	5 sen
brown sugar	5 sen

At the main house: My brother and his wife made sweets together. My mother cut hay for the horse and went out on errands. My brother's wife did the cooking.

According to Mutō's book, *nishikigi* trees [*Euonymus alatus*] are called *tori-tomarazu.*[23] In my village, *megi* trees [*Berberis thunbergii*] are called *torikko*[24] *tomarazu.*

January 19, Sunday

Cloudy but also sunny
Northwest wind
–3°C in the morning, 0°C in the afternoon
Snow accumulation: 2 *sun* (6.1 cm)

Breakfast:	miso soup with dried taro stems in it, *nadazuke,* pickled napa cabbage, *shōyu no mi*
Lunch:	nothing
Supper:	miso soup with green onions (stock made from whale), roasted salted *hatahata, nadazuke*

Today I went to the Wakimoto market to investigate the types of shops and their numbers. I wanted to record the owners' addresses, but they were reluctant to tell me for fear of being charged more for participating in the market. I went home fairly early in the evening and cooked supper, etc. My wife went to the market early this morning to sell burdock roots, but nobody was interested in them—she didn't sell a single bundle.

Today's expenditures:

two pairs of *geta* for the kids	30 sen
sakagasu [*sakekasu*—saké lees]	15 sen
konbu [kelp]	10 sen
whale [blubber]	25 sen
snow burn (*yukiyake*) medicine[25]	10 sen
futoko candy[26]	15 sen
apples	5 sen

sweets	10 sen
shagushi[27]	10 sen
metal screen (*wadashi*)	10 sen

At the main house: My brother minded the sweets shop, and his wife went to the market to do some shopping. She returned early in the evening and cut hay for the horse. My mother cooked and such.

I'll write what I learned about the types and numbers of the shops in the Wakimoto market.

Vegetable shops (*sosai*)	40 shops (9% selling burdock roots)
Cotton and cloth	28 shops
Futoko candy	24 shops (candies that look like human faces—also called *tegoko-ame*—a famous Wakimoto market souvenir)
Eateries	19 shops (*udon, soba,* saké, etc.)
Roasted *mochi*	17 shops (*mochi* made from wheat or rice flour, with sweet bean paste inside, and roasted)
Kelp (*konbu*)	17 shops (finely chopped seaweed for putting in soup)

Figure 11.7. *Kan-mochi,* to be eaten in the spring, drying in the cold air (January 19, in front of the main house).

Metal knives and such	20 shops
General goods	16 shops
Knitting goods	16 shops
Elastic products	16 shops (shoes, clothing, and footwear—basically anything with elastic)
Sweets	10 shops
Salt-preserved foods	12 shops (whale, trout, salmon, etc.)
Fruits	10 shops
Wooden buckets	9 shops
Flowers	9 shops
Geta	8 shops
Fountain pens	9 shops
Ceramics	9 shops
Secondhand kimono, etc.	8 shops
Various small goods/accessories	8 shops
Kitchen pots	7 shops
Butchers	5 shops
Straw products and *kamasu* bags	5 shops
Pine branches and *yujiri*[28] leaves	5 shops (*yujiri* leaves are called *ezonoha* in the village)
Grain winnowers (*mi*)	5 shops
Lacquerwares	5 shops
Hair clippers (*barikan*)	5 shops
Sashimono woodwork	5 shops
Mats, carpets, rugs	4 shops (mainly selling straw *mushiro* mats)
Kimono (specialty shops)	3 shops
Wooden house fittings	3 shops (specialty shops)
Cherry tree bark products	2 shops
Bamboo products	2 shops
Pharmacies	2 shops
Cotton cloth (*momen-gire*)	2 shops
Powder sifters (*kanatoshi*)	2 shops
Kimono, lanyards, collars, etc.	2 shops
Used metal hardware	2 shops
Used Western clothing	2 shops
Glasses	2 shops
Santogo	2 shops (wallet specialty shops)
Skis	2 shops
Hats	1 shop
Roasted sweet potatoes	1 shop
Tabi footwear	1 shop (specialty shop)
Kimono sashes (*obi*)	1 shop (specialty shop)
Laundry hardware	1 shop
Soap	1 shop

Wooden mortars	2 shops
Pencils	2 shops
Electronic goods	2 shops
Scrolls for interior display	2 shops
Fish	1 shop
Secondhand magazines	1 shop
Glass-cutting tools	1 shop
Tools for cleaning pots and pans	1 shop
Chemicals for repairing rubber boots	1 shop

There were far more people at the year-end festival than I've seen in recent years. But even so, the shop owners were very disappointed in the sales figures, which they said were lower than usual.

It was decided that Kanezō, the main house's *wakaze,* will work at the main house again this year. He returned today after resting at his own home for a few days. It's a common custom in the village for those employed on an annual basis to finish their contracts on December 22 or 23 according to the old lunar calendar, and then start again on December 25. Kanezō was employed for eight 4-*to* (72.2 L) *tawara* of rice per year last year at the age of eighteen, but I heard that this year he'll get a raise, and receive 11 *tawara* plus 30 *sen* in cash per month.

January 20, Monday

Snowstorm
Northwest wind
–5°C in the morning, –4°C in the afternoon
Snow accumulation: 2 *sun* (6.1 cm)

Breakfast:	*sakagasu* miso soup with *konbu* in it, *nadazuke,* pickled napa cabbage
Lunch:	nothing
Supper:	green onion and whale [stock] miso soup with *konbu* in it, roasted salted *hatahata,* pickled napa cabbage and *nadazuke*

Early this morning, I carried bundles of green onions on my back over to the main house, and loaded them and the unsold burdock roots from yesterday onto a sled in preparation for taking them to sell along the way to Funagawa. My wife headed out, pulling the sled behind her, and didn't return until evening. I did the post-breakfast cleanup, and washed and bundled green onions the rest of the day, and I cooked and such.

Today's income:

burdock roots (10 *kan* (37.5 kg))	2 yen
68 green onion bundles	1 yen 20 sen

At the main house: Kanezō cut hay for the horse early, and after breakfast he wove ropes for daily use. In the afternoon, he loaded firewood onto the horse-drawn sled at the place where the woodpile is located, and brought it back to the main house. My brother's wife minded the shop. My brother went to Akita City on an errand. My mother washed rice in preparation for making *mochi* tomorrow, and she cooked and such as well.

January 21, Tuesday

Cloudy
West wind
–3°C in the morning, 2°C in the afternoon
Snow accumulation: 1 *sun* (3 cm)

Breakfast:	*sakagasu* miso soup with *konbu* in it, *nadazuke*, pickled napa cabbage, *shōyu no mi*
Lunch:	*mochi, nadazuke*
Supper:	miso soup made with whale stock with dried taro root pieces in it, roasted salted *hatahata, nadazuke*, pickled napa cabbage

Old Yoshida Suekichi died today.

Before breakfast, I pulled out some cabbages I'd stored and prepared them for being taken out to sell. After breakfast, I washed green onions and bundled them, and then went to the main house to pound *mochi*. I pounded twice in order to make *mochi* for my own family, made a pair of straw boots, and then came back home. In the afternoon I pulled out some burdock roots and used a knife (*magiri*) to shave off their skin. Then I put them in water to soak. I collected urine, and then I made supper. My wife went out to sell vegetables, and she returned in the evening.

Today's income:

50 green onion bundles	1 yen
burdock roots (1 *kan* (3.75 kg))	20 sen
cabbages (1 *kan* 400 *momme* (5.25 kg))	25 sen

At the main house: They pounded *mochi* from the early morning. Kanezō did the pounding. My brother's wife lit the fire, steamed the rice, and did *anjō* [turned the *mochi* inside the mortar] and such. My mother kneaded the finished *mochi*, and made individual *mochi* cakes, etc. The *mochi* pounding ended during the morning, and after that Kanezō made more rope for daily use. My brother's wife and my mother got started on their New Year's preparations. In other words, they scraped burdock roots clean and cooked and such. My brother minded the shop.

January 22, Wednesday

Cloudy but also sunny
Northwest wind
–4°C in the morning, 0°C in the afternoon

Breakfast:	*sakagasu* soup with *konbu* in it, side dishes: same as every morning
Lunch:	nothing
Supper:	green onion soup, side dishes: same as yesterday evening

I got up early and pulled out more cabbages and prepared them for selling. After the post-breakfast cleanup, I kneaded the rabbit pelts and then I went to another person's house to collect night soil. In the afternoon I cleaned up inside my house. I lifted all the *mushiro* mats, picked up all the tools that were lying around, dusted off the books and shelves and organized the books, and then dusted the ceiling and swept up all the dust and everything. Then I beat off the *mushiro* mats and put them all back in place on the floor, and put away all the cleaning tools, and that was it for the house cleaning. Later, I kneaded the rabbit pelts twice and nailed them to a board. Then, I cleaned out the rabbit hutch. My wife went out to sell vegetables, and she returned in the evening.

Today's income:

cabbages (4 *kan* (72.2 kg))	80 sen
31 green onion bundles	70 sen

Today's expenditures:

spending money for my wife	1 yen 40 sen
tofu (1 block)	5 sen
soybean curds (*okara*)	10 sen
candies	5 sen

At the main house: Kanezō cut hay for the horse from the early morning, and he spent the rest of the day hauling compost—out to the paddies at Bishamon. My brother's wife and my mother spent the entire day on New Year preparations. My brother was in the shop, making sweets and such.

One recent evening, an old man told me that there are certain customs in the village that differ from household to household. For example, it seems that the people of *Sankurō Magi*[29] clean up all the soot in the house (do *susuhagi*) just after they finish making their New Year *mochi*—but always on that same day. Also, they don't grow sesame. Furthermore, the members of *Sugawara Magi* don't pickle plums (make *umezuke*). Also, at the Takase household,[30] they always give two big round *mochi* (*mejona*) to each Namahage[31] that visits on New Year's Eve, and they don't grow pumpkins. That's what I heard from the old man.

January 23, Thursday

Snowstorm
Strong northwest wind
–6°C in the morning, –6°C in the afternoon
Snow accumulation: 2 *sun* (6.1 cm)

Breakfast:	cabbage miso soup, side dishes: same as every day
Lunch:	nothing
Supper:	I'll explain later.

I went into the village to buy stuff, and then after getting home I cooked rabbit. Then, I cleared some snow off the roof and cleaned off the rabbit pelts that I kneaded yesterday. My wife cooked celebratory New Year's Eve food all day. Tonight's supper: rabbit, tofu and burdock root soup; roasted cod with boiled cabbage and soy sauce for flavor; squid sashimi; *debu; nadazuke,* etc. Also, late, we put *udon* and *soba* in the rabbit and tofu soy sauce soup broth and ate it. The side dishes were the same as for supper.

Today's expenditures:

a bucket	30 sen
cod and squid	30 sen
soy sauce	21 sen
udon noodles	24 sen

At the main house: Kanezō cut hay for the horse from the early morning. And today, he went to serve as a *dami wakaze*[32] for old Suekichi's funeral. He returned in the afternoon, and cut more hay for the horse, cleaned out its stall—removed all the hay and stuff—and washed the horse, and finished early. My brother's wife and my mother continued their New Year food preparations all day. My brother made sweets. In the night, I went to the main house and prepared the *wakamizu oke* (and all the decorations).[33] This means taking care of the bucket itself and all the other necessary items and trimmings, like the ladle and the special white pine branch, the *yuzuri* (*ezonoha*) leaves, dried fish, *mochi, konbu,* charcoal, etc., and setting it all up for the New Year. Normally, the household head does this, but I did it instead.

January 24, Friday

(January 1 according to the old calendar)
Cloudy but also sunny
–4°C in the morning, 0°C in the afternoon
Snow accumulation: 2 *sun* (6.1 cm)

Breakfast:	leftover *soba* and *udon* from last night, *debu,* miso-preserved eggplants
Lunch:	roasted *mochi* with sugar, *nadazuke, debu,* etc.
Supper:	miso soup with rabbit meat and green onions in it, squid sashimi, *debu, nadazuke,* etc.

Since today is New Year's Day (*gantan*) according to the old lunar calendar, I took the whole day off. But I did see to the rabbits in the morning. My wife also took off from the cooking and such.

At the main house: Same as for us. My brother only went to the temple and to the shrine to pray. My mother and my brother's wife cooked and such. In the morning, Kanezō took the horse down to Wakimoto beach, along with other young fellows of the village, to let it run around.

January 25, Saturday

Cloudy, snow later
Northeast wind
–6°C in the morning, 0°C in the afternoon
Snow accumulation: 5 *sun* (15.2 cm)

Breakfast:	*sakagasu* miso soup with *konbu* in it, *nadazuke,* burdock *debu,* pickled eggplants
Lunch:	roasted *mochi* in soy sauce–based soup with rabbit meat in it, side dishes: same as morning
Supper:	We ate rice topped with ground Chinese yam (*tororo meshi*) at the main house.

Yoshida Yasuke's wife gave birth to a baby boy.

I pulled out some cabbages for my wife to take with her to her natal household this morning. Then I went to the main house and made two pairs of straw boots and came home. I took off in the afternoon. My wife went to her natal home today to stay for a while. Today is a day when the wives normally return to their natal homes to visit, so the streets were busy with all the traffic.

At the main house: Kanezō cut hay for the horse early. Later in the morning, he rebuilt the sled for a person to pull, and then he used it to haul soil from the sandy cliff (*mama*) and he dumped that into the hole at the compost pile site. He took off in the afternoon. My brother's wife returned to her natal home. My brother minded the shop. My mother cooked and such.

January 26, Sunday

Sunny but also cloudy
Strong southwest wind
2°C in the morning

Breakfast:	*sakagasu* miso soup with *konbu* in it, pickled napa cabbage, burdock *debu*
Lunch:	roasted *mochi*—eaten with sugar
Supper:	cold rice with hot water poured over it, roasted cod (*tara*) with soy sauce for flavor, burdock *debu,* etc.

I made breakfast and then ate, and after that I roasted *mochi* and put brown sugar inside them—preparation for lunch. Then I went to Funagawa for the Oga Public Oratorical Contest (Oga Minshū Yūben Taikai) to listen to the speakers.

At the main house: Kanezō cut fodder for the horse in the early morning. After that, he wove ropes until about eleven o'clock. Then, he took off and went to Funagawa. My brother was in the shop. My mother went out in the afternoon. My brother's wife returned today and cooked and such.

This year, five households in my village are employing laborers [*wakaze*] on an annual basis. Yoshida Unzō's family's young fellow is twenty years old and is paid ten 4-*to* (72.2 L) *tawara* of rice plus two extra *to* (36.1 L). The payment for my natal household's young fellow [Kanezō] I already explained before. Yoshida Saburobē's family has one who is twenty-three years old and who receives 13 *tawara* and another who is twenty-seven and receives 11 *tawara.* Satō Sonoji's family pays their twenty-year-old young fellow 11 t*awara,* and Satō Shinnosuke's family pays their twenty-one-year-old young fellow ten *tawara* and their seventeen-year-old young fellow five *tawara.* At least, this is what I heard. The young fellows at Unzō's house and at Sonoji's house are both from Momokawa, and so is the young fellow at my natal home. Saburobē's family's young fellows are both from this hamlet—Ōkura, and one at Shinnosuke's house is from Iinomori Village, and the other is from Hizume Village.

January 27, Monday

Snowstorm
Strong northwest wind
–3°C in the morning, –3°C in the afternoon
Snow accumulation: 2 *sun* (6.1 cm)

Breakfast:	*sakagasu* miso soup, burdock *debu, nadazuke,* etc.
Lunch:	roasted *mochi* with sugar, side dishes: same as morning
Supper:	rabbit meat and green onions cooked in soy sauce, side dishes: same as morning

I cooked for myself, and then went out on errands. I went to my old school to hear Nishida Kosakukan speak. There was also a household heads' committee (*koshukai*) meeting today, but I didn't attend.

> At the main house: Kanezō cut hay for the horse early, and wove ropes again after breakfast. He took off in the afternoon. My mother didn't come back yet. My brother's wife cooked and such.

January 28, Tuesday

Sunny, later cloudy, and later, snow
South wind
–3°C in the morning, 8°C in the afternoon

Breakfast:	miso soup with dried taro stems in it—and with *konbu,* too—*nadazuke,* burdock *debu,* etc.
Lunch:	roasted *mochi* with brown sugar, burdock *debu*
Supper:	I ate at the main house.

I cooked and such, and then I prepared a package to send to someone and went to the station to send it. I took off in the afternoon. In the evening, I went to my old school to see a motion picture. The Wakimoto Dekasegi Laborers' Union and the local credit union bought it and showed it for free for their members' entertainment.

Today's expenditures:

fee for mailing package	40 sen
social expenses	10 sen

> At the main house: Kanezō cut hay for the horse early, and wove ropes after breakfast. He took off in the afternoon. He said his throat was sore so he went to bed. My brother's wife sewed while minding the shop. My mother cooked and such.

January 29, Wednesday

Cloudy
Southwest wind
–3°C in the morning, 0°C in the afternoon
Snow accumulation: 1 *sun* (3 cm)

Breakfast:	miso soup with dried taro stems and whale in it, plus some *konbu, nadazuke*

Lunch:	rabbit meat with tofu cooked in soy sauce, with roasted *mochi* in it, *debu*, etc.
Supper:	roasted *mochi* with sugar and with soy sauce, *nadazuke*, *debu*, etc.

I cooked, and then went to the Wakimoto post office to order some seeds and pay for them. Then I picked some *seri* and *taranome*[34] for making *nanakusa jōnimochi*[35] tomorrow. It'll be used at the main house. Then I went to Funakoshi to get a clock fixed. Of course, I fed the rabbits, but I do that every day.

Today's expenditures:

train fare	14 sen
postage stamps	6 sen
green onion seeds (3 *gō* (541.2 ml))	2 yen 40 sen
seed peas (4 *gō* (721.6 ml))	52 sen
burdock seeds (2 *gō* (360.8 ml))	40 sen
tomato seeds (4 *shaku* (72.2 ml))	50 sen
cabbage seeds (2 *shaku* (36.1 ml))	35 sen
daikon seeds (1 *gō* (180.4 ml))	50 sen
postal order fee	10 sen

At the main house: Kanezō cut hay for the horse early. After breakfast, he wove a 2.5 *hiro* (3.75 to 4.5 m) *tatenawa* rope for [making] rice *tawara*. He also made straw boots, etc. My brother's wife minded the shop, and she and my mother cooked together, etc.

January 30, Thursday

Snowstorm
Northwest wind
–6°C in the morning, –6°C in the afternoon

Breakfast:	potato miso soup with *konbu* in it, *nadazuke*, burdock *debu*
Lunch:	rabbit meat and tofu cooked in a soy sauce–based soup with roasted *mochi* in it [*ozōni*], *nadazuke*, burdock *debu*, etc.
Supper:	leftover soup from lunch and *nadazuke* and burdock *debu*

Yoshida Shinichirō received a bride (*yome*) today.[36]

I cooked and such, and during the morning I wove ropes for daily use. In the afternoon, a friend came over to visit, so I stayed at home and read over my manuscript some.

Today's expenditures:

tofu (1 block)	6 sen
kerosene (1 *shō* (1.8 L))	30 sen

At the main house: Kanezō cut hay for the horse in the early morning, and wove more *tatenawa* ropes like yesterday after breakfast. He took off in the afternoon. My brother's wife sewed and such while minding the shop and my mother cooked, etc.

January 31, Friday

Blizzard
Northwest wind
–8°C in the morning, –7°C in the afternoon

Breakfast:	*sakagasu* miso soup with *konbu* in it, *nadazuke,* burdock root *debu*
Lunch:	nothing
Supper:	rabbit meat and *konnyaku* cooked in soy sauce, side dishes: same as morning

I did the cooking and such, and then I worked on the rabbit pelts that I put in water to soak the other day—from the *urauchi* stage to the alum-soaking stage. I also shoveled snow—there's so much it seemed like the house would get buried, so I really worked up a sweat clearing everything.

Today's expenditures:

konnyaku (1 block)	2 sen

At the main house: Kanezō cut fodder for the horse early, then he spent the rest of the day making the straw end covers (*sadara*) for *tawara.* My brother's wife sewed in the morning, and went to the school for a Mothers' Association[37] meeting in the afternoon. Mrs. Wazaki Haru[38] came and spoke. My brother's wife also cooked and such. My mother went to my younger sister's house.

You make the end covers (*sadara*) for *tawara* on top of a stone *fugushi* mortar. They're the perfect size for that. You can make one with a bundle of straw that's just about the limit for a man to grab with one hand. You start by pounding the straw some in the middle to make it soft, and then you tie it (at the middle) and push it down over the lower part of the mortar, centered on the little stick in the center (*hojo*), and flatten out the lower ends of the straws so they cover the surface of the mortar. And you make sure the part you tied fits onto the *hojo.* So, you flatten out the lower parts of the straws, and then you do the same for the upper parts (see Fig. 11.8).

Once you've spread and flattened the straws well, you can get started making the *sadara.* First, you stand on it, placing both of your heels on the *hojo* at the center of the mortar. Then you start making it by working with the straws beyond your toes. Like I wrote above, it will be about the size of the stone mortar, so you grab five or six straws with each hand and twist them two times, and,

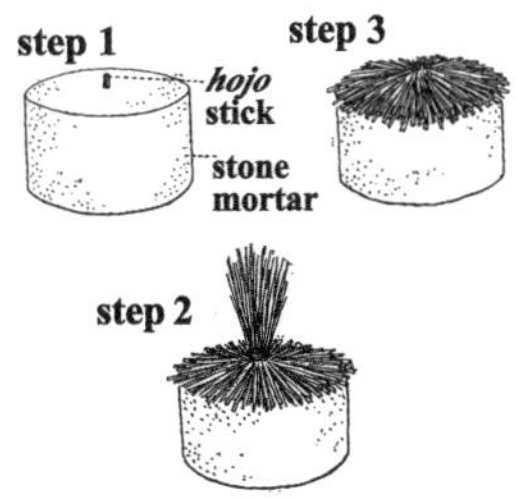

Figure 11.8. How to use a mortar to make a *tawara* end cover (*sadara*).

leaving three or four straws sticking out straight, you fold the straws in your right hand over the straws in your left hand, and hold them down with your toes. Then you grab the next little bunch of straws with your right hand and take them between the ring finger and pinkie of your left hand and then pass the straws you were holding between the thumb and forefinger of your left hand to your right hand and fold those over and hold them with the toes of your right foot. And you just go all the way around like this. Finally, when you get to the end you take the straw you left sticking out straight and use that to bind up the straw you have in your hand at the end.

Chapter 12: February 1936

February 1, Saturday

Snowstorm
Strong northwest wind
–8°C in the morning, –6°C in the afternoon
Snow accumulation: 5 *sun* (15.2 cm)

Breakfast:	burdock root miso soup with *konbu* in it, *nadazuke*
Lunch:	nothing
Supper:	same soup as morning, roasted *hatahata, nadazuke,* pickled napa cabbage, burdock *debu,* etc.

Again, today, I stayed indoors and read because of the snowstorm. I went [to the main house] to fetch Fusae in the evening, because my wife came back home a bit late. She brought many *mochi* that she'd received with her. I did all the cooking again today.

At the main house: Kanezō cut hay for the horse early, and again spent the day making *sadara.* My brother's wife cooked and minded the shop. My brother went out and about to play. My mother didn't return today, either—my sister is in bed with a cold, so it seems that my mother did the cooking at her home in place of her.

Today, a new bride (*yome*) arrived at the household of Sugawara Sukezō. She is for his second son, who is generally known as "Onchako." She's from Futto Village.

February 2, Sunday

Snowstorm, later just cloudy
Strong northwest wind
–2°C in the morning, 2°C in the afternoon
Snow accumulation: 7 *sun* (21.2 cm)

Breakfast:	saké lees (*sakagasu*) miso soup with *konbu* in it, *nadazuke,* pickled napa cabbage, burdock *debu*

Lunch:	roasted *mochi* with brown sugar, side dishes: same as morning
Supper:	miso soup with daikon and a cod head cooked in it, side dishes: same as morning

I started clearing snow early this morning. We've had heavy snowstorms since the day before yesterday, and it was still the same this morning. It seemed like this little shack might get buried in the snow. I thought there was no room for complacency, so I gave it everything I had, and by nine o'clock I had finally cleared a path. This is the snowiest winter in my seven years of mountain life, and the storms are really powerful. Around nine o'clock, the storm started to let up, so I went out to fetch a *tara* [an angelica tree: *Aralia elata*]. The tree isn't for us—it's to be used at the main house. Tomorrow is January 11 according to the old lunar calendar. It's the day to cut a *tara* tree, some pine for the *kadomatsu*[1] and willow and such for *wakagi mukai*[2] for *koshōgatsu*.[3] But in recent years, *tara* trees have become kind of scarce in the village, and without them people will have a problem. So, instead of following the custom of cutting these on January 11, most households now go out and compete with one another to cut them a day or two early. And nowadays, some families have even stopped using *tara* trees for *koshōgatsu*. Not only this, but some houses have even stopped setting up the traditional mortar-bucket-pine arrangements on New Year's Eve. In other words, you could say that villagers are gradually starting to cast aside traditional village ways and mimic the lifestyles of city folk. Some locals say that even if they follow these time-consuming traditions and customs, nothing really changes anyway, and many say there's no point because they just remain poor all the same. But it can't be helped—this is just the reality of farming village life today. So, this morning, I cut a *tara* tree and took it to the main house. In the afternoon, I pulled *migo* out of rice straws at the main house. These will be used for weaving a *kera*.

To pull the *migo* out of the straws, you slide the straw you tied the bundle with down toward the ends a bit and lay the bundle down on the dirt floor of the workspace. Then, you fold the upper parts of the straws, where the *migo* are, all the way back over the tied part, just above the point where you tied it. Then you put your right foot on the tied-up base part and pull very hard with both hands on the *migo*, and once they've come out you put your left foot down on the straws that have had the *migo* pulled out, and then fold the next part back and hold the base parts down with your right foot again and pull the *migo* out with both hands, and you just keep going like this. This is the way it's usually done these days. But my father, and the older generations, did it a totally different way. They used to tie up the top parts—where the *migo* are—with straw, and lay a board on those parts, and stand on the board with both feet and pull little by little on the tied-up straw bundle base. So, the *migo* parts would remain on the ground at their feet.

Anyway, once I had pulled out all the *migo*, I pounded them to make them soft. Then I wove those *migo* into strings in preparation for making a *kera*. And then I made the *amae*, which is the neck and shoulder part [of the *kera*]—the part you always make first. To do this, you start by weaving a rope with straw, and you bend it over in the center. In local speech, people say *nihonka ni suru*.[4] Then, at a point 5 or 6 *sun* (approx. 16.7 cm) from the fold, you tie the two parts in a knot called *emomusubi*. And then, finally, you weave it with *migo*. First, you put the big toe of your right foot through the loop you've made by tying the rope in an *emomusubi* knot. Then you weave *migo* through the two parts of the rope, moving down away from the loop, from the knot. You grab a *migo* with your right hand and pull both parts of ropes with your left hand.

Then, you put the *migo* behind the two parts of the rope so that the base part sticks out 3–4 *sun* (approx. 10.6 cm) toward your left hand. Then, you pinch the *migo* and rope parts with the thumb and forefinger of your left hand and hold them tight. Then, you fold the end part of the *migo* with your right hand over the rope parts leftward. While doing that, you hold everything in place with your left thumb, and at the same time, quickly let go of the base part of the *migo* that you were holding with your left thumb. Then, with your right hand, you grab the base part of the *migo* you have been holding with your left thumb and fold it back over the part you've been holding with your left thumb to the right and run it between the two ends of the rope. Then, you let go of the *migo* with the right hand and pull the base part of the *migo* that you have just put through the two parts of the rope hard from under the rope toward your toe. You keep going just like this until you've reached a length of about 1 *shaku* 8 *sun* (54.5 cm). Once you've done that, you tie another *emomusubi* knot to hold it all together. This is how to make the *kera* piece that's known as *amae* (see Fig. 12.01). Next, you move on to making the actual *kera*.

Next, I'll explain about how to tie an *emomusubi* knot.

First, you hold the two parts of the rope with your left hand and pull. You put the left one on top of the right one and hold them. Then, you hold the left one with your right hand and make a loop around the right one and pinch it with the thumb and forefinger of your left hand. Then you grab the right one, which is going through the loop, with your right hand, and bring it under your left hand, and let it stick out leftward, and then grab it again with your right hand. Next, you stick that part back through the loop, and let go of it with your right hand, and again grab that part that's now going through the loop with your right hand and pull hard on it, and then hold it down with the middle finger of your left hand. And finally, if you pull nice and hard on the end of the rope part that you used to make the loop in the first place, it will make a good, tight knot (see Fig. 12.02).

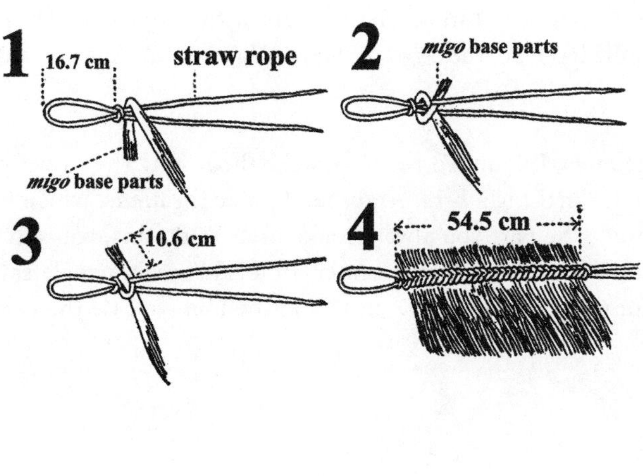

Figure 12.01. How to make the *amae* part of a *kera*.

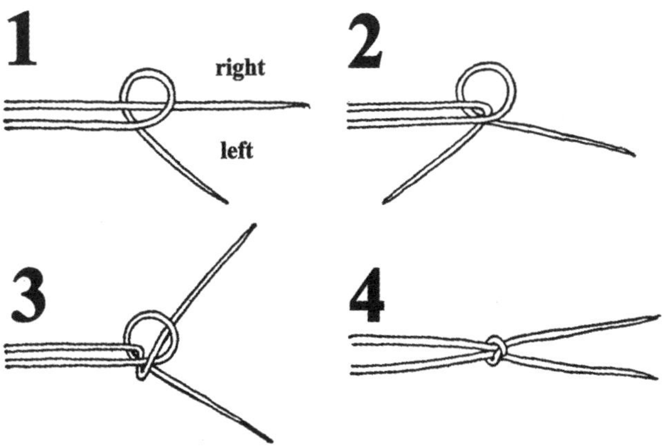

Figure 12.02. How to tie an *emomusubi* knot.

My wife pulled some daikon out of storage and made soup. In other words, she cooked pretty much all day.

Today's expenditures:

brown sugar	10 sen

At the main house: Kanezō cut hay for the horse early. After breakfast, he went up into the hills for *wakagi mukai* preparations. I had already cut a *tara* tree, but he still needed pine for the *kadomatsu,* and some willow, so he cut those and took them back to the house. Typically, villagers will cut an entire small pine tree or the upper part of a larger pine tree. And, if someone with no mountain land cuts a tree on someone else's property without permission, the landholder will make a big fuss. In the afternoon, he wove the main rope (*motozuna*) for a horse saddle and made straw boots. In the past, a *wakaze* like Kanezō would have been expected to make straw boots for himself to wear during his personal downtime (*shigoto no agama*)—that is, after supper, during his own free time—but today they generally do this during their regular work time. My mother didn't return today. My brother went out on errands, and his wife minded the shop and cooked and such.

February 3, Monday

Cloudy
West wind
–2°C in the morning, 2°C in the afternoon

Breakfast:	daikon miso soup, side dishes: same as yesterday morning
Lunch:	roasted *mochi* eaten with brown sugar, roasted *hatahata,* other dishes: same as morning
Supper:	miso soup with dried taro stems and whale, etc., in it, side dishes: same as lunch

Figure 12.03. Young men at the Ōkura meeting place (February 3). From Yoshida's personal collection: substituted for similar photo used in 1938 book.

I cleared snow off the roof and around the house early. In the afternoon I didn't work—I went out to take some photos. My wife did laundry and cut *mochi* into thin pieces and tied them up with straw for making *hoshi mochi.* She also did some sewing.

On the afternoon of January 11 according to the old lunar calendar, during *koshōgatsu,* young men of the village always gather and make Namahage masks. The fellows in the photo belong to section (*ku*) one of the hamlet (see Fig. 12.03).

Today's expenditures:

salt	10 sen

At the main house: Kanezō cut hay for the horse early, as always. After breakfast, he wove a *toshina* rope for the *koshōgatsu* celebration. He took off in the afternoon. The villagers always take off from work starting on the afternoon of January 11 according to the old lunar calendar. Since the old days, they have taken off this day especially for making Namahage masks. In other words, people have always told children that the adults are going to fetch the Namahage on this day. My brother went to a communal labor meeting and to a personal group meeting. His wife cooked while minding the shop.

February 4, Tuesday

Sunny
–3°C in the morning, 4°C in the afternoon

Breakfast:	saké lees (*sakagasu*) miso soup with kelp (*konbu*) in it, side dishes: same as every morning
Lunch:	same as yesterday
Supper:	same as yesterday evening

In the morning, I worked on making the *kera*. In the afternoon, I went to pick up the clock I got repaired the other day. My wife sewed all day, but also ran errands, and cleared some snow and dug up burdock roots and shaved off their skin.

Today's expenditures:

clock repair	80 sen
train fare	14 sen

At the main house: Kanezō did his usual morning work. And again, he worked on weaving a *motozuna*. My brother's wife prepared for making *mochi* again tomorrow, washed rice, and watched the shop and cooked. My brother went out on errands.

My *kera*-making work for today took me pretty much up to making the body of the *kera* itself (*hon-zukuri*). First, I wrapped the *amae* part I made the other day around a specially shaped wooden board, as the diagram shows (see Fig. 12.04).

The board is 6 or 7 *bu* (approx. 2 cm) thick, and I carved a groove running around the edge of the big, curved part that's 2 or 3 *bu* (approx. 8 mm) deep. The *amae* fits into that to prevent it from moving around. So, if you wrap the *amae* around the board and tie it tightly, the end parts and base parts of the *migo* become separated. I kept the end parts of the *migo* on top and fluffed them up a bit. Then I wove some *irimigo* into the *amae* in the parts of the wooden board that stick out. In other words, I wove more *migo* into the *amae* in those spots.

To do this, you separate three or four *migo* from the others with your right hand and grab them with your left finger. Then you grab one of the *amitsura* [weaving ropes] with your right fingers and put the *migo* you're holding with your left fingers in, and then grab five or six new *migo* with your

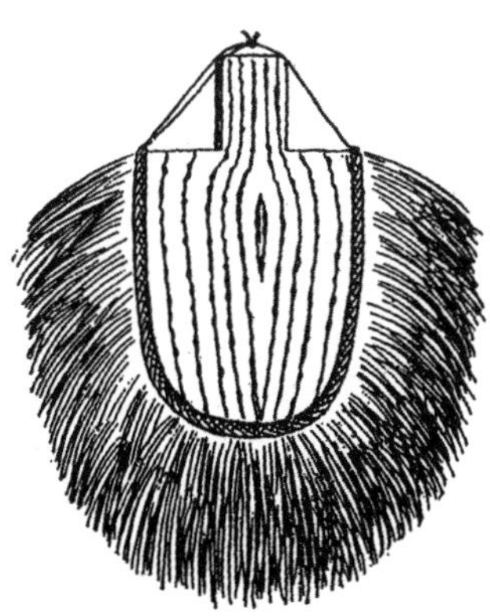

Figure 12.04. The *amae* part of the *kera*, wrapped around a wooden board.

right hand and weave them in, too, and then, the *amitsura:* you move the right one to your left and pull the left one over to your right. Then you tie them up tightly with the fingers of both hands. When you do it like this, the base parts of the *irimigo* (the *migo* you are adding) are pointing away from you and the end parts are pointing toward you. And, as before, you weave the base parts so that 3 or 4 *sun* (approx. 10.6 cm) are sticking out. You do this twice. The third time, you fold the base part of the *migo* you added first under the base part of the *migo* you added second and weave it together with the one you added third. Then you add more *irimigo* and keep going. So, the base part of the second *irimigo* is woven together with the base of the fourth, and the base of the third goes with the base of the fifth, etc., until you reach the ending point. And once you've finished with the *irimigo,* you go back around the shape of the wooden board, leaving about 2 *sun* (6.1 cm) at the base of each *migo.*

Next, you flip the entire board over, so that the part you've woven faces downward. In other words, the end parts of the *migo* of the *amae* are facing upward. Then you grab the base parts of the *migo* and weave *irimigo* in as I explained above. And you weave the *irimigo* in at a point about 1 *sun* (3 cm) from the base of the *migo* in the same way as you did before. If you do this, you'll have two distinct layers of *migo* (on the *amae*). If the *migo* on the *amitsura* get too short, you just add more and weave them in, and you must completely remove the stamens and pistils from all of the straws and cut and discard about 6 *sun* (18.2 cm) of the base of each straw, and pound them until they're soft. Then, you prepare a thin bamboo stick, 4 or 5 *sun* (approx. 13.6 cm) long and 1 *bu* 5 *rin* (4.5 mm) in diameter. You have to cut it at an angle from a point about 2 *sun* (6.1 cm) from the end. Now you're finally ready to start making the body of the *kera* (ready for *honzukuri*).

Today, the head of Akita Library,[5] Ono Yuzuru, came to speak at the invitation of the Wakimoto Young Men's Association.

February 5, Wednesday

Occasional snow
North wind
–3°C in the morning, 2°C in the afternoon
Snow accumulation: 3 *sun* (9.1 cm)

Breakfast:	miso soup with salted vegetables we received from my wife's natal household in it, side dishes: same as every day
Lunch:	same as yesterday
Supper:	miso soup with dried taro stems and whale in it, side dishes: same as yesterday evening

Today I stayed in bed all day because I caught a cold. My wife sewed and cooked and such.

Today's expenditures:

brown sugar	20 sen

Figure 12.05. Pounding *mochi* at Yoshida Kintarō's house (February 4, 10:00 a.m.).

At the main house: They started pounding *mochi* for the *koshōgatsu* celebration early this morning (see Fig. 12.05). My brother's wife and my mother and Kanezō did the pounding together. My mother got home last night, but she left again in the afternoon. My brother's wife cooked and such. Kanezō made a pair of straw boots after finishing the *mochi,* and then he wove more *motozuna* ropes. My brother went out on errands again in the afternoon.

February 6, Thursday

Cloudy with some sunshine
Northwest wind
–1°C in the morning, –1°C in the afternoon
Snow accumulation: 1 *sun* (3 cm)

Breakfast:	daikon miso soup, side dishes: same as every day
Lunch:	nothing

Supper:	miso soup with dried taro stems and whale in it, side dishes: same as always

I was in bed all day again. I finally managed to get up in the evening, and smeared an egg yolk/ starch mixture on the rabbit pelts, etc. My wife sewed, cooked and such all day.

Today's expenditures:

apples	5 sen

At the main house: Kanezō cut hay for the horse early, and spent the rest of the day making straw horseshoes, making leg gaiters (*kyahan*), and weaving ropes for the horse.

February 7, Friday

(January 15 according to the old calendar)
Cloudy with some sunshine
–1°C in the morning, 4°C in the afternoon

Breakfast:	same as yesterday morning
Lunch:	side dishes: same as always[6]
Supper:	I'll explain below.

I felt a little better today, so I got up. I put fresh dried grass and weeds in the rabbit hutch, shoveled some snow, and dug up some daikon for feeding the rabbits. Then, I went into the village to buy some things for the *koshōgatsu* celebration. On the way home, I stopped and made two pairs of straw boots at the main house. My wife sewed, did laundry, and cooked and such. Today is January 15, the Little New Year's Eve (*koshōgatu no toshitori*), according to the old lunar calendar. It's the night for Namahage to visit people's houses (see Fig. 12.06). I went to the main house to have a look.[7]

For supper we ate miso soup with *shinjo*[8] and tofu in it. There also was flounder *sashimi,* and roasted halibut and boiled cabbage, both of which we ate with soy sauce for flavor. We also ate *hatahata-zushi*[9] that we received from the main house. And we ate fried tofu that we got from my wife's brother's household, with soy sauce. There was also burdock *debu,* etc.

Later, we had a soy sauce–based soup with *shinjo* and tofu in it, and we poured the broth over *udon* noodles and ate them. Everything else was the same as supper.

Regarding the New Year customs and the Namahage, they're discussed in detail in my [1935] book, *Kanpūsanroku Nōmin Shuki.*

Figure 12.06. Two Ōkura-style Namahage. From Yoshida's personal collection: not included in 1938 book. The September 1934 Attic Museum trip to Wakimoto, during which this photo was taken by Takahashi Buntarō, was documented on motion picture film by Shibusawa Keizō. Upon viewing the film in 2022, Saburō's daughter Fusae surmised that the men dressed as Namahage were members of the Attic Museum group.

Today's expenditures:

udon	23 sen
fish (*shinjo,* halibut, flounder)	25 sen
soy sauce (2 *gō* (360.1 ml))	6 sen
tofu (1 block)	6 sen

At the main house: Kanezō made straw boots. After breakfast, he cleared all the hay and dung out of the stable and piled it up outside. He also cut a 5 or 6 *sun* (approx. 16.7 cm) piece of *tara* tree and split it down the center. One piece he shaved and cut to make one end pointed, and on it he wrote "the former thirteenth-month auspicious day"[10] with brush and ink. In normal years, one would write twelve—not thirteen—but this a leap year. Then he cut a 1 *shaku* 5–6 *sun* (47 cm) piece of *tara* tree and shaved the bark off one end of it, and repeated the process, to make a pair of chopsticks for handling charcoal and pieces of wood. One pair is needed at each hearth in the house. Also, he got the *toshina* rope he made the other day and shoved some white pine leaves, false daphne leaves (*yuzuriha*), charcoal, *mochi*, a *minjashi*,[11] kelp (*konbu*) and such—all cut small—into the rope itself. Then he took some willow and the split piece of *tara* wood with writing on it and stuck those into the rope as well (see Fig. 12.07).

After that, Kanezō made a *kadomatsu* decoration. And he made *mochi* offerings to all the important tools throughout the house, and wrapped a length of *toshina* rope, complete with the above decorations, around each such tool or important item (see Figs. 12.08, 12.09, 12.10, and 12.11). He also cut horse fodder and such. He went to the agricultural warehouse to buy some charcoal. My brother minded the shop. My brother's wife and my mother did the New Year preparations. And, since it was the night for Namahage to come, my mother cooked *kenoko jiru*. *Kenoko jiru* is a soup containing *sashidori* [Japanese knotweed], *enyo* [*Angelica ursina*], *warabi* [*Pteridium aquilinum*],

Figure 12.07. A stone mortar, with the *toshina* tied around it, and with *mochi* on top (February 8, at the main house). The piece of wood on which Kanezō wrote, and other items, have been stuck into the *toshina*.

Figure 12.08. An old-time *nawakagi* with a *toshina* and *mochi* hanging from it (February 8, at the main house). [A *nawakagi* was a device for hanging a kettle or pot over an open hearth (*irori*).]

fuki [butterbur], and eggplant and such, which have all been preserved in salt, and it also contains tofu and *utsumame* and such all cooked together. *Utsumame* are cooked lumps of ground raw soybean powder.[12]

February 8, Saturday

Sunny, later cloudy
–5°C in the morning, 3°C in the afternoon

Figure 12.09. A *toshina* and a bucket of water adorn a stone mortar for the annual ritual (February 8, at the main house).

Breakfast:	*kenoko jiru* we received from the main house, side dishes: same as every day
Lunch:	roasted *mochi* with brown sugar, etc.
Supper:	same as morning

Today I didn't work. In the evening I went for the *hido*[13] at the shrine.

Today's expenditures:

tabi	35 sen
sweets	5 sen

At the main house: Everyone took off and rested. My mother went to pray at the temple. My brother's wife cooked and such.

Figure 12.10. A large, overturned iron pot, with a *tsukenawa* and a *mochi* on top (February 8, at the main house).

February 9, Sunday

Rain, sunny later, then cloudy
Southwest wind
2°C in the morning, 4°C in the afternoon

Breakfast:	same as yesterday morning
Lunch:	same as yesterday
Supper:	miso soup with whale and dried taro plant stems in it, roasted *hatahata,* side dishes: same as lunch

In the morning I went to the main house to help them pack up some stuff for sending somewhere, and then I took the package to the station and posted it right away. I took off in the afternoon. My wife, as usual, spent the entire day sewing and cooking. All four of us are troubled now by a cold and a nasty cough (*sabugi*).

Today's expenditures:

postcards	15 sen
apples and *mikan*	5 sen

Figure 12.11. A hoe, and a three-pronged hoe, adorned with *mochi* and a *toshina* (February 8, at the main house).

At the main house: Kanezō cut fodder for the horse early, and after breakfast he cleared snow around the house. In the afternoon, he took off. My brother's wife minded the shop, and my brother went out on some errand. My mother cooked.

February 10, Monday

Cloudy, later sunny
West wind
0°C in the morning

Breakfast:	same as yesterday morning
Lunch:	same as yesterday's lunch
Supper:	basically, the same as yesterday evening

Figure 12.12. The *kadomatsu* displayed on January 16, according to the old calendar, with a *toshina* (February 8, at the main house).

Figure 12.13. A *toshina* and candles, presented to the local New Year Kōshin (Okano) deities (February 8, in front of Mishima Shrine).

I haven't recovered from the cold, so I spent the day in bed. My wife fed the rabbits, did laundry, sewed, and cooked.

At the main house: Kanezō cut hay for the horse and wove ropes. In the afternoon, he took off; he went to my old school to hear the leader of the 16th brigade[14] speak. My brother went to the general assembly of the Agricultural Support Association (Shinōkai). My sister minded the shop, and my mother cooked, etc.

February 11, Tuesday

Sunny, later snowy
East wind
–6°C in the morning, 0°C in the afternoon
Snow accumulation: 5 *sun* (15.2 cm)

Breakfast, lunch, supper: no change from yesterday

In the [early] morning I kneaded the rabbit pelts, and I went into the village on an errand. In the afternoon, I went to a big party hosted by the Young Men's Association (Seinenkai[15])—for the fellows who participated in the Namahage ritual this time, and for those who've come back from *dekasegi* work in other places (see Fig. 12.14). I took some photos there, and I spoke for about ten minutes. My wife cooked all day. In the afternoon, we hosted my mother and my wife's mother, and my wife's grandmother (of her natal household) at our house for a little party in their honor.

Today's expenditures:

fish	50 sen
tofu	6 sen
dried roasted sardines	5 sen
udon	33 sen

Figure 12.14. Young men of Ōkura at the hamlet meeting place. They were celebrating the merging of the hamlet's three Namahage groups, the receipt of gifts on that occasion, and Seinenkai financial support. The party was also for thanking members for their efforts, and for honoring all those who had engaged in *dekasegi* work (February 11).

soy sauce	15 sen
konnyaku	5 sen
saké	20 sen
envelopes	10 sen

At the main house: Kanezō cut hay for the horse in the early morning, and after breakfast he went to Nagane Village on an errand. In the afternoon, he attended the Young Men's Association party. My brother minded the shop. His wife was sick in bed. My mother cooked, and she came over to my house.

February 12, Wednesday

Sunny, later intermittent snow
Northwest wind
–2°C in the morning, 4°C in the afternoon
Snow accumulation: 2 *sun* (6.1 cm)

Breakfast, lunch, supper: same as yesterday

In the morning I cleared snow, and then my crazy neighbor and I cut each other's hair. After that, I went down to the main part of the village to mail something from the post office. I took off in the afternoon. My wife sewed all day.

Today's expenditures:

postal fee	10 sen

At the main house: Kanezō cut hay for the horse in the early morning, as usual, and after breakfast he wove ropes for making a *mokko.* He took off in the afternoon. My brother's wife was sick in bed again. My brother minded the shop. My mother cooked and such.

February 13, Thursday

Cloudy
Northwest wind
–4°C in the morning, 1°C in the afternoon

Breakfast:	miso soup with daikon and tofu in it, daikon *ipponzuke, debu,* pickled eggplants, etc.
Lunch:	roasted *mochi* with brown sugar, burdock *debu,* daikon *ipponzuke,* etc.
Supper:	daikon miso soup, roasted *hatahata,* everything else: same as morning

Early this morning I cleared snow and later in the morning I cleaned rabbit pelts. I prepared lunch early and went over to the main house after eating and did more work on the *kera* I've been making. My wife sewed and cooked and such.

Today I finally got into making the main part of the *kera*. I'll explain the process. So, I already put the *amae* piece on a board and wove both sides. I prepared a piece of bamboo. The rice straws (*wara*) were already pounded and ready, so the task for today was to weave the base parts of the *migo* at a point about 1 *sun* (3.03 cm) from the ends.

I shoved the piece of bamboo between every other *migo* I wove, from the right hand over to the left hand. Then, I grabbed about three rice straws in my right hand and passed them to my left hand and stuck them into the piece of bamboo and pulled the bamboo through the woven rice straws. And so, because the bamboo had rice straws in it, the base ends slid out as I went along. I pressed down on the straws with my hand, and I could pull with my right hand, and the bamboo just slipped out. So, the rice straws and the bamboo separated from each other. And then, I pulled the straws out about 5–6 *sun* (approx. 16.7 cm) and folded them around a strand of *migo* in the center, so the top ends and base ends both hung down. I just kept passing the piece of bamboo through the *amae*, two strands of *migo* at a time, and each time I'd stick several more strands into the bamboo. I'll draw a diagram (see Fig. 12.15).

If you keep going like this, stuffing straws [into the bamboo], the straws go through the part you wove twice each time. That is, the base parts and the end parts of the rice straw strands will be sticking out, making for a total of six strands in each spot. And so, you just weave the straws as you go around the wooden board and weave all the straw ends that are sticking out at the 2–2.5 *sun* (6.8 cm) point. The thing you really have to be careful about here is not to weave the new straw parts together with any of the straw you already wove before. Once you've finished with this weaving, you do the same thing again with the bamboo and straws all the way around again. If you run out of straws, you have to pound more to soften them and use those. Then, you continue weaving. If you weave outward from the base part of the *amae* like this several times, you'll have 5–6 *sun* (approx. 16.7 cm) of the woven part. Next, you pass a piece of bamboo through the strands of straw just as before, at a point about 5–6 *sun* (approx. 16.7 cm) away from the previous spot. Once you reach the end, you again put the bamboo through, with straws, 5 *sun* (15.2 cm) away from the end, and then you stop. After that, you weave straws through it again. Of course, you should start at the part where you last put the bamboo through. But this time, you don't weave the straws through

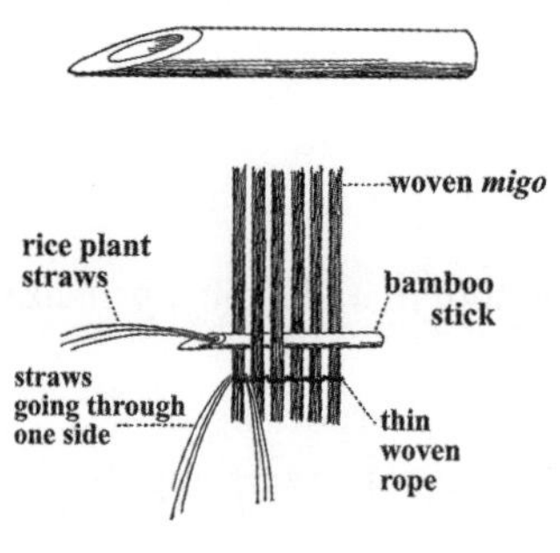

Figure 12.15. Weaving straws and *migo*, with a piece of bamboo, to make a *kera*.

with the same spacing—you have to pass the bamboo through at intervals about 2 *sun* (6.1 cm) apart. So, if you do it like this, you'll weave both sides at about 5 *sun* (15.2 cm) intervals, and your weaving will get tighter and tighter.

If you do it this way, moving toward the parts where you inserted strands of straw, you'll be weaving the straws you wove before twice. I did this to the front and back of the *kera,* weaving the straws while leaving about 5 *sun* (15.2 cm), centering on the places where I'd passed the bamboo through, tightening it on both sides, and the sun went down while I was doing this work.

Today, the Wakimoto Women's Association (Fujinkai) had a speaker; Principal Numata of the Akita Women's High School[16] came and gave a lecture for the association members' training study session.

At the main house: Kanezō cut hay for the horse early, and after breakfast he used a horse-drawn sled to haul sand from the sandpile at Kiyozō's house over to a point along the edge of the road beneath the main house. In the afternoon he used a different sled, which he pulled behind him, to carry that sand out to the nursery paddies (*nashiro*). He spread a layer of sand across each paddy—on top of the snow, of course. My brother went out on some errand. His wife seems to have caught a cold, so she just minded the shop. My mother cooked.

Here, we generally call pots and scythes and other items sold and repaired by tinsmiths *doya.* And we call lewd and lustful people [men] *doya* as well.

February 14, Friday

Cloudy with occasional snow
Northwest wind
-2°C in the morning, 3°C in the afternoon

Breakfast:	miso soup with *egeshi* in it, side dishes: same as every day
Lunch:	same as yesterday
Supper:	daikon miso soup, side dishes: same as yesterday evening

My wife and I did the same things as yesterday, and I finally finished making the *kera.*

Yesterday I tightened it up a bit, by about 5 *sun* (15.2 cm), on each side, and wove it twice. I did it again today. Then I tightened it up more on each side by about 3 *sun* (9.1 cm) and put more straws into it [using the bamboo]. Then I took the *kera* off the board I use for making them. The part I put more straws into has been tightened so much that is now measures 1 *shaku* 4–5 *sun* (approx. 44 cm). Next, I wove a rope about 5 *shaku* (1.5 m) long with several strands of straw. I wove this rope by holding the straws between my hands—centered—and twisting them. Once that was done, I folded it over in half and hooked it around the big toe of my right foot at the midpoint, and pulled it tightly with both hands, and twisted it very tightly (*yori o kaketa*). Then I rubbed it with a wad of *migo* in order to make it smooth and soft (*kokuri o kaketa*). Then I inserted this cord into the *kera* where I'd put straws earlier and wove it through. That is, I put it in there among the straws and wove it in. Then I wove it at a length of about 2 *sun* 5 *bu* (7.5 cm) like before. I didn't try to tighten it up anymore after that—I got it to a size of about 1 *shaku* 4–5 *sun* (approx. 44 cm) wide and then wove straws through it about five more times, leaving 2 *sun* 4–5 *bu* (approx. 7.4 cm) between each pass. And finally, I wove the straws I'd already inserted again twice, at 2 *sun* (6.1 cm) intervals. And I left about that amount of length at the last part I wove and cut all that off with a big cleaver (*nata*) and then I had a nice complete *kera*. (I used a total of about seven bundles of straw to make the *kera*.)

At the main house: Kanezō did exactly the same work as yesterday—that is, he hauled more sand out to the nursery paddies (*nashiro*) and spread it over the surfaces. He also used the sled to haul *godomori*[17] back to the house (from the paddies), but it was frozen hard, so he had to use an axe for splitting firewood to break it up and get it onto the sled. That was the day's work for Kanezō. My brother's wife minded the shop. My mother cooked. My brother was also in the shop today.

February 15, Saturday

Cloudy
Northwest wind
-3°C in the morning, 2°C in the afternoon

Breakfast:	same as yesterday morning
Lunch:	roasted *hatahata, nadazuke,* pickled napa cabbage, burdock *debu*
Supper:	miso soup with taro stems, potato and whale in it, side dishes: same as lunch

I spent the whole day weaving *habaki*.[18] For my wife, it was the same as yesterday.

Today's income:

rabbit pelt tanning fee	25 sen

Today's expenditures:

twine	5 sen
candies	3 sen

At the main house: Kanezō cut hay for the horse early, and after breakfast he collected night soil and spread that across the nursery paddies (*nashiro*). In the afternoon he made paths in the snow for hauling sand out to the Shinden paddies. In other words, he went out and stomped the snow down. My brother went out on some errand, and his wife minded the little shop. My mother cooked, and went out to pick daikon and potatoes, and she made soup with those, etc.

In the old days, people wouldn't make *habaki* during the "master's time" (i.e., the official working hours of the day). Instead, they'd weave them on their own time—primarily in the evenings, after supper—and they'd make many of them, pretty much all winter long. But I'd say that over a decade or so people gradually came to make *habaki* during the "master's time"—during the daytime, that is. After all, *habaki* have always been regular work gear, necessary for working for one's master, so it's stupid to make them only at night. People should proudly make them during the daytime if they need them, some people started saying.

In this way, this position became ingrained in the minds of the young fellows of the village, and the masters (*oyakata*) gave the practice their tacit approval, so the young fellows came to boldly make *habaki* during the daytime on workdays and dedicate themselves to playing on days off and in the evenings. But this didn't last long—you never see anyone making *habaki* during the daytime today. Pretty soon they'll probably just completely disappear from the farming villages. In my case, *habaki* still serve a purpose, so I started making a pair today. I'll explain how it's done.

In the old days, people would use hemp twine, which they always had on hand, year-round, and which came from a plant called *itobo*. These plants grew in people's fields, so they'd just get the twine from the plants. But today, those plants and that twine are nowhere to be seen in the villages. So instead, people now buy a kind of twine made from tree bark at a store and use that.

I already had a bunch of *migo* that I'd pulled out of rice plant stalks, so I just had to pound that a bit to soften it. Then I went and bought some twine to use.

The next thing I did was weave a rope about 5 *shaku* 5 *sun* (1.67 m) long with that twine. As for the method, I must note that there are two types of such ropes: left-woven and right-woven. The way you make the latter is the usual way: when you weave rope with your left hand below and your right hand above. That is, as you weave you clearly differentiate between the straws in your left palm and those you hold in your right hand and bring the straw in your right hand over to your left wrist, and open your right hand, release the straw you're holding down with your right thumb, and rub your right palm across your left, toward your left hand's fingertips, wrapping the straws around one another. And if the straw in your right hand, which is on top, comes close to your left thumb, you hold it down with that thumb. Then, you grab the straw you rolled toward your left fingertips with the right hand and roll that toward your left wrist. That's basically how it's done. If you do it too slowly you can't make rope; you must work as fast as you can in order to get some nice rope out of it.

The counterpart of the rope you can make like this is *hidari nawa* (left-handed rope), and of course it's done in the opposite way—with the right hand down and the left hand up.

Okay, so I made a left-woven rope with two lengths of twine, about 5 *sun* (15.3 cm) long. Then I looped it over the big toe of my left foot, at the center, and twisted it tightly with both hands. Once that was done, I split the end part into two strands and began to weave again—adding twine—because I wanted to make a rope about 5 *shaku* 5 *sun* (1.67 m) long. Once that was done, I used a wad of straws with pistils and stamens to polish the rope by rubbing it up and down—to smooth it out [i.e., to do *kokuri* to it]. This is necessary. Then I looped it over my right big toe, to get started, and stretched it out hard and twisted it again with both palms. When you twist a right-woven rope like this, you put your left hand beneath your right hand and then you put your palms together with your left fingertips leading, and you roll the straw toward your left wrist. Once it reaches your wrist, you push it toward the fingertips of your left hand with your right hand, and then further—toward your left wrist—and you must do all of this very quickly. And for a left-woven rope, you do the opposite. In other words, you push it from the left wrist toward the fingertips of that hand. Next, you crush a bunch of straws in your hand and rub that wad of straws up and down the length of the rope quickly, to make it soft and smooth [do *kokuri*]. Once I had two ropes done, I used the piece of bamboo to pull the ropes nice and hard. Then, I started weaving more ropes with *migo*. In local speech, this is called *habaki no nogitsuke*. You must put the *migo* in a winnower (*mi*), with the end parts pointing toward yourself. And when you weave, you start with the ones on your left, using your left hand.

First, you need to make some more nice, tight ropes. You take seven or eight *migo* by their bases each time and make two right-woven ropes and two left-woven ropes, to make for a total of four ropes. Then you twist each one tightly and rub each with crushed straws [to make them smooth]. Then you separate the left-woven ropes and the right-woven ropes so that you have one of each per *habaki* and prepare to attach their ends—where you started and where you stopped weaving them—to the *habaki*. The lengths will be different for each person because everyone has different legs—thicker, thinner, etc. Once that's done, you can start weaving the *habaki*. The technique is the same as for the *kera*. So, first you've got to weave tight-tension ropes, which you can start with the twine. These are called *nogitsuke* ropes. After you've woven two of these, you grab seven or eight of the *migo* by the tips with your right hand and then bring the end of the rope to those with your left hand. Then you put the rope on top of those and hold the *migo* in place at their centers with your thumb. You have to be sure to not let go of the twine you used to start the rope. Then, you carefully pull the *migo* toward yourself with your right hand. You make sure that the base parts of the *migo* you pulled out are right at the center of the rope with thread at the end and pull it about 2 *sun* (6.1 cm) toward yourself. Then you grab the base parts of those *migo* with your right hand and pass them under the rope with thread, and then hold with your left hand. So, the *migo* were above the rope with thread at first, but now they're below it. Then, you bend the *migo* around the rope, and start weaving them. You go as far as you can, and then about 1 *sun* (3 cm) going the other way. The *migo* I used today were thick and fairly short, so I used seven to eight each time. Normally four to five would be enough.

Once you've done all this, you won't really need to use any more *migo*. Now you just weave the *migo* tips a certain length—some people go with 1 *sun* (3 cm) and some with more like 1.5 *sun* (4.5 cm). But you'll need to start aiming for longer length because as you weave toward the tips of the *migo* you'll run out and find that you can only go so far, and you want to get the right width. Then at the end you weave the other way once, or maybe twice.

If you do it like this, you'll have wrinkles in your *habaki*—that's a sign of a well-made one. They say that if you find *habaki* with wrinkles, you had better buy them. But a *kera* with wrinkles should be tossed out.

That's how to make *habaki*. Today I could only get about two-thirds of the way through. But this means that I can finish them early tomorrow.

February 16, Sunday

Sunny
–5°C in the morning, 4°C in the afternoon

Breakfast:	no change at all from yesterday
Lunch:	roasted *mochi* with sugar, roasted cod (*tara*) with soy sauce, *debu*, etc.
Supper:	miso soup with daikon, whale, and kelp (*konbu*) in it, side dishes: same as yesterday evening

Early this morning I shoveled snow and pulled cabbages out from winter storage. My wife went out to sell them. After breakfast I cleaned up and prepared for lunch and wove *habaki*—finished those today. My wife got home around three o'clock and got ready for supper.

Today's income:

cabbages (4 *kan* 300 *momme* (16.1 kg))	85 sen

Today's expenditures:

cotton cloth	2 yen 63 sen
tofu dregs (*kasu*)	6 sen
kelp (*konbu*)	5 sen
candies	3 sen

At the main house: Kanezō wove thin ropes starting in the early morning. These are for Minami Iso Village's giant fish trap (*daibōami*). They're the ropes for weaving the *tawara,* for the trap. At the main house, they took on that job (this year). You have to attach the *kakari* ropes to the *tawara,* and also attach two lids [*sadara*], and each *tawara* is considered to be worth 6 sen. You use your own straw to make them. It's not much money, but people say it's better (to make these and get some cash for them) than to simply goof around. In our village alone, thousands will be made, people say. My brother's wife wove thick ropes—these will be sold to the agricultural cooperative's warehouse. My mother cooked and such.

Old Papa Suekichi,[19] who died recently, is one person who will not soon be forgotten here in the village. I'll write about a particular episode. Now, this fellow's balls weren't originally especially large or anything, but as he aged, they got bigger and bigger, so at the time of his death he was the owner of a huge pair of testicles. A man with gigantic balls like that is called a *dobin* here in the village.

People also call such a man *ōhaba fudoshi*.[20] There's also *mata ippe*,[21] but *dobin* is the usual term, and these are both represented in a little riddle that goes like this:

> Take *satomochi*.[22] (*Satomochi to kakete nanto toku?*)
> And take a *dobin*. (*Dobin to toku.*)
> What do they have in common? (*Sono kokoro wa?*)
> Both are delicious. Another drink! (*Amari umakute, mata ippe!*)[23]

Even just as far back as I can remember, there was another such fellow besides Papa Suekichi in the village—the *dobin* of Kihachi's family. He's deceased, too—died some years ago.

Old Papa Suekichi worked hard all his life, since he was young. And he loved to drink saké. He always said that he had been saving up money for buying saké since he was young—in anticipation of becoming unable to work enough to buy saké for himself in his later years. But he doesn't really seem to have been able to save very much. As he got older, and became less able to work, his energy level and health declined surprisingly fast, but when he did drink, he'd get all his strength and energy back—a hundred times—and he'd be able to take on just about any sort of work. But sadly, he died. Shrewd villagers kept wary eyes on him for years, thinking, "His balls will probably return to normal size after he dies, anyway."

This fellow, when he was young, received some residential land from his own main [natal] house, and he founded a branch household of his own. But the thing is, he couldn't read or write, so he just left the matter of the written agreement up to the people of his main house. Well, as it turns out, at one point—some decades ago—the main house started getting charged huge amounts of money for rent by the local landlord. You can't blame him for being surprised when he learned that the landlord was able to simply seize his residential property once the compound interest had been calculated, even though he thought he had received the land. So, he was shocked, and he got someone to look into the matter for him, and it was found that his agreement over the land separation with the main house had only been made orally, and that after the separation the main house had begun to experience financial hardships and had at one point sold off that particular land, along with the rice paddy land it had owned. And this fellow in question had been totally clueless. And the landlord hadn't realized that someone had a house in the middle of those rice paddies. It wasn't until some decades had passed that the landlord learned about this and started demanding rent payments, according to the story. I've heard before that this old fellow often complained that his house and land had been taken away by the landlord with someone else's help, and that although the landlord had forgiven the debt, he had started charging rent from that point.

Yoshida Umetarō's wife gave birth to a baby boy today.

February 17, Monday

Cloudy and rainy
South wind
2°C in the morning, 8°C in the afternoon

Breakfast:	miso soup with dried daikon leaf stems and kelp in it, side dishes: same as every morning
Lunch:	roasted salted *hatahata,* all else: same as morning
Supper:	miso soup with dried taro stems cooked with whale, and with kelp (*konbu*) in it, side dishes: same as lunch

I went to my natal home early in the morning and got some night soil and urine, and got more from other people's homes after breakfast. Then I started making a *senaka-ade.*[24] My wife sewed all day.

I'll write about how to make a *senaka-ade,* the way I did it today.

You pound some straws and weave five thin ropes, each about 1 *jō* (3 m) long, using five or six straws at a time. Then you tightly twist them and rub each with a wad of straw to make them smooth. Next, you set up a ladder and tie them to the rung that is closest to 4 *shaku* (1.2 m) from the ground. And you loop them over the rung and tie them so that you basically have two lengths of rope per single rope—ten lengths of rope. And you make the middle rope go down lower than the others, and the two next to it go down almost as low, like the diagram shows (see Fig. 12.16).

You can make these with a variety of materials. People often use weeds. I sometimes see such ropes made of a weed called *gama.* When I plant seedlings in the rice paddies in May, I'll often pick the extra seedlings that have grown to a height of 7–8 *sun* (approx. 22.7 cm) and let them dry in the sunshine. The lower stem parts turn blue, and the root parts turn red. I make *senaka-ade* with them. They look real pretty if you make them with those seedling straws. And the roots are as soft as cotton, so it doesn't hurt your back much.

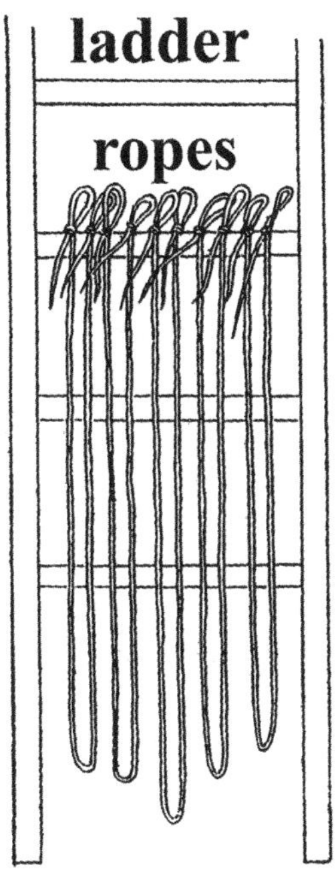

Figure 12.16. The first stage of making a *senaka-ade.*

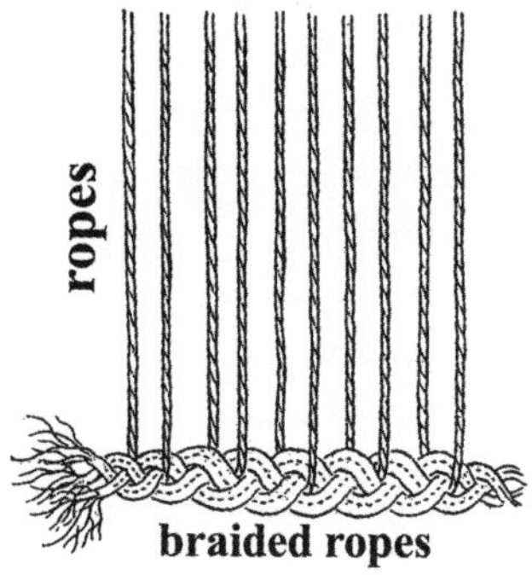

Figure 12.17. Braided ropes woven through the loops of the vertical ropes.

Anyway, after you've tied the ropes on the ladder as shown in the diagram, you braid three ropes made with 14 or 15 straws, and weave them through your five long looped ropes (see Fig. 12.17).

Once that long braided part is done, you attach each end to both sides of your ropes, and you should weave more straws into it as you go.

At this point, you're ready to start the main work. First, you take five or six rice plant straws and stick them 5–6 *sun* (approx. 16.7 cm) into the rope on the left. And you do the same to the side pointing toward you. But you must make sure the base ends of the straws go between the second and third ropes from the left, and that the base ends are pointing toward you. And you need to twist the ropes tightly as you go along. So, you have the second and third ropes going through from the back side and pointing toward you, and the base parts on the back side, and the tips pointing toward you. In other words, you twist the tips (pointing toward you) tightly and keep the second rope in the center and weave between that and the third rope. Next, you twist tightly as you weave and go between the third and fourth ropes, with the third one in the center, and so the straws are going through between the third and fourth ropes. You just continue in this order, and twist the straws tightly, going up and down, with the (vertical) ropes in the center and the two ends of the (twisted) straw parts on either side of the rope on the right. Next, you want to make the base ends of the straws point to the right and go leftward with them. Then, you make the base ends of the straws you're using point to the left, and move rightward with them as you go, and then both ways, so that you have three layers of straw. Then, you want to weave straws through the second and third ropes to the left, and stick the root [base] ends of the straws through to the back, and then you do the same for the third and fourth ropes, going between them. As for the amount of straw to use, this differs from person to person . . . at least a little, anyway.

Next, you hold the straws you've stuck through between the third and fourth ropes with your right fingers and twist them with your left hand, and then stick them through between the third and second ropes. Then, you want to take those straws at the third rope and shove them through to the back on the other side of the second rope on the left. Next, you twist the straws that are sticking out at the fourth rope and going in between the third and second ropes, and shove them through, toward yourself, between the second and leftmost ropes. Do it like this, and you'll have four layers of straw tips sticking out on the left side. You want to make sure that the ends of the straws where you started—that is, the bottom—are wrapped with straws, and basically sticking out the back.

You want to stick them out toward yourself while pulling them up, and hold them with the fingers of your left hand and then pull them toward yourself with your right hand. Then you hold the straws at the starting point—the bottom—with the fingers of your left hand and push the straws downward, and point them up toward the top and put all the tips of the straws together, so that the rope and straws on the left are inside. Then you twist those once and separate them, keeping the second rope in the center, and push the ends of the straws through the back and toward yourself, and then you weave in some thin, young, straws.

You then need to point the base ends of the straws to the back, and take the ends of the straws you separated, which are pointing toward yourself, and polish them with a wad of crushed straws, and shove them through between the second and third ropes and then back through again between the third and fourth ropes. Then you weave in some more thin straws. As you go along like this, you need to add thin straws, and polish them with a wad of crushed straws. So, if all the straw tips are sticking out on the right side, you'll have four layers of them on that side. And then, you work on the straws on the bottom right, just as you did for the other side, and stick the tips out toward the left, and then work from the left and stick the tips out toward the left, and then work from the left and stick them out on the right, and then go back with them to the left, and go on like this until you've reached a height of about 1 *shaku* 5 *sun* (45.5 cm). And the width at the bottom will be about 5 *sun* (15.2 cm). And if you have a width at the widest part of about 8 *sun* (24.2 cm) then you're in good shape. In other words, you want the width to increase as you go from the bottom to the top. Once you've reached a height of about 1 *shaku* 5 *sun* (45.5 cm), you had better stop, untie the two center ropes from the frame [ladder], and then tie them together tightly. I did up to this point today and quit (see Fig. 12.18).

At the main house: Kanezō cut hay for the horse early and spent the rest of the day weaving big fish trap (*daibōami*) *tawara* (see Fig. 12.19). My brother's wife minded the shop. My brother went out on some errand, and my mother did some cooking while helping with the *tawara* work by making thin ropes. It's a great advantage for a person to be able to make these *tawara* for the big fish traps. Basically, it takes three bunches of rice plant straws (*wara*)—which currently cost somewhere between 9 *rin*[25] and 1 sen each—to make a *tawara*. But that's only for a very skilled person. You can basically say that it's 3 sen for the straws and 3 sen for the labor, per *tawara*. In one day, three people can make 20–25 *tawara*, working as hard as they can. Let's say they make 25. That means they earn 75 sen. This is pretty good for the winter off-season. But Kanezō—at the main house—isn't very experienced, so he needs an extra bunch of straws to make each *tawara*. In fact, once he has finished one *tawara* he has probably gone through an extra bunch and a half of straw. That's 4 sen 5 *rin* per *tawara*, just for the raw materials. And he gets 1 sen 5 *rin* per *tawara*. So, for 25 *tawara* it's 37 sen 5 *rin* each for three people. The pay difference between the highly skilled and the unskilled really is shocking.

February 18, Tuesday

Rain and clouds, later sunny
Southwest wind
0°C in the morning, 2°C in the afternoon

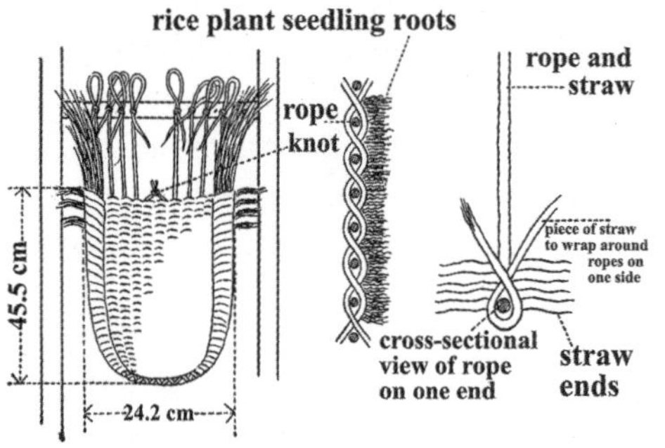

Figure 12.18. Views of a *kera* and its parts, in progress.

Figure 12.19. Kanezō making *tawara* lids (*sadara*) (February 17). The *senaka-ade* that Yoshida started making on that day is visible in the background.

Breakfast, lunch, and supper: exactly the same as yesterday

My work for today was just the same as yesterday, and I finished making the *senaka-ade*. My wife sewed and cooked, as always.

Now, yesterday I made the *senaka-ade* up to a height of about 1 *shaku* 5 *sun* (45.5 cm) and did up to the point where I tied the central ropes, so today I mainly concentrated on working from that point—continuing by completing the two sides. That is to say, I made the *senaka-ade* so that the sides would go over the shoulders. First, I took some colored strips of cotton cloth—about 1 *shaku* 5 *sun* (45.5 cm) long—and stuck them through the body of the *senaka-ade* in the back at a point roughly 3 *sun* (9.1 cm) below the knot I tied yesterday [in the center]. Using colored cloth is good.

Then I did the same with two more strips of cloth at the 7.5 *sun* (22.7 cm) point and at the 1 *sun* (3 cm) point. That made for three layers of cloth strips sticking through the *senaka-ade* [on each side]. I wanted to start working on the right side, so I untied the knot I made yesterday with my left hand. Then I focused on the four ropes on the right side. So, since there already were some straws attached to the rope on the far right, I started attaching some to the ropes to the left—little by little. About one-half of the amount on the rightmost rope is enough. Then I continued making the right edge, working with the fine straw tips that were already sticking out there, just like before. And I stuck the tips of the straws toward the left. Then, I wove the bottommost two strips of cotton through the straw on the right side, like before—going from right to left—and then started on the second strips of cloth on the right side and went to the left. Then I did the same with the third cotton strips. Next, I pushed the tied-up straws on the right side upward, and cut off the extra parts with a knife, to make the right side match the left. So, after making about 3 *sun* (9.1 cm) of it like this, I tied a new knot just about 2 *sun* (6.1 cm) to the right of the knot I tied at the center yesterday, after weaving the middle ropes to the ropes to the right of the center. Then I held one of those ropes and entwined it with about 3 *sun* (9.1 cm) of what I'd made by twisting straws on the right side. What I mean is, I separated the rope into two parts and wove it together with the four ropes on the right.

After that, I did the same things I'd done to the right side to the left side, and once I'd made about 7 *sun* (21.2 cm), I attached all the ropes and straws and cut off the excess straw with a knife and wove the ropes together using about 3 *sun* (9.1 cm) [of the remainder] and wove in extra straws where needed and then tied it all up. Then I proceeded to make the rest of it [the two sides that go over the shoulders] with soft, thin straws (*nae*), and then once they were all gone, I cut off the excess. After that, I used some strips of cotton cloth and four [leftover] ropes and made some sandals (*waraji*). I'll explain how to make these later. Then I made some 1 *shaku* 5 *sun* (45.5 cm) long ropes that get thinner going toward one end. I put these ropes together, divided into three parts, and wove with them. That is, I made *mikko nawa*—ropes made by adding them to pre-existing ropes.[26] I made them so they grow thinner going downward. At that point, one side of the *senaka-ade* was done. Next, I finished the left side. I didn't have three layers of straw sticking out on that side, so I added straws at the center point and after doing that it was almost exactly as the right side had been. Once that was done, I trimmed off all the fuzzy rough parts from the smaller thinner straws, and attached the ropes to the *senaka-ade* at points about 5 *sun* (15.2 cm) from the bottom on each side. In this way, I made a real nice *senaka-ade*.

At the main house: Kanezō wove *tawara* from the early morning. My brother's wife wove ropes to sell. My brother manned the shop. My mother did the same as yesterday.

February 19, Wednesday

Cloudy with snowstorms
Strong west wind
–4°C in the morning, –2°C in the afternoon

Breakfast:	miso soup with cabbage and kelp in it, side dishes: same as every morning
Lunch and supper:	same as yesterday

This morning I carried seven pieces of wood that had been lying around down to the main house's property, one by one, because they need to be taken over to the lumber mill. In the afternoon I ran some errands. My wife did her usual work, which included making a kimono for my eldest daughter because she will start elementary school soon, and cooked and such.

Today's expenditures:

candies	5 sen

At the main house: Kanezō cut hay for the horse early, and spent the rest of the day making *tawara*. My brother minded the shop. His wife wove more rope for selling. My mother wove rope for *tawara* and cooked and such.

Today I'll write about the changes over time in sleds. In the old days, people carried fertilizer out to the paddies on sleds they pulled behind themselves (*hito-sori*). They say that once snow started falling, adult men and women—and even kids who'd help them—would be out pulling sleds from before sunrise until sunset every day. But with modernization, the sleds began to change rapidly, and people could avoid the bone-breaking work of pulling sleds themselves and instead start relying on horses to pull them. In other words, horse-drawn sleds (*uma-sori*) [*ba-shori* in local dialect] appeared on the scene.

The first sleds had runners that were made from a single piece of wood—about 5 *shaku* (1.5 m) long. These were split into two pieces and made to curve up a bit at the front. And there were two toothlike things sticking up farther back, each with a hole in it. These sleds were called *hito-sori,* and people still use them to pull stuff today. There are also long sleds for people to pull stuff with. The runners for these are the same as those for regular sleds but they generally range from 2 *shaku* (60.1 cm) to 2 *shaku* 5 *sun* (75.75 cm) in length. And when you pull one of these using a rod and rope, you attach a special, thick, upward-curved wooden pole to the front. These are mainly used to haul large and heavy loads down from the mountains. And a person will grab the end of this pole and walk along, to steer the sled (*kajitori*). Probably, attaching such a pole to one of those sleds is a vestige from the one-runner sleds (*ippon-sori*) that people used in the old days—at least that's what I think.

But nowadays you can't find a one-runner sled anywhere, even if you go looking for one. And I think that the horse-drawn sleds of today are basically variants of these long sleds that people pull. At some point, people started making the runners so that both ends curve upward. But before that, they were just like the long sleds that people pulled in that only the front ends curved upward. And then, there came another change: people started attaching strips of iron about 5 *bu* (15.2 mm) wide—cut to fit—on the bottoms of the sled runners to make the sleds slide more smoothly. Apparently, this doubled the maximum load weight limit. But more recently, materials

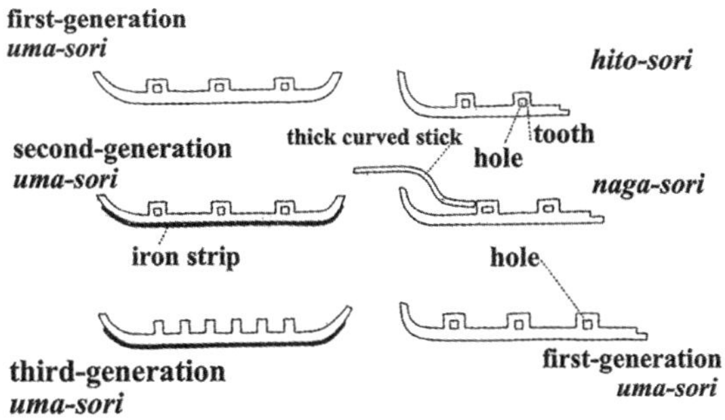

Figure 12.20. Sled runners: Changes over time.

for making sleds have become scarcer, and even though people still know the techniques, they started cranking out multiple sleds from the same big pieces of wood. And they even stopped making runners differently—not carving them out of single pieces of wood but making the "tooth" parts with holes separately and attaching them. In other words, the sleds of today are not as strong as the sleds of old. Oak is said to be the best wood for making sleds, but sometimes I spot sleds made of cherrywood. I'll try drawing some diagrams of sled runners (see Fig. 12.20).

I've just remembered some fun experiences I had with sleds in the past. I'll write about them. When I was a kid, the third-phase horse-drawn sleds had already begun to appear. The center of the village is the highest spot, so the roads slope downward on both sides.

After eating supper, we kids would often gather at those sloped roads. Right away, we'd find a house with a horse-drawn sled sitting outside and we'd quietly lift it and take it over to the sloped road. Then we'd take it up to the highest spot and we'd all climb on—more than ten kids at a time—and then slide all the way down on that sled. That was our fantastic nighttime fun then. And every night we'd get chased around by the owner of whichever sled we'd used. The sled owner would angrily shout, "You little dogs! You've stolen my sled again! Whoever you are, I'll kill you when I catch you!" Chased by someone so angry, shouting like this, we truly had to run for our lives. We'd run into the paddies, on the ridges, and hide out there, or crawl deep into someone's shed—whatever it took to escape. Or sometimes we'd get a sled that slid really well, and even with the very powerful [kid named] Manpei at the rudder, we'd fly off the road and crash into some house, breaking a door or a window, and then the owner would call the cops and all ten or so of us would get into all kinds of trouble. We'd really be screwed then because we'd get it from three sides: the cops, the schoolteachers, and the parents. But the thing is, we never gave up. Instead, we made huge circular ropes with rice plant straws (*wara*) to use as brakes, so anytime we felt that we were about to fly off the road and lose control of the sled we'd throw those down under the sled to stop it. In that way we avoided disaster many times.

February 20, Thursday

Occasional snowstorms
Northwest wind
–5°C in the morning, –3°C in the afternoon

Breakfast:	miso soup with kelp (*konbu*) in it, *ipponzuke,* miso-preserved eggplants
Lunch:	boiled potatoes, *ipponzuke* and such
Supper:	miso soup with daikon and whale in it, roasted *hatahata, ipponzuke*

This morning I had to run down to the post office in the main part of the village, and while there I voted in a local election. In the afternoon I made a pair of straw boots, which I gave to my mother. My wife did the same things as yesterday.

Today's expenditures:

postage stamps	2 sen

At the main house: Kanezō wove *tawara* again from the early morning, and my brother's wife made ropes for selling during the morning and minded the shop in the afternoon. My mother did the same things as yesterday. My brother went to vote in the election.

Today I made a pair of boots with straw, so I'll write in detail about how it's done.

First, you remove all the pistils and stamens from a bunch of straws, and then you pound them starting from a point about 2–3 *sun* (7.6 cm) from their bases [to soften them]. Then you grab and hold some with your left hand, right at the dividing line between the pounded and un-pounded parts, and pull two or three straws out from the bunch just above where you're holding it and bring those down and hold them with your thumb and then wrap them around the bunch twice, and then when you release your thumb you pull those straws downward. But you don't pull too hard—just lightly. Then you separate the straws into two halves and hold one half with your left hand and then with your right hand you wrap one half back around, away from yourself. If you do this, then when you release the bunch of straw it will naturally split into two (equal) parts. Once you've done that, you grab the straw hard with your left hand—this time without pushing it down with your thumb—you . . . the upper part of what you're holding . . . to put it another way, you take six or seven straws and wrap them very tightly around the spot where you just fastened everything together. Then you pull on those straws with your right hand and quickly grab the rest with your left hand and separate them into two parts again.

Then, you take the part you were holding and pulling with your right hand and shove it through to the back, in between the separated parts, with some force. Then you grab the left side of the straws that are separated with your left hand and take eight or nine straws from there with your right hand and shove them through between the divided parts, out through the back. Then you stick the straws on the right through to the left rear. You don't want to stick the straws through more than once. If you do it so that all the straws on the left go through to the back on the right side, and that all the straws on the right go through to the back on the left side, then with the 4–5 *sun* (13.3 cm) of straw left over, you'll have the boot's sole—that is, you want to leave that unwoven straw sticking out on the right and left sides, pointing toward the back. This forms the bottom—the sole (see Fig. 12.21).

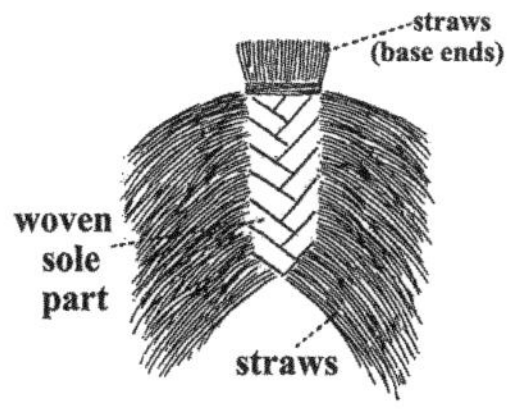

Figure 12.21. Making a straw boot: Early stage.

Figure 12.22. How to weave a straw *irigo*.

Once you have that, you set it down and then you'll put something called an *irigo* inside, and then you make the upper parts. You want the heel of the foot to rest on the base of the straws. And the reason why you need the *irigo* is because if you don't make the boots big enough, it will be hard to put them on and take them off. In other words, if you make the boots big and loose, you'll have to make the soles thick, and to do that you put *irigo* inside, under the soles of the feet. They're made of straw. To make one, you loosely grab a bunch of straws and separate them into two halves, and then you [take both halves] and bend them in the middle. Then you separate those into four parts. You weave them in almost the same way as how you did the three-part weaving I mentioned before (see Fig. 12.22).

This is how they're made. And when you have about reached the 5 *sun* (15.2 cm) point, you stop and tie it all up, leaving about 2 *sun* (6.1 cm), which you cut off, and that's it. All you need to do from that point is pound it some to soften it, and then you can use it. You put a foot down on top of it and take five or six straws from the left with both hands and twist them tightly and then put that between your big toe and the next toe, and then do the same for the right side, and then you take five or six straws on the left and weave them into a small rope, and then do the same on the right.

Then you take the lengths of straws you twisted and held with your big toe and hold them—without letting them unravel. Then you hold the right-side straws with the thumb and forefinger of your left hand, and at the same time you hold the left-side straws with the pinky and ring finger of your left hand, and then pull toward the right. Then you quickly pass the right-side straws you're holding with your left hand over to your right hand and pinch that with your thumb and forefinger. Then you separate five or six straws from the rest of the left-side straws you're holding with your left hand and move those over to your right thumb, and just as you take these between your right thumb and forefinger you let go of the straws you've been holding with your right hand and pass them to your left hand and grip them between the pinky and ring finger and pull them. Then you combine the two different lengths of straws you're holding with your right hand, and twist them together just a little, and pass them to your left hand and pinch them between your thumb and forefinger.

You use your middle, ring, and pinky fingers for hooking straws, and your thumb and index finger for holding straws, and keep them pointing toward yourself. Next, you separate some right-side straws from the others with your right hand and move them over to your left hand and move what you have been holding with your thumb over to a place in between your middle and ring fingers. Then you pinch the straws you are holding with your right hand between your thumb and index finger. After that, you use your right hand to further separate the two lengths of straw you're holding with your left thumb and index finger and move what you've been holding the longest with your left hand to your right hand, and you take one of those between your pinky and ring finger and pinch them and pull. Then you combine the straws you've had hooked over your left pinky and the straws you've been holding with your thumb and hold them between the thumb and index finger of your right hand and repeat the process.

If you do this multiple times, the straws that are separated both ways are called *tsu-tsu.* If you make it until three of these *tsu-tsu* stick out on each side, you can do what people call *shieru,* which means taking two of the *tsu-tsu* on each side between your thumb and index finger and then taking the remaining one—that is, the first one you separated—and then with your right hand pulling them toward yourself. Then you grab the straws sticking out to the side, and take the ones that you just folded toward the front—the *tsu-tsu* you pulled—and hold them downward and take them between your thumb and index finger, and then divide the straws between your middle finger and the one next to it, and further divide them, and leave one of the former and hold it in place with your right pinky finger. Then you take one cluster of straws that you're holding with your left pinky and combine it with the one you're holding with your thumb and catch it between the middle finger of your right hand and the one next to it. Then you grab the front two *tsu-tsu* on the left side with your right thumb and hold them, and then bend one forward with your left hand. Then you let go of the two *tsu-tsu* you're holding with your right thumb and pull both toward yourself with your left hand. Then you further separate the straws on the left side with your left hand, and bend downward the *tsu-tsu* that you've already bent forward, and take them with the thumb of your right hand and then take one of the two *tsu-tsu* you've got between the middle finger and the one next to it, leaving one behind, and pinch it between your right pinky and ring finger, and then take one you're holding with your left pinky and one you're holding with your thumb and combine them and then catch that between your left middle finger and the one next to it.

You keep doing this until all loose straws on both sides are gone. Once that's done, you forcefully pull apart the straws of the *tsu-tsu* that you bent forward earlier. And if you're holding all the straws with your right hand you use the left, and you take the three leftover *tsu-tsu* and tuck them under the straws that you bent toward the front and pull them hard to the left, and you switch hands as you work. You'll have used up all the unbent straws, so you work holding the straws of the *tsu-tsu* you already bent forward, and you do the same thing on the other side, repeating the process. Once you've gone 7–8 *sun* (approx. 22.7 cm) from the starting point, you stop. At that point, if you still have straws in your right hand (or your left), you let go of what's being held in place by your middle finger and mix it in with the unused straws on the left side. Then you put the straws you're holding with your pinky together with the straws you're holding with your thumb and hold all that with the thumb of your right hand. If you do this, you'll have four different *tsu-tsu*. In other words, counting the *tsu-tsu* that have been bent forward and those that have not been bent forward on both sides, you'll have a total of four. What you'll do then is point these four *tsu-tsu* toward the front and use what you're holding in your left hand as the base and make a straw sandal (*waraji*).

If you find yourself running out of *tsu-tsu* straws along the way, you just add some as you go, and after adding a little you stop adding them and start adding different straws. And as you go along like this, you make it so that the sandal gets narrower. Then you put it under your foot and see how it fits—if the end is like the part where you started making it, then you stop. Then, you take the four separate *tsu-tsu* straws you had left over from before and put them together to make two, and then add some straws to them and weave them into two separate ropes. Once you have these, you twist them well (*yori o kakeru*) and bring the left one to the spot where you started making the sandal and stick it through the top, out through the bottom, and over to the right, and hold the end of it, and run it through the one you stuck through the sandal, and bring it toward yourself, and then run it back toward the right. Then you stick the one on the right through and do the same as you did with the one on the left (see Fig. 12.23).

Once you've reached this point and attached these ropes, you bend the boot from the inside, and then there's the back-end part, or the heel. This is called the boot's tail. You turn it so this is sideways, and you separate the straws by a ratio of 1:2. That is, you make it so that one-third point

Figure 12.23. A nearly completed straw boot.

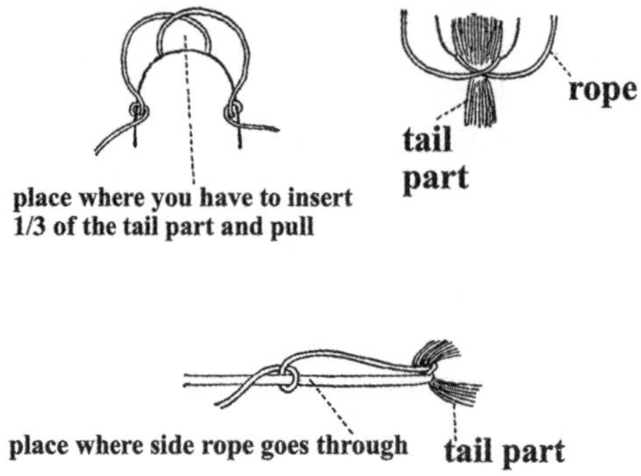

Figure 12.24. Ropes and tail part of a straw boot.

downward and two-thirds point upward. And you put the [upper] one-third into the ropes you attached to the front of the boot and pull hard on both ropes at their centers (see Fig. 12.24). Do it like this, and the tail part will be nicely tied up, and the tip part will fit right into the tail part.

After pulling both ropes nice and tight, you pull on the ropes that go through the sole part. Then you tie both to the inside of the boot. Next, you make another left-woven rope and twist it tightly (*yori o kakeru*) and tie it to the boot at the same place. So, now you have those two ropes from when you started making the boot—one on each side—running through the tail and up to the center, just inside of the outer side ropes. You wrap them once around the outer ropes, and back to the inside again, and then back through the split tail part, and through the ropes you twisted tightly when you started making the boot, and then out the top, through the *tsu-tsu* you bent forward earlier, and then out the front. If you complete this for both sides, and then weave the ends of these ropes as left-woven ropes and tie them to the boot right there, you're basically done making a boot. To make a left boot, you put your left foot in when making it and to make a right boot you put your right foot in while making it.

February 21, Friday

Sunny at first, but some snow later
Northwest wind
–5°C in the morning, 0°C in the afternoon
Snow accumulation: 1 *sun* (3 cm)

Breakfast:	miso soup with burdock root and daikon in it, *ipponzuke,* miso-preserved eggplants
Lunch:	cold rice with hot water poured over it, roasted salted *hatahata, ipponzuke*
Supper:	same soup as morning, side dishes: same as lunch

Early this morning I pulled cabbages out from winter storage and gave them to my wife so she could take them out and sell them. I cleaned up after breakfast and went to the main house to pound *mochi,* and then made lunch. In the afternoon I looked at some books and goofed around a bit. My wife returned in the evening; she cooked and such.

Today's income:
cabbages (2 *kan* (7.5 kg)) 38 sen

Today's expenditures:
tofu dregs (*okara*) 10 sen

At the main house: Early this morning Kanezō and my brother's wife and my brother pounded *mochi*. Today is the February first New Year (*shōgatsu*) celebration—called *mikka-shōgatsu*—and, of course, people prepare for it in the usual way. They finished the breakfast preparations. Kanezō wove ropes. My brother's wife minded the shop. My mother did New Year preparations—prepared burdock root and such—and cooked, etc.

Today I'll write about how to make *tawara*.

Until recently, it was generally accepted that a person needed 9 *hiro* (13.6 m) of *migonawa* rope to make one *tawara*, but today people use a 10 *hiro* (15.2 m) rope to make one. If you cut a 10 *hiro* rope into four equal parts you'll have eight rope ends, and eight lengths [from their center points]. You put a knee down on each piece and run the ends through holes you've made in small round pieces of wood and tie them on and wrap them around the wood. These pieces of wood are called *komozutsu* (see Fig. 12.25).

The way to wrap the rope around a *komozutsu* is as follows. First, you stick the end of the rope through a hole in the *komozutsu* and tie the rope. Then, you hold the *komozutsu* in your left hand and bring the rope to your knee—under it. Then, you raise the *komozutsu* up and hold the rope with your thumb. Next, with your right hand, you run the rope up and then over the top and down, going under itself. Once you've done all eight like this, you drape them over a bar that's specifically made for this (see Fig. 12.26).

Next: the weaving. You take 15 or 16 straws that have had the stamens and pistils completely removed and start weaving them through the ropes from the leftmost *komozutsu* rope on one side,

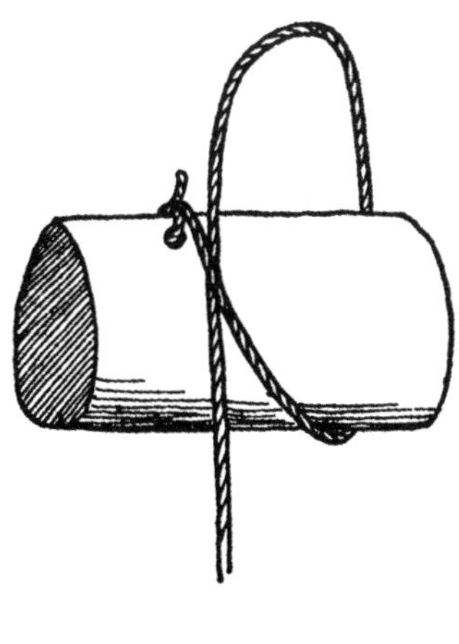

Figure 12.25. A *komozutsu*, for making *tawara*.

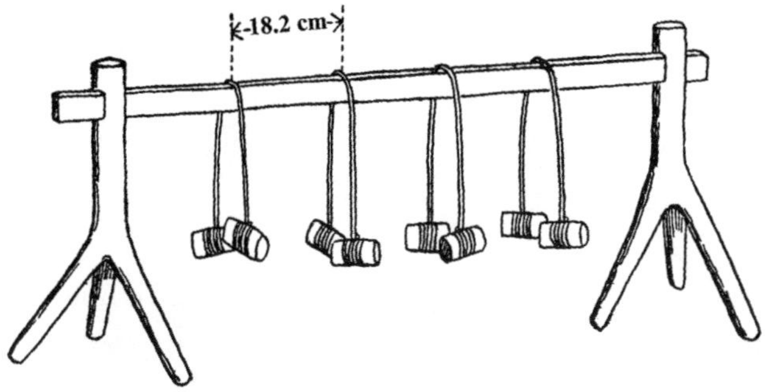

Figure 12.26. Bar and *komozutsu* device for making *tawara*.

go through 5–6 *sun* (approx. 16.7 cm) and then skip one rope and go on to the second *komozutsu* from the right, and go through 5–6 *sun* again. And you must make sure that the base ends of the straws stick out beyond the ropes, and that each time you weave the straws you bend them upward. You weave on both sides, but you only weave through the middle two ropes once each time you do it. And when you weave, you go from the *komozutsu* on the far left on your side, and you hold the ones on your side to your right. Going along like this, you progress about 1 *shaku* (30.3 cm) and then you bring it up so that it's wrapped around the bar and tie it tightly around the center. And as you go, you sometimes loosen the ropes wrapped around the *komozutsu*. You go on like this until the *komozutsu* ropes have all been used up. Then you untie the *tawara*—which is wrapped around the horizontal bar—and untie all the *komozutsu* ropes and take the ropes on both sides and stick them upward through the *tawara* at the points in the middle of where you started weaving first and second. Then you tie once where all the rope ends come together, and then make two left-woven ropes and stick them through the *tawara* from the outside. And if you do the same at the other three places, you'll have finished making a *tawara* (see Fig. 11.2).

February 22, Saturday

Cloudy sometimes
Northwest wind
–4°C in the morning, 0°C in the afternoon
Snow accumulation: 1 *sun* (3 cm)

Breakfast:	same as yesterday morning
Lunch:	nothing
Supper:	I'll explain below.

During the New Year (*shōgatsu*) time, we pound and eat *mochi,* and don't usually do very strenuous work, so we have eaten 6 *gō* (1.1 L) of rice mixed with 2 *gō* (360.8 ml) of oats (*mugi*). And because we consumed a lot of soy sauce, we only used about one-half of the usual amount of miso. We finished all the *nadazuke* and the pickled napa cabbage, and recently started eating the *ipponzuke*. We bought more *hatahata* (sandfish) this time than we do in most years, so we didn't have to buy any salt-preserved salmon for the New Year celebration. My wife hasn't been eating much *hatahata* this year, but the kids have been eating a lot. Over three New Year celebrations, we've pounded about 1 *to* (18 L) of rice into *mochi*.

Today my wife spent the whole day on food preparation because it's the New Year's Eve of February. This morning I hauled all the lumber I carried down to the main house the other day over to the mill, and on the way back I did some shopping. In the afternoon I made a pair of *ashinaka* and ate supper pretty early, and then went to bathe [at the main house].

Today's expenditures:

fish	20 sen
tofu	6 sen
soy sauce	6 sen
sugar	8 sen

At the main house: Early this morning Kanezō pulled barn manure out from the stable and hauled it, and later in the morning he made more *tawara.* In the afternoon he cut hay for the horse. My brother went out on errands, and his wife minded the shop and warmed up the bath.[27] My mother wove ropes during the morning, and in the afternoon she and my brother's wife did New Year preparations.

Today I'll write about how to make *ashinaka.*[28]

In this village, you can still find people who can make *ashinaka,* but not anyone who can make straw sandals (*waraji*). Straw *ashinaka* are becoming increasingly scarce—they'll probably just disappear eventually.

Straws for making *ashinaka* and *waraji* need to be thoroughly pounded to make them very soft. Once ready, you weave two thin ropes, each with five or six straws, about 2 *shaku* (60.1 cm) long. Then you overlap their base ends about 2 *sun* (6.1 cm) and weave them together there. Then you hook the rope over the big toe of your right foot and twist it hard and rub it with a wad of straw to make it smooth. Next, you make a diamond shape with your legs and loop the rope over both big toes and pull nice and hard to prevent it from unraveling.

You hold the part where you attached the two original lengths of rope together with your left hand, and with your right hand you hold the ends of the rope and hook the rope over both big toes. If you do this, the two lengths coming back toward you will be together, parallel, in the center. The distance between the part you were holding with your left hand [before you hooked the rope over your toes] and the points where the rope loops over your toes will differ from person to person, but for me it's about 8 *sun* (24.2 cm). You must make sure that the ends of the rope are in the middle. I'll draw a diagram to illustrate this (see Fig. 12.27).

Once you've prepared for making *ashinaka* like this, you're ready to get started. You have to keep the rope hooked over your toes to prevent the two end lengths from becoming untwisted, and you grab three or four straws with your right hand and bring them over and wrap them tightly around the rope where you've been holding it with your left hand. Then you hold the part where you wrapped the rope with straw again with your left hand, and catch the two lengths of

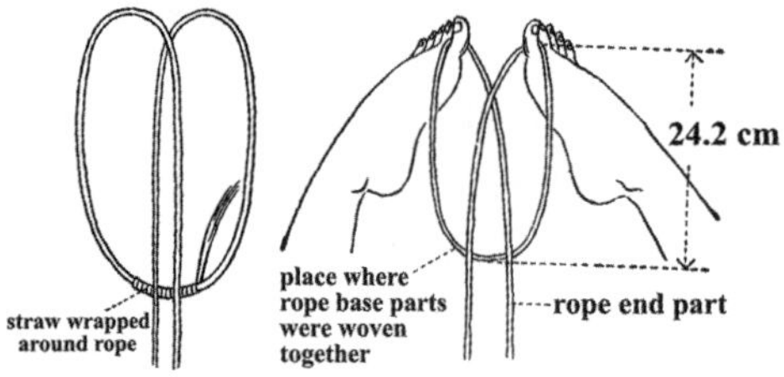

Figure 12.27. How to use the feet to make an *ashinaka*.

rope in the center with your right thumb and run them up and over the other parts, lined up together.

Once you've evenly aligned the rope lengths, you hold them in place with your left thumb. Then, you grab the leftover part of the straw you wrapped around the rope earlier with your right hand. It should be on the inside of the rope part on one side, and should be sticking up from below, and needs to be sticking through the rope on the other side. While hanging on to that straw, you bring the ropes together and stick that through the loop on one side from above, and then outward. Next, you wrap the rope on one side once and run it beneath both straight rope lengths and then up through the loop from beneath.

If you do this three times, alternating left-right-left, then next you weave the *ashinaka* using all four rope lengths separately. And anytime you find yourself short on straws, you just add some, working from the center. For example, say that the straw is sticking out from beneath the rope on the left side. What you'd do is take that straw and stick it through to the back side, with the leftmost rope at the center, between that rope and the second one to the left, and then upward between the second and third ropes. Then you bring it over the third rope and then back through, toward the back side, and make it stick out from under the rightmost rope length. Then you center the right rope length and stick the straws through near the third rope, and then back to the top between the third and second rope lengths. Then you stick the straws through again, between the leftmost rope and the second one, so that it sticks out from under the leftmost rope length. Here in the village, people call this process *waraji-zukuri.*

At this point, you put the five fingers of your right hand together and measure the width and length. And if you get that all set, you have to attach the side strings. When you make sandals (*waraji*) you have to stick all the fingers of one hand through the ropes and pull hard toward yourself to tighten everything from time to time. You need three strings to make and attach the side strings, and they must be well-twisted and rubbed with a wad of straws to make them soft and smooth. You take one of these strings and cut it in half—but you must keep them twisted—and weave one piece together with the second rope and the other with the third rope on the left side, but you don't run these over to the right side; you just let them go down to the back side. Then you bring them back over close to the point where you started them. After that, you repeat the process for the right side. Once that's done, you take four or five straws from the center and weave them from the left side, coming out in front of the strings, and then back to the main body and through and then back to the outside, and then do the right side in the same way. And then, pulling tightly as you go, you'll eventually see the *ashinaka* coming together (see Fig. 12.28). You only want to

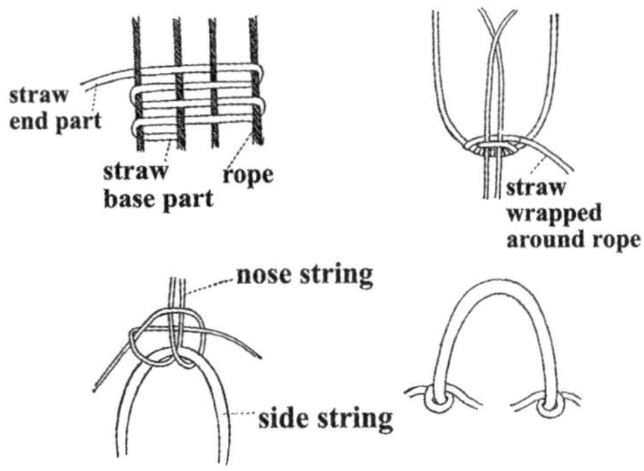

Figure 12.28. Making an *ashinaka*: Parts and views.

make them so that they're a little narrower than your foot, measured from the big toe to the little toe, and roughly as long.[29]

So then, you pull hard on the ends of the two ropes I mentioned above—one by one—and they'll basically be finished. Then you just shave off the extra fluff and such on the bottom and polish the upper surface with a wad of crushed straws, and then do the front center ("nose") string. You weave two ropes and fold them in half and attach them at the front of each *ashinaka,* where they should be so that they come out between your big toe and the next toe, and attach them where the side strings are attached. You tie them so that the one on the left goes down under the side strings and then up, and the other so that it goes over the side strings and then down on the right. Then you grab that and hold it with a left finger. Next, you grab the rope on the right with your right hand and make a circle, and move the end of that over to the left, where you're still holding a rope with a finger. Then you take the left upper rope and run it under that looped part and then up through the rope at the very front, and through the circle on the right and out. Then you hold that with your right finger and make a loop on the left with your left hand, and then if you pull very hard on the looped rope it will all be tied up very nicely. If you have about 1 *sun* (3 cm) of extra straw on the sides, you just trim all that off and then you're done.

For supper tonight we had miso soup with flounder, tofu, and burdock root [pieces] in it. We also had some roasted flounder and boiled cabbage with soy sauce for flavor. There also was octopus sashimi and a burdock root *debu* dish. The flounder and octopus were both caught off the shore of Oga.

February 23, Sunday

Cloudy with occasional snow later
Northeast wind
-3°C in the morning, 3°C in the afternoon
Snow accumulation: 1 *sun* (3 cm)

Breakfast:	miso soup with tofu, saké lees (*sakagasu*), and kelp (*konbu*) in it, and leftover dishes from last night

Lunch:	roasted *mochi* eaten with sugar, *ipponzuke, debu,* etc.
Supper:	same soup as morning, *ipponzuke, debu,* etc.

Today I went to Akita City to buy some fertilizer, and I stayed overnight at a friend's home. My wife sewed all day.

Today's expenditures:

fertilizer	17 yen 5 sen

itemized:

superphosphate of lime (1 *kamasu*[30])	1 yen 75 sen
Hōnen[31] soybean meal (2 *kamasu*)	6 yen 50 sen
Sanriku[32] sardine bits (1 *kamasu*)	5 yen 40 sen
oil cake (1 *kamasu*)	3 yen 20 sen
ground sardine powder	20 sen
saké lees (*sakagasu*)	10 sen
kelp (*konbu*)	15 sen

At the main house: Kanezō made *tawara* from the early morning. My brother's wife minded the shop and cooked and such. My brother went to a meeting at the agricultural warehouse, and my mother went over to a relative's house to help prepare for a celebration because the father of that family turns forty-two this year. Of course, she drank celebratory saké there before going home.

February 24, Monday

Occasional light snow
–3°C in the morning, 4°C in the afternoon
Snow accumulation: 1 *sun* (3 cm)

Breakfast:	saké lees (*sakagasu*) miso soup with kelp (*konbu*) in it, *ipponzuke,* burdock root *debu*
Lunch:	same as yesterday's lunch
Supper:	taro stem and potato miso soup with kelp (*konbu*) in it, octopus sashimi, *ipponzuke, debu,* etc.

Today I returned from Akita City, and immediately went to see the village young men's *tawara* rope competition (see Fig. 12.29). My wife—as usual—sewed and cooked.

Today's expenditures:

train fare	1 yen
gift (*omiyage*) for my friend	65 sen
sweets for the kids	5 sen

Figure 12.29. The Young Men's Association straw products competition (February 24, 1936, at the hamlet meeting hall). [Saburō's elder brother is at far left.]

Figure 12.30. The straw products competition.

At the main house: Kanezō cut hay for the horse early, and after breakfast he made *tawara*. He took off in the afternoon; he also went to see the straw products and rope competition. My mother went to her natal home to pray before the *butsudan*. For my brother's wife, it was the same as yesterday. My brother served as a judge for the village straw products competition in the afternoon (see Fig. 12.30).

February 25, Tuesday

Occasional snowstorms
Strong west wind
–4°C in the morning, –1°C in the afternoon
Snow accumulation: 6 *sun* (18.2 cm)

Breakfast:	burdock root and potato miso soup with kelp in it, *ipponzuke,* burdock root *debu*
Lunch:	roasted *mochi,* roasted salted *hatahata,* the rest: same as morning
Supper:	cold rice with hot water poured over it, everything else: same as lunch

Yoshida Toyosaburō's wife gave birth to a baby boy today.

My house was pretty much buried in snow this morning because of last night's snowstorm. I cleared all that early, and then made straw boots. My wife sewed and cooked, as usual.

Today's expenditures:

ink	35 sen

At the main house: Kanezō made *tawara* this morning, starting early, and took off in the afternoon. My brother went out on errands, and his wife minded the shop and cooked and such. My mother didn't return today.

February 26, Wednesday

Cloudy
Northwest wind
–3°C in the morning, 0°C in the afternoon

Breakfast:	daikon and saké lees (*sakagasu*) miso soup with kelp in it, *ipponzuke,* burdock root *debu*
Lunch:	roasted *mochi* that we ate with sugar, salted *hatahata* simmered in water, *ipponzuke,* burdock root *debu,* etc.
Supper:	taro stem and whale miso soup with kelp in it, roasted salted *hatahata,* all else: same as morning

This morning I hauled fertilizer home from the station. In the afternoon, I went to the lumber mill and had the wood I carried over there the other day cut into boards, and then I carried those back with me. My wife did the usual things, like every day.

Today's expenditures:

fee for having lumber processed	80 sen

At the main house: Kanezō cut hay for the horse early, and spent the rest of the day making *tawara*. Everyone else did the same things as yesterday.

February 27, Thursday

Snowy, later just cloudy
Northwest wind
–4°C in the morning, 1°C in the afternoon

Breakfast:	exactly the same as yesterday morning
Lunch:	same as yesterday
Supper:	same as last night

This morning I put dry *mochi* in a box and then [took them out and] strung them up in places where they can dry further in the wind. I also made more *ashinaka*. In the afternoon I worked more on making *ashinaka*. My wife went over to her brother's place this morning on an errand, and she sewed and cooked and such in the afternoon.

At the main house: Kanezō made *tawara* from the early morning. He made seventeen or eighteen today. My brother's wife made rope for *tawara*. My brother manned the shop. My mother cooked and such.

February 28, Friday

Cloudy, later snowy
Northeast wind
–6°C in the morning, 2°C in the afternoon

Breakfast:	seaweed (*shigana*)[33] and potato miso soup, other dishes: same as usual
Lunch:	roasted salted *hatahata, ipponzuke,* burdock root *debu,* etc.
Supper:	taro stem and whale miso soup with kelp (*konbu*) in it, *ipponzuke,* burdock root *debu,* etc.

The *shigana* seaweed we put into the miso soup this morning was from the Minami Iso Village shore; my wife received it at her brother's house and brought it back with her. They got it from relatives in Minami Iso Village.

This morning I cleared snow, and then I pulled out some miscanthus I'd stored and started making *sudare* screens with that. I'll stand them up around my spring onion beds to protect the onions. The way to make these is a little different from the way in which *tawara* are woven. As I explained

Figure 12.31. The *toshina* rope and *kadomatsu* components and such from the New Year celebrations tied to a tree branch (February 28, 1936, at the main house).

before, when you weave *tawara* you alternate between two places at the center, but when you make *sudare* you weave all spots every time, and you make some thin sticks from tree branches, and whereas such sticks aren't woven into *tawara,* when you make *sudare* you weave them into it when it gets long, and then tie it in place. I used about 16 *hiro* (24.2 m) of rope to make one *sudare.* They're 4 *shaku* 5 *sun* (136.4 cm) wide, and roughly 6 *shaku* (181.8 cm) tall. My wife sewed and cooked and such all day.

Today's expenditures:

magazine(s)	1 yen 10 sen
postal fee for mailing book	10 sen
candies and sweets	5 sen

At the main house: Kanezō cut hay for the horse early. After breakfast he wove *kagari* ropes through the *tawara* he recently made [in early January]. Then he made more *tawara* in the afternoon. My brother's wife was in the shop all day, and my brother went to a local Agriculture Association (Nōkai) gathering. My mother cooked and such.

My natal household's main house—that's the big main house—is currently headed by Yoshida Kinjirō. He recently reorganized the household property. He sold about 7 *danbu* (6,942 m^2) [of paddy land] to Yoshida Takichi, of this village, for the high price of 550 yen per *danbu.* They say that Takichi intends to farm the land himself. And, at the same time, Kinjirō's younger brother Kintarō[34] separated from the household and set up a new branch household of his own.

He received a mere *chōbu* (0.99 ha) of paddy land: just a trifle. It seems, though, that he plans to continue living in the family house for the foreseeable future. And Kinjirō and his son will open a shop (*mise*) and continue living there as well.[35]

February 29, Saturday

Sunny with occasional snow
Northwest wind
–3°C in the morning, 3°C in the afternoon

Breakfast:	same as yesterday morning
Lunch:	roasted salted *hatahata, ipponzuke,* burdock root *debu,* etc.
Supper:	miso soup with pieces of pickled daikon stems and kelp in it, side dishes: same as lunch

Yoshida Masakichi's wife gave birth to a baby girl today.

Today I cleared snow and wove more *sudare,* just like yesterday. My wife did the same things as yesterday.

At the main house: Kanezō made *sadara* [*tawara* lids] from the early morning. My brother's wife did the same things as yesterday. My brother went out. My mother cooked, and during breaks she wove rope for *tawara.*

Chapter 13: March 1936

March 1, Sunday

Occasional snow
–6°C in the morning, 2°C in the afternoon

Breakfast:	miso soup with salted vegetables we received from my wife's brother's family and some kelp (*konbu*) in it, side dishes: same as every morning
Lunch:	roasted salted *hatahata,* everything else: same as morning
Supper:	taro stem and whale miso soup with kelp in it, side dishes: same as morning

Today I cleared snow and made more *sudare* screens. My wife did the same things as yesterday.

Today's expenditures:

candies	2 sen
fee for telephone call	4 sen

At the main house: My brother manned the shop, and his wife wove ropes for selling. My mother did what she did yesterday. Kanezō cut hay for the horse early, and made *tawara* the rest of the day.

March 2, Monday

(February 9 according to the old calendar)
Sunny
–6°C in the morning, 4°C in the afternoon
Snow accumulation: 2 *sun* (6.1 cm)

Breakfast:	saké lees (*sakagasu*) miso soup with kelp (*konbu*) in it, miso-preserved eggplants, *ipponzuke,* etc.
Lunch:	salted *hatahata* simmered in water, *ipponzuke*
Supper:	same soup as last night, side dishes: same as morning
Snack:	dried *mochi*

Figure 13.1. People on their way to the agricultural warehouse to sell ropes (March 2, 1936, in front of the main house).

Figure 13.2. Rice (in *tawara*) being taken to sell on a horse-drawn sled (March 2, 1936, in front of the main house).

Today I cleared snow. Since I'd heard that someone wanted to have some rabbit pelts tanned, I went down into the village and returned with seven, which I put into water to soak. In the afternoon I went to observe a meeting of the local agriculture improvement association.[1] My wife did the same things she does every day.

Today's expenditures:
postage stamps and postcards 36 sen

At the main house: Kanezō wove *tawara* from the early morning, but he took off in the afternoon. Every year on the ninth of February, according to the old lunar calendar, people roast *Ogano mochi* and eat them. And folks say you're not supposed to let anyone else eat yours. My brother's wife was in the shop all day. My brother went out on errands. My mother didn't rest—she did the same work as yesterday.

March 3, Tuesday

Cloudy
South wind
–15°C in the morning, 5°C in the afternoon

Breakfast:	miso soup with salted fish and kelp (*konbu*) in it, side dishes: same as yesterday
Lunch:	roasted salted *hatahata, ipponzuke,* etc.
Supper:	same soup as yesterday evening but with *arame* seaweed in it, *ipponzuke,* etc. We received the *arame* from the main house; it came from the Wakimoto shore.
Snack:	dried *mochi*

Today I made a *fugo.* First, I wove a *migo* rope the same thickness as what you need to make a *tawara.* I already had pounded *migo* on hand, so I didn't have to pound any before getting started. Since I was planning to make a small *fugo,* I only needed a 7 *hiro* (10.6 m) rope. The weaving part is done exactly like making a *tawara,* the only exception being that you use pounded straws for the starting point and for the ending point; you just weave your rope in the normal way, but with those pounded straws. And you must twist the rope and polish it with crushed straws to make it smooth. Actually, the rope weaving method varies some from person to person. I use four or five straws. And the spacing of the rope strands also varies, according to the intended use of the *fugo.* If you're making a big *fugo,* you'll want it to be wide from the start, but if you're making a small *fugo,* you can just go narrow. Today I wove with a spacing of about 3 *sun* 2 *bu* (3.7 cm). That was because I was aiming for a width of about 1 *shaku* (30.3 cm). And I wanted it to be about 1 *shaku* 2 *sun* (36.4 cm) tall, and roughly 5 *sun* (15.2 cm) deep. And I wanted a lid, too. That meant I needed to weave a piece about 3 *shaku* 4–5 *sun* (1.1 m) long. At the end, when you're done, you need to weave it tightly and then wrap your ropes around the end part twice and tie it up real tight and trim off the excess end parts. You want to leave 5 *sun* (15.2 cm) of straw on each side.

Next, you bend the straws on both sides inward a bit to mark the point where the *fugo* will be bent, about 1 *shaku* 2 *sun* (36.4 cm) from where you started. Then you do the same at a point 5 *sun* (15.2 cm) further down. Next, you start weaving from both sides at those points, with *migo* rope, lining up the tips of the straws you use to weave and starting at the center. And you weave so that the tips of the straws stay on the inside. You want to go to a height of about 1 *shaku* 2 *sun* (36.4 cm) and then stop and wrap the rope around once and tie it up real tight. That's one side, so next you do the same on the other side.

At this point, you pretty much have the *fugo* shape, so what you do next is just finish it off. You already have the bottom part—about 5 *sun* (15.2 cm) wide, with the tips of the straws sticking out—so you leave about 1 *sun* 5 *bu* (4.5 cm) of that which extends beyond the rope you used to weave, and cut off the rest. And you do the same for the top, where the lid will go, at the 5 *sun* (15.2 cm) point. Next, you make a *migo* rope, about 2 *sun* (6.1 cm), for the center of the lid, and twist it and polish it with straw. I attached a fairly thick [three-rope braided] *mitsuko* rope at a point about 2 *sun* (6.1 cm) from the point where I started weaving, and then I stuck another rope that I'd twisted well and polished with crushed straws through to the back in two places with the weaving rope at the center, and I fastened that well—just a regular knot. Then I turned it around and wove the two *migo* ropes to a length of about 3 *shaku* (90.1 cm). And then I tied up the weaving rope at the right corner of the bottom part, about 1 *shaku* 2 *sun* (36.4 cm) from the top part, and I verified the corners of the lid part, and tied the weaving ropes running from each corner to the *fugo* body at points 6 *sun* (18.2 cm) from the bottom corners, so that was that for the side ropes of the *fugo*. At that point, the ropes for carrying the *fugo* were also done, so all I had to do was shave off the extra straw stuff and I was finished (see Fig. 13.3).

My wife did the same things today as yesterday.

Today's expenditures:

candies	2 sen

At the main house: Kanezō cut hay for the horse early, and he spent the rest of the day weaving *tawara* again. My brother's wife went to the market at Nagane to buy stuff. My brother manned the shop and my mother cooked and such.

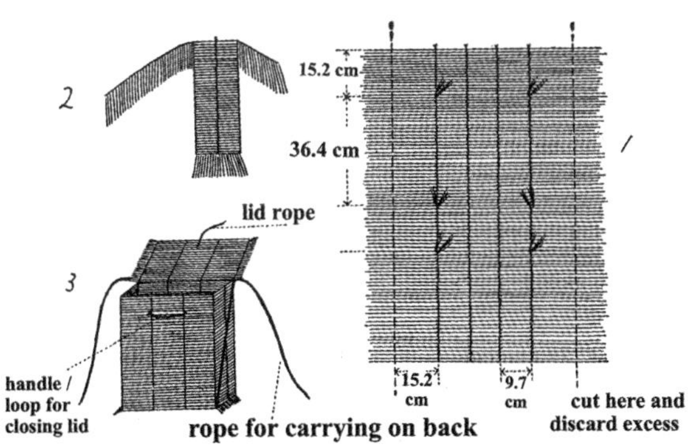

Figure 13.3. A *fugo*: Parts and the finished product.

March 4, Wednesday

Sunny
Strong west wind
–4°C in the morning, 2°C in the afternoon

Breakfast:	saké lees (*sakagasu*) miso soup with *arame* seaweed in it, *ipponzuke,* etc.
Lunch:	salted *hatahata* simmered in water, *ipponzuke,* etc.
Supper:	same soup as yesterday evening
Snack:	dried *mochi*

Yoshida Ryōzō's[2] wife gave birth to a baby boy today.

The house was buried in snow again this morning because of last night's snowstorm. I cleared all that during the morning. And I went to pay my respects to a cousin because he has been sick, and in the afternoon, I took a shot at making two pairs of straw sandals (*waraji*). You can hardly find anyone in the village who can make straw sandals today.

First, you beat the straws thoroughly to soften them, and then make two ropes, both 1 *hiro* (1.5 m) long, using four or five straws at a time, going in both directions. In other words, you make one 2 *hiro* (3 m) rope. Then you bend it at the center to make for two lengths and hook it over the big toe of your right foot and twist it thoroughly and polish it with crushed straws. Then you weave it in exactly the same way as when you make *ashinaka,* except that you don't need the middle two ropes when you make *ashinaka.* And, of course, you need to make them about as wide as five fingers, and as with *ashinaka,* you'll have *tsu-tsu* on the *waraji.* You take two or three straws and twist and polish them thoroughly and bend them and tie that tightly to the rope frame of the *waraji* 3–4 *sun* (10.6 cm) from the base part of the *waraji,* and just leave that for the time being. Then you weave as you do for *ashinaka,* splitting the straws in two parts as you go, making a rope. Then you bring the straw you have woven to the center and weave like you do for *ashinaka* and make *tsu-tsu* on the sides. Next, you take three or four straws and weave them in to make a base or cushion, attaching them to the *tsu-tsu* as you would for *ashinaka,* and keep weaving.

Once you have the width of several fingers you go on and make the second *tsu-tsu.* Then you continue until you have the width of four fingers and stop there. Oh, and I forgot to write this, but you need a rope about 1 *shaku* 2 *sun* (36.4 cm) long to make a straw sandal. When you're done, you must have at least 3 *sun* (9.1 cm) left over. Then, once it's done, you attach a rope to each side near where you started and run the rope you had hooked over your right big toe through the loop created by having the other part hooked over your left big toe, and hook that over your left toe, and hook the one that was over your left toe over your right toe. Then you grab the rope you stuck into the center of the sandal with your left hand and grab a side rope with your right hand and pull toward yourself. At the same time, you also push hard with your legs. In this way, you can finish by tightening it up well. Then, you pull the rope you stuck through the sandal out. And then, if you spread open your big toes and pull outward with both hands and push down, then the heel of the sandal will be just right, and it will fit well. Then you take the ropes off your toes and hold them in both hands and pull the left rope up, to bend the sandal upward. Then, you stick the two ropes on the right through again and perform the same routine. And then, finally, you untwist the end part of the rope on the left some and stick the two on the right in where that is. This forms the "ear" of the sandal. Then, you put the two central ropes together once and then separate them and run them through the *tsu-tsu* and through the ear part, and then trim the extra fluff and such, and then you're done. You can make many straw sandals in this way.

My wife went into the village during the morning to buy some stuff, and she did the usual things after lunch.

Today's expenditures:

kerosene (1 *shō* (1.8 L))	30 sen
mandarin oranges (*mikan*)	3 sen
charcoal (8 *kan* (30 kg))	1 yen 60 sen

At the main house: Kanezō wove *tawara* starting early. He made about 15 and finished that pretty soon, so he pulled *migo* out from rice plant straws. My brother and his wife made sweets during the morning, and in the afternoon my brother went to the school for a student guardian conference.[3] My brother's wife minded the shop [in the afternoon]. My mother wove rope and cooked and such.

March 5, Thursday

Sunny with occasional snowstorms
West wind
−4°C in the morning, 2°C in the afternoon
Snow accumulation: 6 *sun* (18.2 cm)

Breakfast:	miso soup containing salted vegetables and *arame* seaweed, *ipponzuke*, pickled eggplants
Lunch:	salted *hatahata* simmered in water, *ipponzuke*
Supper:	cold rice with hot water poured over it, octopus tentacle slices with soy sauce, *ipponzuke*, etc.

I wove more *noma* mats for putting on my hotboxes, and made a pair of straw boots in the evening. My wife took our eldest daughter, Fusae, to my old school. It was for a physical exam to make sure she can be a student. She's going to start the first grade this year.

Today's expenditures:

dried alum (*yaki myōban*)	13 sen
octopus tentacle	15 sen

At the main house: Kanezō cut hay for the horse early, and after breakfast he wove *tawara*. During lunchtime one *waraniyo* collapsed, so he had to haul all that straw to the storage shed and put it away in the afternoon. After that, he was sent to another village on an errand. My brother was in the shop during the morning, and he went out again in the afternoon. His wife took a child[4] to my old school during the morning, and she minded the shop in the afternoon. My mother wove rope and cooked and such.

March 6, Friday

Occasional snowstorms, but sometimes sunny and sometimes just cloudy
Northwest wind
–3°C in the morning, 2°C in the afternoon

Breakfast:	same miso soup as yesterday morning but with kelp in it, *ipponzuke*, pickled eggplants
Lunch:	octopus tentacle slices with soy sauce, *ipponzuke*, pickled eggplants, etc.
Supper:	exactly the same as yesterday evening
Snack:	dried *mochi*

About those rabbit pelts I put into water to soak three days ago, today I pulled them out to have a look and found that the fur was coming off, so I gave up on tanning them. I don't know if that happened because they stayed in the water for two extra days or what—it's a mystery.

Today's expenditures:

tofu dregs (*okara*)	6 sen
postage stamps	3 sen

Today I'll explain how to make *agutomaki*.[5]

First, you take three or four straws that you've already beaten to soften and start weaving from a point about 6 *sun* (18.2 cm) from their bases. You weave about twelve times and then stop. Then you separate all the unwoven straws with both hands and then put two groups together all the way across to make for six groups. Then you lap the five or six groups on the left over the five or six groups on the right and put the three or four on the right atop the five or six on the left, and then the three or four on the left under the five or six on the right, and then above the three or four on the right. Then you run the two or three groups on the left under the five or six groups on the right, and then atop the three or four on the left (see Fig. 13.4).

Next, the one or two groups on the left go atop the five or six on the right, and then under the three or four groups on the right. And that's all you need to do to make the body of one of these. Then you grab it with both hands on the sides and pull very hard. Next, you hook it over your big toes at

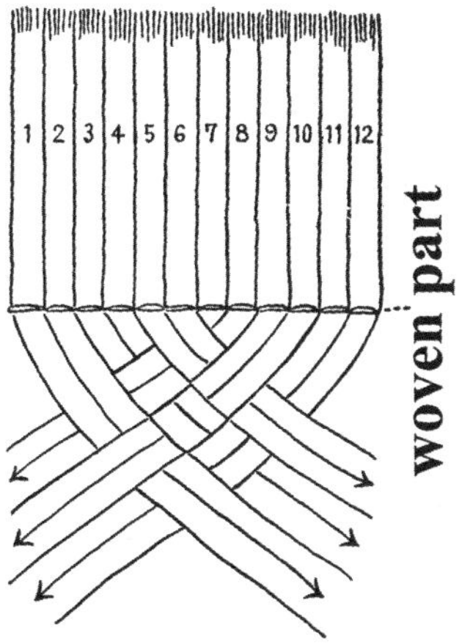

Figure 13.4. How to weave straws to make an *agutomaki*.

the middle point, where you started putting the body together by overlapping straw groups, so that all the straws cover your toes. Then you pull your toes out and put the straws—from the lower parts to the tips—under your knees and weave again. This time you want to weave all the straws, starting from a point about 1 *sun* (3 cm) from where you started weaving earlier. In other words, you make it so that the tips of the straws you wove are positioned in the center of the bases.

Next, you weave the straws in the middle two times. From the second spot, you want to give it about 3 *sun* (9.1 cm), and next you weave from about the 1 *sun* (3 cm) mark. Then, you separate the straws on both sides into groups of three and then weave them together, and in the end weave them into ropes. That's for both sides. And that's basically it for making an *agutomaki* (see Fig. 13.5). If you make another like this, you'll have a pair. I never see anyone making or wearing these in the village today.

At the main house: Kanezō cut hay for the horse early, then wove *tawara*. My brother's wife wove ropes for selling, and cooked. My mother went to pay a *mimai* visit to my cousin, who has been sick. Ichisuke minded the shop. My brother attended a gathering at the agricultural warehouse.

March 7, Saturday

Sometimes sunny, sometimes cloudy, sometimes a bit of snow
Strong west wind
–2°C in the morning, 3°C in the afternoon

Breakfast:	saké lees (*sakagasu*) miso soup with kelp in it, *ipponzuke*, pickled eggplants
Lunch:	salted *hatahata* simmered in water, *ipponzuke*, pickled eggplants

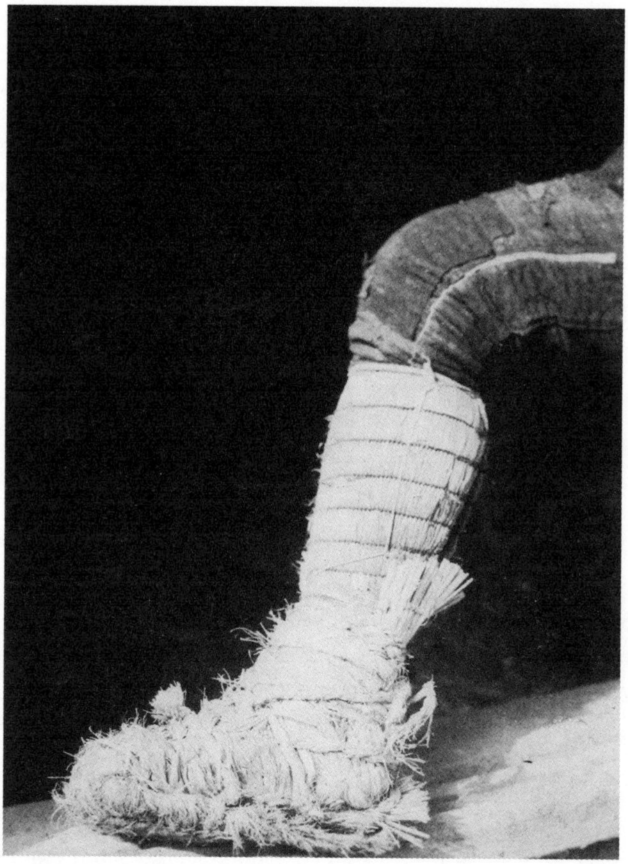

Figure 13.5. Winter footwear and legwear for working outdoors: *waragutsu, agutomaki, habaki.* Photo credit: Ōnishi Goichi. From Yoshida's personal collection: not included in 1938 book.

Supper:	*udon* noodles in soup made from dried sardines (*iwashi*) with some soy sauce, side dishes: same as morning
Snack:	roasted dried *mochi*

This morning I made a pair of *fumi dawara.*[6] They're for using next year. These *fumi dawara* are yet another tool (*dōgu*) that have pretty much vanished from this village. They're not hard to make. And I already had a *migo* rope, so I just used that to make them. You need 7 *hiro* (10.6 m) of rope, and you attach the pieces to *komozutsu,* just like when you make any *tawara.* And you want about a 4 *sun* (12.1 cm) distance, so I went with lengths of about 1 *shaku* 2 *sun* (36.4 cm). And I lightly pounded the base parts of the straws I used up to about 1 *shaku* 5–6 *sun* (approx. 47 cm) from the ends and used those. So, I faced the *komozutsu* and arranged the straws so that the bases were sticking out to my left and wove them through the ropes like that—with the tips to the right. You only need a few straws, maybe five or six each time.

So, as you go, after weaving twice, you move it down and then you bend the tips of the straws you wove first upward to the point where the third straws will go and touch them to that spot and hold them in place with your right hand. Then you take five or six straws in your right hand and put them in there and weave those straws in, through the ropes, always being certain to keep the pounded base parts of the straws on the right side. If you do that, about 4 *sun* (12.1 cm) of the base ends will remain on the left side each time. And each time you weave straws in, you must put them through all four ropes. So, you do the third straws like that, and then move them down, and then take the tips of the second straws and bend them upward to where you'll weave the fourth straws, and then do the same as you did with the third straws and go on and on like that until there is no more rope left to weave.

Next, you tie the part where you started to the part where you finished, just like when you make a *tawara* for rice. Then, you weave some thin rope (*kagari nawa*) and make a lid (*sadara*) to fit the *tawara*. You then fold the bottom edge of the *tawara* body inward a bit and fit the lid on and attach it with the rope you wove, and then wrap the rope around the body of the *tawara* in several places. Then, you weave two relatively thin but hard ropes about 2 *shaku* (60.1 cm) long, and bend one at the midpoint and put the two ends into the *tawara*, so that they're about 2 *sun* (6.1 cm) from the side and attach each to a place where you attached the lid with the rope, and then attach both ends to the *tawara*. Then, you do the same with the other rope, on the opposite side of the *tawara*. If you pull hard on the ropes [and they don't come loose] that's fine. That's how to make a *fumi dawara*, and if you make another, you'll have a pair (see Fig. 13.6).

Figure 13.6. A pair of *fumi dawara* in use. Photo credit: Ōnishi Goichi (detail). From Yoshida's personal collection: not included in 1938 book.

This morning I went to another person's house to get night soil, and I also collected the urine at my own place, and I had to clear a bunch of snow around the cesspit. My wife went out to buy some tofu dregs again today. In the afternoon she pulled some daikon out [from a winter storage place] and then she used them to make some dishes, and she did some other cooking.

Today's expenditures:

tofu dregs (*okara*)	9 sen
tofu (4 blocks)	24 sen (We gave the tofu to my wife's younger brother's family, in celebration of the birth [three days ago].)
udon	28 sen
dried roasted anchovies	8 sen

At the main house: Kanezō cut hay for the horse early, and after breakfast he wove more *tawara*. In the afternoon he used the horse to haul firewood from the wood shack to the house with a sled. My brother manned the shop, and his wife wove more rope for selling—a usual activity of hers lately. Tonight, my mother was preparing for what people call *nehan* [Nirvana Day] here, which means that she cooked food for offering to the ancestors at the household *butsudan*. She cooked red bean rice (*sekihan*) and a burdock *debu* dish, and she made a dish with dried leaves (*nonoha*) and dried royal fern (*zenmai*) and sesame miso, and such.

The main house's wood shack collapsed today under the weight of the snow.

March 8, Sunday

Cloudy
–6°C in the morning, 4°C in the afternoon

Breakfast:	daikon miso soup with kelp in it, *ipponzuke*, pickled eggplants
Lunch:	roasted salted *hatahata*, *ipponzuke*
Supper:	taro stem and whale miso soup with kelp in it, side dishes: same as morning

No matter how long I wait for the snow to melt away naturally, it shows no signs of doing that, and it is nearly time for me to start my hotbox preparations, so I have to get some of the snow out of the way. I started on that today. I borrowed a sled from the main house and used that to clear away some snow. In the afternoon, since today was a day for villagers to rest, I followed suit and took it easy for a bit. This morning my wife went to her brother's place to get some straw, and she stayed there and rested in the afternoon and returned in the evening.

At the main house: Kanezō cut hay for the horse early, and after breakfast he went out to Shinden to stomp down the snow to make paths. My brother went to an Agriculture Committee (Shinōkai) meeting. My mother went to the [Buddhist] temple to pray.

March 9, Monday

Sunny
–2°C in the morning, 6°C in the afternoon
Snow accumulation: 3 *sun* (9.1 cm)

Breakfast:	daikon miso soup with kelp in it, *ipponzuke,* pickled eggplants
Lunch:	salted *hatahata* and grated daikon and such cooked in water, side dishes: same as morning
Supper:	daikon and whale miso soup with kelp in it, side dishes: same as morning

Matsushita Junrō's wife gave birth to a baby boy today.

I cleared snow again. Starting today, my wife will take her lunch with her to the main house and make ropes for selling every day. I made lunch and ate with the kids.

At the main house: Kanezō cut hay for the horse early, and after breakfast he used the horse to haul a load of sand out to Shinden with the big sled, but there was so much snow that the horse was in danger of getting stuck and not being able to walk anymore—despite the snow-stomping Kanezō did yesterday—so he gave up and took the horse home. Instead, he wove lids (*sadara*) for *tawara.* My brother's wife wove more rope for selling. My brother manned the shop during the morning but had to go out on errands in the afternoon. The kids minded the shop in the afternoon. My mother went to another village on an errand, and she returned in the evening, and then she cooked supper and such.

March 10, Tuesday

Sometimes sunny, sometimes cloudy
Strong northwest wind
–2°C in the morning, 3°C in the afternoon
Snow accumulation: 2 *sun* (6.1 cm)

Breakfast and lunch:	exactly the same as yesterday
Supper:	dried taro stems and whale miso soup with *aosa* seaweed in it, roasted salted *hatahata, ipponzuke*
Snack:	roasted dried *mochi*

The *aosa* we ate today was from the Minami Iso coast; we got it from my sister-in-law.

Today my wife and I did exactly the same work as yesterday.

At the main house: Kanezō cut hay for the horse early, and later he affixed *kagari* ropes to *tawara.* My brother's wife minded the shop in the morning and wove ropes in the afternoon.

My brother did the same as yesterday. My mother cooked and such. The kids watched the shop in the afternoon.

March 11, Wednesday

Sunny
Northwest wind
–5°C in the morning, 5°C in the afternoon

Breakfast, lunch, and supper:	same as yesterday
Snack:	roasted dried *mochi*

Today it was the same as yesterday for both my wife and me. I just used the small sled for clearing snow today. I pulled it myself.

Today's expenditures:

candies	3 sen

At the main house: Kanezō made *tawara* from the early morning, and he affixed ropes to the *tawara* he'd made during the afternoon. Also, since he needs to start hauling *tawara* tomorrow, he prepared the horse-drawn sled and put the poles on it and loaded up some *tawara* and tied them down and such. My brother's wife wove ropes for selling again today. My brother was at home during the morning but went out again in the afternoon. The kids minded the shop again [during the afternoon]. My mother cooked and such.

Soon after the second-largest landowner in this village, Amano Gen'ichi, became mayor, the village was designated by the prefecture government as one of its leading "economic rehabilitation villages."[7] Most of the village's paddies and dry fields are located inside the village. I'm not sure if that has anything to do with it, but the mayor became the head of the [local] farmland reorganization association, and then he set about making new water reservoirs and fixing up the old ones, so even paddies that were always poorly irrigated [in the past] became beautiful almost overnight. But at the same time, the farmland prices went up, they say. I don't know exactly what the mayor is thinking right now, but I hear that he's been trying to convince smaller-scale farmers to sell their rice paddy land. And it seems that there are some farmers here who have taken advantage of the low-interest government loans to buy paddy land. And I've heard of another way—some who are unable to buy paddies outright have paid half the money and then had the titles transferred and then mortgaged the land to cover the rest of the cost.

One such farmer said: "We don't really want to buy paddy land so badly that we'd be willing to go into debt, but once you start buying like that, if you don't buy certain paddies, they'll be bought

Figure 13.7. The home of the village mayor, Amano Gen'ichi. From Yoshida's personal collection: not included in 1938 book.

up by someone else who has enough money, and then we won't be able to survive. When you think about it this way, we can't just focus on the future—we just have to borrow money and buy those paddies now."

But even if they build a bunch of reservoirs, the cost isn't borne only by the landlords. It puts a tremendous labor burden on the farmers. Some people are saying, "The association head just sells his own paddies, and then soon he'll initiate some paddy land reorganization measures."[8]

This mayor really is ambitious. This village has two schools, but he's trying to merge them into one, and public support for that is rising. What's more, he's trying to organize a merger between this village and two adjacent municipalities—Futto Village and Funakoshi Town—to make a new town called Higashi Oga Town. He's really pushing that—gradually aligning the various involved groups of people with his vision.[9]

March 12, Thursday

Cloudy, later sunny
Southeast wind
-1°C in the morning, 8°C in the afternoon

Breakfast:	miso soup with potato and *aosa* seaweed in it, *ipponzuke,* pickled eggplants
Lunch:	roasted salted *hatahata, ipponzuke,* etc.
Supper:	dried taro stems and whale miso soup, *ipponzuke,* etc.

Figure 13.8. Preparing the luggage of a man who is going away for *dekasegi* work (March 12, 1936, in front of Wakimoto Station).

Today I cleared more snow—the stuff covering my *waraniyo*[10] and what was surrounding my hotbox tools and parts. I pretty much finished clearing all the snow today. My wife wove rope for selling again.

At the main house: Kanezō loaded up the horse-drawn sled with *tawara* for the big fish trap (*daibōami*) early this morning. Then, after breakfast, he hauled the load out to the hamlet of Onagawa in Minami Iso Village.[11] He got back around two o'clock, and after having lunch he wove ropes for selling, at the main house. My brother's wife minded the shop in the morning, and wove more ropes for selling during the afternoon. As for my brother, it was the same as yesterday. The kids minded the shop [in the afternoon] and my mother cooked and such.

A fellow named Yoshida Saburobē sold some of his cedar and pine forest land, which faces the east side of a place called Hachimori, near Mt. Kanpū, and the trees were cut and hauled away. And prior to this, the same thing happened with pine forest land owned by Yoshida Yasuke at a place called Ishikura.

I have a friend named Yoshida Shūzō. When this young fellow was working as a stone mason, a big stone fell on his right foot, and bones were broken.[12] Consequently, he used up all his father's savings [for treatment]. Moreover, since that wasn't enough, he wound up going deeply into debt, and becoming unable to pay the rent on his residential property, and therefore getting kicked out of his house. Now, he's building a little tiny house alongside a prefectural road located some distance from the village.

APPENDIX 1: *MIKO* TALES

My mother went to the home of some of her father's relatives in another village[1] to hear what a *miko*[2] had to say. About five or six years ago, a daughter of this family went to Tokyo to work in a textile factory, but she got very sick and had to return to her village, where she soon met a miserable end. She was only seventeen.

When my mother arrived, the *miko* was going strong. As soon as my mother sat down, the *miko* turned her head and spoke.

"There's someone here—a relative of mine[3] on my father's side, who passed over the mountain to get here. I'd like to receive a cup of fresh water from that person."

The *miko* said this, so the people searched a bit among the group, and eventually it was decided that the *miko* was talking about my mother, so my mother filled a cup with fresh water and placed it on the household's *butsudan*. Then, the *miko* spoke again.

"Yes, yes, I really do appreciate that. I'm so grateful to you for coming all the way over the mountain to care for me by paying your respects like this—to everyone, for gathering and giving me fresh water—I really do appreciate it.[4] Thanks to you, I could finally join the ranks of 'normal' people. Really—I truly do appreciate it."

Everyone there was in tears by this time, I heard. And so, the *miko* continued speaking.

"There's someone here whose spouse is deceased and who is not well. If this person offers some *konnyaku* to the Ogane-sama[5] in Yabase they'll certainly recover, so you should believe me, and go make the offering."

After another search, it was determined that again, the *miko* was talking about my mother. Indeed, she has been complaining a lot lately of persistent headaches. She wept, saying that the *miko* was correct. Yabase is a reference to a certain place—there's an Ogane-sama in Terauchi Village in Yabase, South Akita County.[6] Ogane-sama is the same as Kōshin-sama.

But it seems that my mother has no intention of making an offering to Ogane-sama in Yabase. "I don't care what some *miko* says," she scoffed. Even though she was crying while the *miko* spoke, afterward she just blew it off like that.

There also was a boy who drowned when the *Eitoku Maru* went down. The boy, Yoshida Kenji, was from my hamlet. In order to let Kenji know about how things were going in the world of the living, the *miko* held a séance called *nanakurayose*. My mother also went for that. I'll write what I can, according to my mother's memory.

The *miko* spoke: "Ah, if only I'd stayed on my own family's boat, I wouldn't have wound up like this. But I was consumed by greed, so I rode on other boats all the time, and ended up like this. Heavy rain, strong wind, huge waves . . . like filling a pot with water and letting it sink in a well—the boat I was on went down just like that. We held on to each other's hands and cried out for help. Those with living parents called out their father's name, or their mother's name, and those with wives or children called out their wife's name, or the names of their children—firstborn, second-born, and so on. As for me, my father and my mother are already gone, and I am alone, with no spouse or child. The only person whose name I could call out was that of my elder brother. Thinking about it, it's really sad—pathetic—but even though I lost my body, I'm sure that eventually a big sum of money will come to this house, where my parents' spirits dwell. The household will become wealthy, and then its members will be envied by the neighbors."

The *miko* continued: "When I think now of my good friend, I really feel pathetic. Before I rode on that boat, I always shared everything I had with him—such a good friend . . . always played together. That friend of mine, he traveled far over the mountains and now he's in the city of Kyō,[7] living happily. Ah . . . oh . . . I'm so envious. That good buddy of mine left the place where he was before and moved away."

Again, the people tried to figure out who she was referring to, and eventually it was decided that she was talking about my younger brother, Shirō, who is now in Tokyo. After all, Kenji and Shirō were inseparable best friends.

The *miko* continued: "Burning the gods of various countries of the world is not acceptable. Even here, in this village, someone recently set a shrine on fire. Fortunately, someone passing by noticed, and the fire was put out before the shrine was destroyed. If this happens again, the gods are going to be very upset. The people of the village should pay very close attention to this situation."

She was referring, of course, to the "crazy neighbor" of mine who lives down below my fields. And as a matter of fact, I'm the one who noticed him trying to set fire to that shrine and stopped him.

Finally, I think I'll make note of some of the vocabulary that the *miko* used.

yomidori	father
heradori	mother
shigizu	relatives
narabi	couple [spouse]
tagara	child [children]
hadeda	to die [to be dead]
hobe	friend
hanamizu	fresh water
kamado	wealth [wealthy person]

Appendix 2: Maps

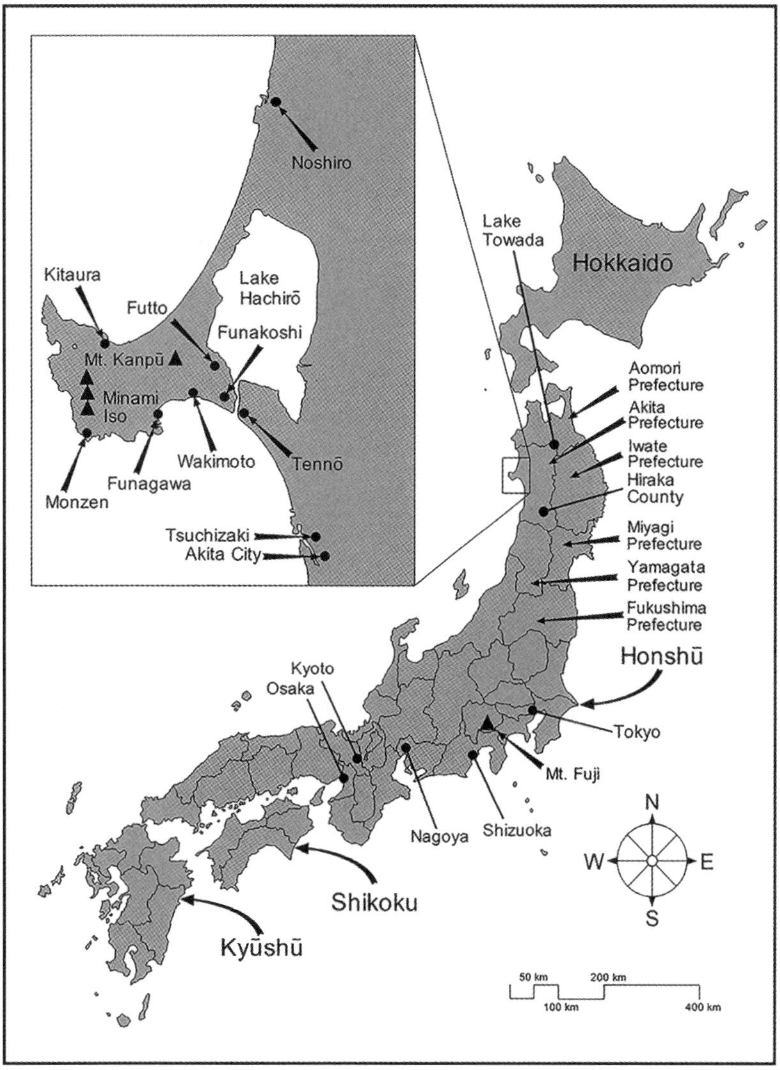

Map 1. Japan (without Okinawa and some islands), showing the Oga Peninsula and surrounding area in the 1930s. Illustration credit: Jack Walsh.

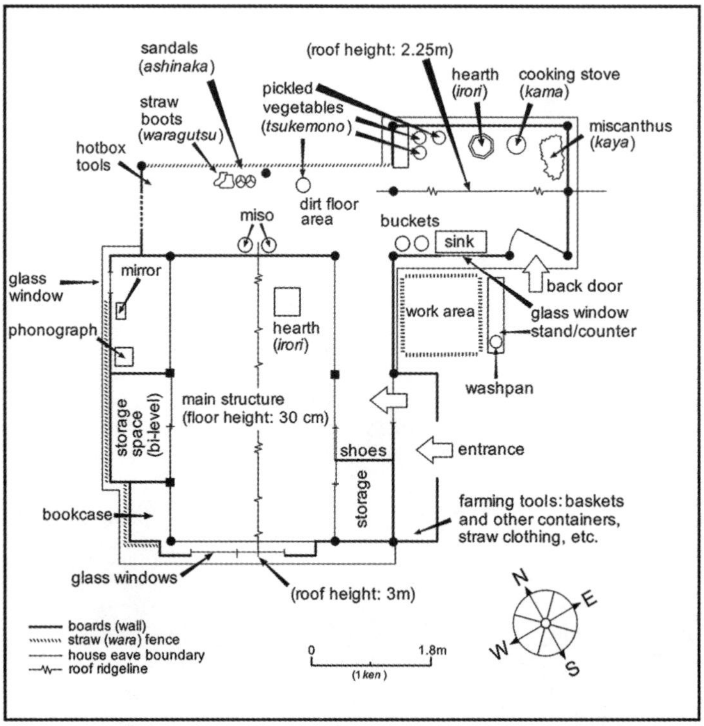

Map 2. House diagram, from original by Ogawa Tōru. Illustration credit: Jack Walsh.

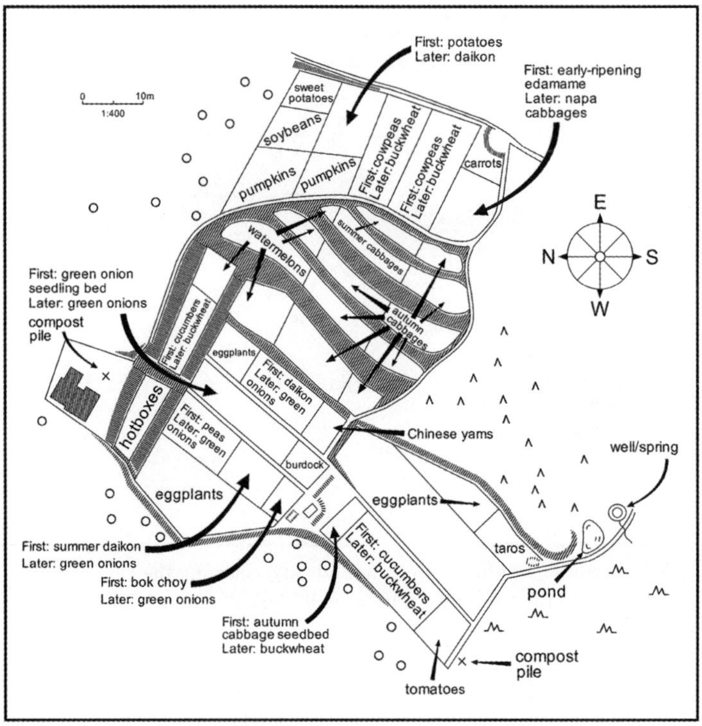

Map 3. Field use arrangement, 1930. Illustration credit: Jack Walsh.

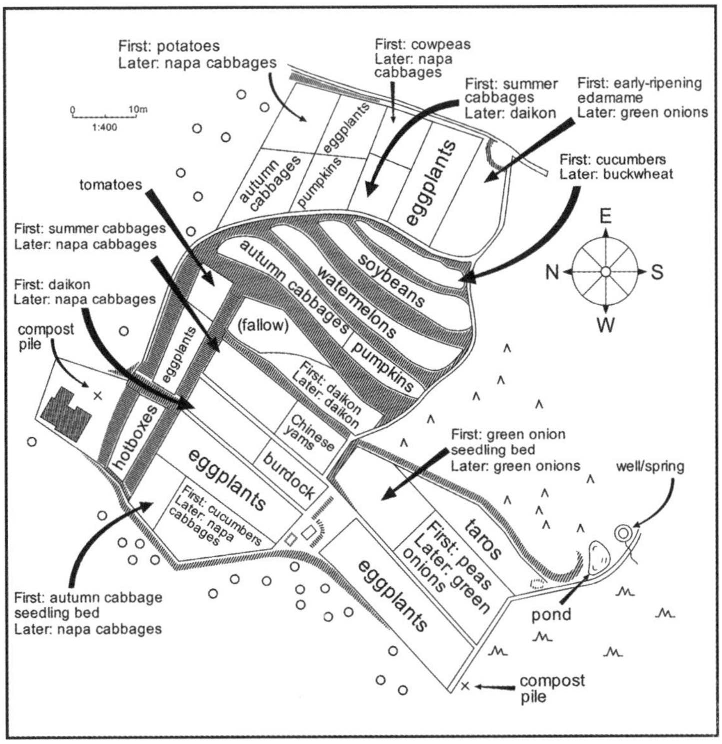

Map 4. Field use arrangement, 1931. Illustration credit: Jack Walsh.

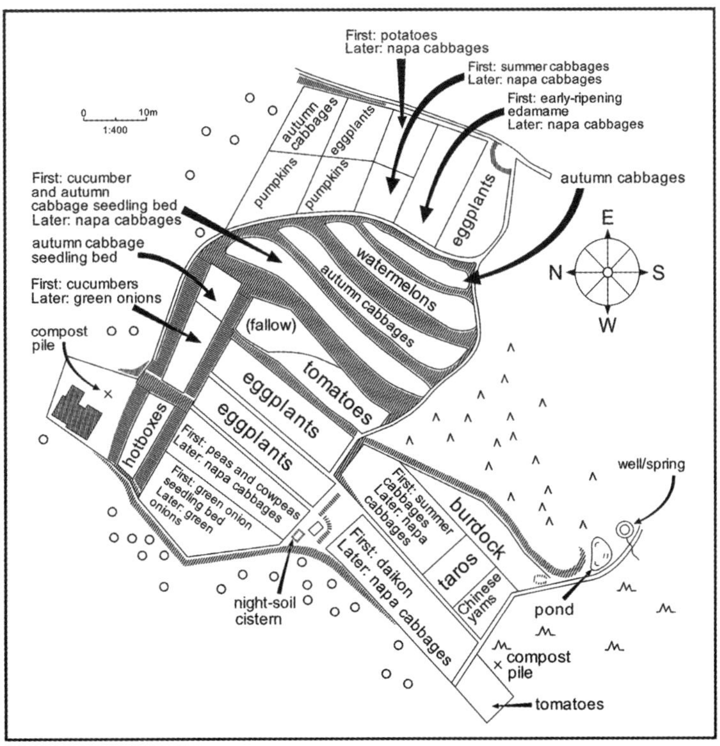

Map 5. Field use arrangement, 1932. Illustration credit: Jack Walsh.

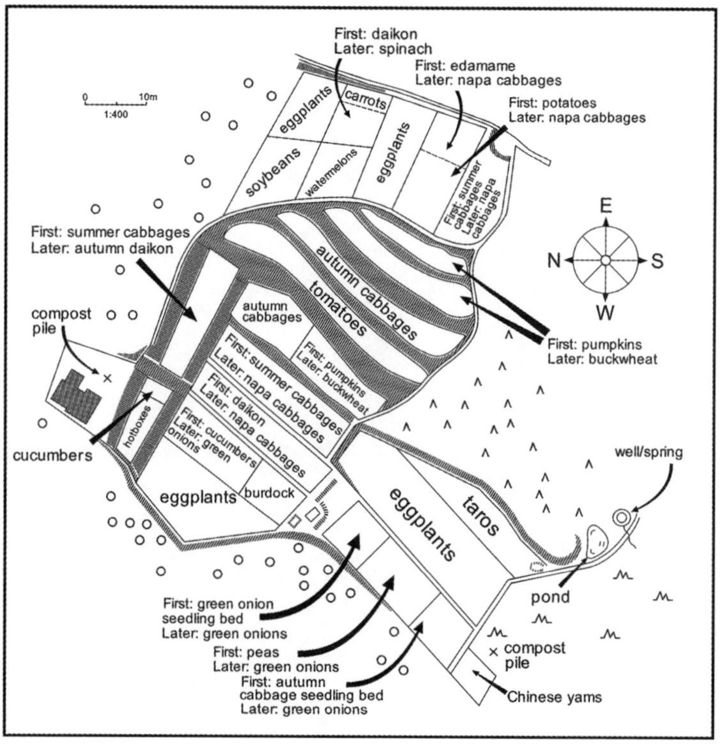

Map 6. Field use arrangement, 1933. Illustration credit: Jack Walsh.

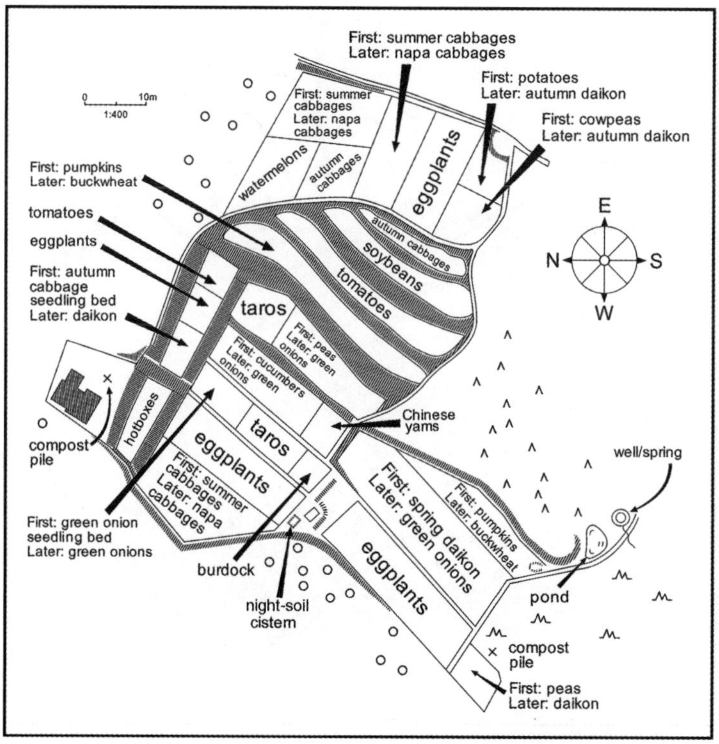

Map 7. Field use arrangement, 1934. Illustration credit: Jack Walsh.

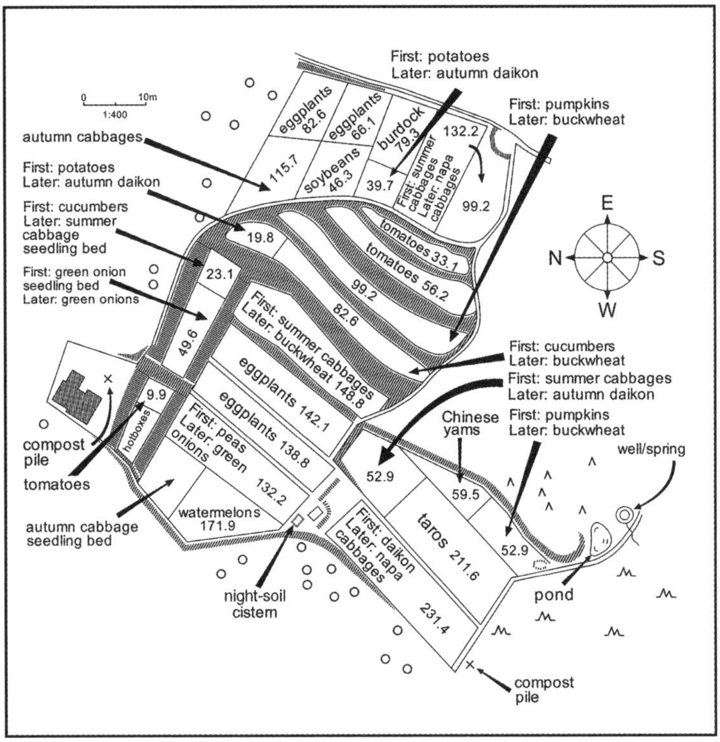

Map 8. Field use arrangement, 1935, with numbers indicating areas in square meters. Illustration credit: Jack Walsh.

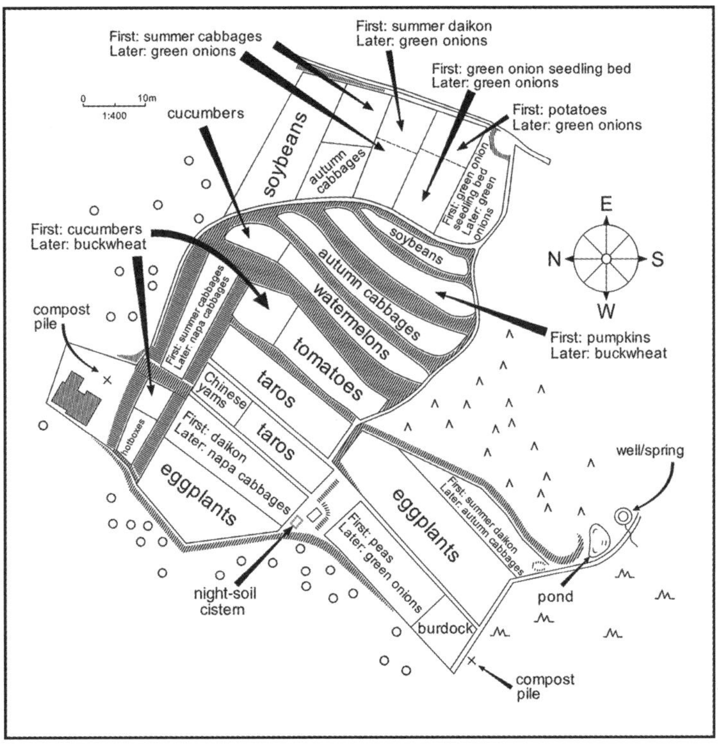

Map 9. Field use arrangement, 1936. Illustration credit: Jack Walsh.

Appendix 3: Financial Data for March 13, 1935 to March 12, 1936

Date	Item	Quantity	Income (in yen)	Expenditures (in yen)	Difference (in yen)
Mar. 15	stamps			0.30	
	seeds			1.64	
	postal fee for book			0.40	
16	kerosene	721.6 ml		0.12	
	a new glass cover for the lantern			0.10	
	ink from Maruzen			0.30	
	straw			1.00	
17	straw			2.00	
	candies			0.05	
18	soy sauce	541.2 ml		0.08	
	sugar			0.10	
22	fertilizer			12.50	
	brown sugar			0.10	
	candies			0.06	
	alum			0.04	
25	postcards	10		0.15	
	soy sauce	361 ml		0.06	
	candy	30		0.10	
27	candies			0.05	
28	cold medicine (antipyrine)			0.20	
	apples	2		0.05	
29	*shiroyo* (*shirauo*)	180.4 ml		0.10	
	tofu	1 block		0.06	
	white sugar			0.05	
	green tea			0.05	
	soy sauce	360.8 ml		0.10	
30	nails			0.05	
	Subtotal (March 13–31, 1935)		**0.00**	**19.81**	**–19.81**
Apr. 1	rabbit pelt tanning fee	4 pelts	0.50		
	candies			0.03	
	fee for mailing photo negatives			0.10	
5	postcards	10		0.15	
	candies for the kids	6		0.02	
	herring	4 fish		0.10	
7	Lion toothbrushing powder	1 bag		0.09	
	wire			0.07	
	alum			0.05	
	candies for the kids	4		0.02	

Date	Item	Quantity	Income (in yen)	Expenditures (in yen)	Difference (in yen)
8	salted herring	4 fish		0.10	
9	sweets for the kids			0.03	
11	brown sugar			0.10	
	sweets			0.05	
13	soy sauce	541.2 ml		0.08	
	seeds			0.50	
	herring	11 fish		0.20	
	rubber boots for the kids	2 pairs		0.60	
	stamps	10		0.30	
	postage for mailing copy of book			0.10	
	apples, candies	3, 6		0.12	
14	profits from selling daikon seeds		0.05		
15	wire			0.05	
	candies, sweet potatoes			0.05	
16	kids' haircuts			0.10	
	candies			0.02	
17	taro seeds			3.30	
	rubber hose for sprayer			1.60	
	train fare			1.00	
	gifts for the kids			0.20	
	summer cabbage seedlings			2.00	
20	nails			0.05	
	kids' allowances			0.07	
21	candies			0.05	
23	candies			0.03	
25	rabbit pelt tanning fee		0.50		
	cash given to friend			5.00	
	postage for photo film			0.23	
	postcards			0.06	
	udon	6 bunches		0.16	
	soy sauce	902 ml		0.13	
	treats for the kids			0.07	
26	salted herring	75 fish		1.50	
27	salt	3.6 L		0.20	
	kids' allowances			0.12	
29	kerosene	721.6 ml		0.12	
	mothballs			0.05	
	candies and treats			0.05	
	rabbit pelt tanning fee		0.30		
30	candies			0.05	
	Subtotal (April 1935)		**1.35**	**19.07**	**–17.72**
May 1	payment for some taros		2.00		
3	train fare			1.00	
	pens and envelopes			0.10	
	one hammer			0.20	
	tobacco powder	1.5 kg		0.25	
	bananas	375 g		0.25	
	toys for the kids			0.40	
	treats for the kids			0.25	

Date	Item	Quantity	Income (in yen)	Expenditures (in yen)	Difference (in yen)
4	*nanadogo gasa*	2		0.20	
	treats			0.05	
5	tobacco (cigarettes)	1 pack?		0.08	
	postcards			0.15	
	candies/treats			0.05	
	polished rice for helping	1.8 L	0.30		
8	salt	19.8 L		1.00	
	soybeans	18 L		1.80	
	brown sugar and candies			0.08	
9	harmonica		0.40		
11	salted herrings	35 fish		0.50	
	loss (reimbursement for ball)			0.05	
14	soy sauce	541.2 ml		0.09	
	ink			0.05	
	tofu	1 block		0.06	
	sweets, flour, cloth, etc.			0.10	
15	*konnyaku*	1 block		0.02	
	sweets			0.02	
	one red cord			0.01	
18	postage stamps			0.61	
	money for kids to buy food			0.10	
20	tomato seedlings	60	1.20		
	salt			0.10	
	sweets			0.05	
21	tomato seedlings	22	0.55		
22	tomato seedlings	42	0.90		
	sweets			0.05	
	soy sauce	541.2 ml		0.10	
23	rolled oats	1 bag		2.40	
	copper sulfate	1.9 kg		0.75	
	black cotton cloth			0.10	
	sweets			0.05	
24	tomato and eggplant seedlings	10, 50	0.50		
26	eggplant seedlings	100	0.57		
	sweets			0.05	
28	brown sugar			0.05	
	Subtotal (May 1935)		**6.42**	**11.17**	**–4.75**
Jun. 1	cash for the kids			0.05	
3	white sugar			0.10	
	soy sauce	361 ml		0.10	
6	sweets			0.10	
8	brown sugar			0.05	
	candies			0.05	
12	haircuts for the kids			0.10	
	sweets			0.10	
13	green peas	937.5 g	0.25		
	stamps and postcards			0.40	
	postal fee for mailing photo film			0.10	
	sweets			0.05	

Date	Item	Quantity	Income (in yen)	Expenditures (in yen)	Difference (in yen)
14	green peas		0.25		
	eggplant seedlings		0.40		
17	sugar			0.08	
	soy sauce			0.03	
	udon			0.10	
	konnyaku			0.06	
18	green peas		0.45		
	daikon		0.24		
19	green peas		0.15		
	daikon		0.55		
21	daikon		0.60		
	green peas		0.28		
	snacks			0.10	
22	green peas		0.13		
	daikon		0.10		
23	daikon		0.60		
	green peas		0.42		
	jikatabi	2		0.75	
	secondhand shirts	2		0.40	
	snacks			0.14	
24	daikon		0.60		
	green peas		0.36		
	postage on package to Tokyo			0.75	
	fish for mother			0.10	
25	daikon		0.60		
	green peas		0.40		
	kerosene			0.10	
	candies			0.10	
26	daikon		0.50		
	green peas		0.20		
	soy sauce	541.2 ml		0.09	
	snacks			0.05	
27	daikon		0.60		
	green peas		0.05		
	used kimonos for the kids			0.15	
	mackerel	2 fish		0.12	
28	daikon		0.60		
	green peas		0.20		
30	snacks			0.02	
	nails			0.05	
	Subtotal (June 1935)		**8.53**	**4.39**	**4.14**
Jul. 1	daikon		0.60		
	udon	8 bunches		0.20	
	candies			0.02	
2	daikon		0.35		
	summer cabbages	4.9 kg	0.45		
	green peas		0.03		
	sweets			0.05	

Date	Item	Quantity	Income (in yen)	Expenditures (in yen)	Difference (in yen)
3	daikon		0.25		
	green peas		0.10		
	cabbages	7.5 kg	0.55		
	postage stamps			0.08	
4	daikon		0.20		
	cabbages	15 kg	1.05		
	envelopes			0.05	
	sweets and candies			0.05	
5	daikon		0.20		
	cabbages	16.1 kg	1.05		
	snacks and candies			0.05	
6	daikon		0.20		
	green peas		0.10		
	cabbages	24.4 kg	1.87		
7	daikon		0.20		
	green peas		0.05		
	cabbages	17.25 kg	1.25		
8	daikon		0.10		
	cabbages	21.75 kg	1.32		
9	daikon		0.10		
	cabbages	24.75 kg	1.50		
	postage stamps			0.03	
	agricultural chemicals			0.75	
10	daikon		0.12		
	cabbages	25.1 kg	1.20		
	money for selling a dove chick		0.30		
	postage stamps			0.10	
11	daikon		0.10		
	cabbages	24 kg	1.10		
	eggplants		0.10		
	a *koshimaki* for my wife			0.50	
	deworming medicine			0.05	
	candies			0.05	
12	daikon		0.10		
	eggplants		0.40		
	cucumbers		0.40		
	cabbages	30 kg	1.65		
	snacks			0.05	
13	daikon		0.10		
	cucumbers		0.10		
	cabbages	67.5 kg	3.20		
	scrap iron (sold)		0.05		
	soy sauce	360 ml		0.06	
	sweets			0.05	
14	daikon		0.10		
	eggplants		0.30		
	cucumbers		0.30		
	cabbages	14.25 kg	0.72		
	various sweets			0.10	

Date	Item	Quantity	Income (in yen)	Expenditures (in yen)	Difference (in yen)
15	eggplants	24	0.35		
	cucumbers	24	0.30		
	cabbages	14.25 kg	0.60		
	seeds			1.25	
	stamps and postcards			0.25	
	snacks			0.03	
16	daikon		0.13		
	cucumbers		0.50		
	eggplants		0.50		
	cabbages	21 kg	1.00		
	sweets			0.05	
18	daikon		0.10		
	cucumbers		0.36		
	eggplants		0.60		
	cabbages	23.25 kg	1.25		
	paper for mounting photos			0.10	
	sweets			0.05	
19	daikon	12	0.10		
	eggplants	32	0.50		
	cabbages	13.1 kg	0.57		
20	daikon		0.10		
	eggplants	32	0.50		
	cabbages	18.75 kg	0.78		
	sweets			0.10	
	salted trout	1 fish		0.30	
21	eggplants	32	0.57		
	cucumbers	25	0.25		
	cabbages	18.75 kg	0.90		
22	eggplants	50	0.90		
	cucumbers	25	0.25		
	cabbages	9.4 kg	0.40		
	white cloth			0.08	
	candies			0.02	
23	eggplants	60	1.10		
	cucumbers	30	0.22		
	cabbages	15.75 kg	0.65		
	mosquito repellent			0.06	
	candies, etc.			0.04	
24	eggplants	60	1.00		
	cucumbers	25	0.20		
	cabbages	12 kg	0.50		
	soy sauce			0.09	
	candies			0.03	
25	eggplants		0.77		
	cucumbers		0.30		
	cabbages	14.25 kg	0.75		
	postage stamps			0.06	
	black cotton cloth			0.10	
	candies			0.03	

Date	Item	Quantity	Income (in yen)	Expenditures (in yen)	Difference (in yen)
26	eggplants	45	0.65		
	cucumbers	30	0.25		
	cabbages	12.75 kg	0.50		
	kerosene	720 ml		0.12	
	roasted dried anchovies			0.05	
	lamp wick			0.01	
	white sugar			0.05	
	candies			0.05	
27	eggplants	60	0.50		
	cucumbers	33	0.25		
	tomatoes	1.1 kg	0.15		
	cabbages	9.4 kg	0.40		
	salt			0.10	
28	eggplants	50	0.70		
	cucumbers	33	0.25		
	tomatoes	37 (3.75 kg)	0.65		
	cabbages	12 kg	0.60		
29	eggplants	111	1.30		
	cucumbers	28	0.20		
	tomatoes	40	0.50		
	cabbages	10.5 kg	0.35		
	postage for mailing a manuscript			0.15	
	candies			0.05	
30	eggplants	65	0.84		
	cucumbers	27	0.20		
	tomatoes	37 (3.75 kg)	0.50		
	cabbages	9.75 kg	0.40		
	udon	3 kg		0.40	
	soy sauce	900 ml		0.14	
	white and brown sugar			0.10	
	konnyaku and *fu*			0.05	
	fish			0.15	
	oil			0.10	
	postage stamps			0.09	
	soap			0.05	
31	eggplants	80	1.00		
	tomatoes	60 (6.75 g)	0.90		
	cabbages	12.4 kg	0.60		
	Subtotal (July 1935)		**50.55**	**6.54**	**44.01**
Aug. 1	eggplants	61	0.70		
	cucumbers	34	0.13		
	tomatoes	46 (6.4 kg)	0.75		
	cabbages	6 kg	0.25		
	snacks			0.10	
2	eggplants	61	0.73		
	tomatoes	60 (7.5 kg)	0.70		
	cabbages	6 kg	0.30		

Date	Item	Quantity	Income (in yen)	Expenditures (in yen)	Difference (in yen)
3	eggplants	113	1.15		
	tomatoes	85 (12.4 kg)	1.00		
	postage stamps			0.06	
	candies			0.04	
4	eggplants	80	0.95		
	tomatoes	70 (11.3 kg)	1.00		
	cabbages	5.6 kg	0.25		
	rice	4.5 L		6.50	
	soap			0.05	
	candies			0.03	
5	eggplants	74	0.70		
	tomatoes	80 (10.5 kg)	0.80		
	cabbages	6 kg	0.25		
6	eggplants	80	0.75		
	tomatoes	52 (5.6 kg)	0.62		
	cabbages	9 kg	0.40		
	tax			0.76	
	sweets made from candy			0.05	
7	eggplants	140	1.18		
	tomatoes	56 (6.4 kg)	0.55		
	cucumbers	30	0.15		
	pumpkins	2.3 kg	0.15		
	candies			0.02	
8	eggplants	90	0.70		
	tomatoes	75 (10.5 kg)	0.75		
	cabbages	11.5 kg	0.48		
	mochi made from wheat flour			0.05	
9	eggplants	104	1.00		
	tomatoes	67 (8.25 kg)	0.60		
	pumpkins	2.6 kg	0.15		
	cabbages	3.75 kg	0.15		
10	eggplants	170	1.20		
	tomatoes	24 (3 kg)	0.24		
	pumpkins	2 (2.6 kg)	0.15		
	hat	1		0.45	
	spending money for my wife			2.00	
	kimono cloth			1.50	
	salted trout	1 fish		0.50	
	whale			0.15	
	summer *geta* for kids	2 pairs		0.30	
	scrubbing brushes (*tawashi*)	2		0.05	
	teacups	3		0.10	
	agar-agar	375 g		0.25	
11	eggplants	280	2.03		
	tomatoes	102 (12.8 kg)	0.95		
	pumpkins	10 (10.5 kg)	0.50		
	cabbages	38.25 kg	1.70		
	soy sauce	1.8 L		0.30	
	white and brown sugar			0.10	

Date	Item	Quantity	Income (in yen)	Expenditures (in yen)	Difference (in yen)
12	eggplants	72	0.60		
	tomatoes	16	0.15		
	cabbages	6 kg	0.32		
	cooking oil			0.05	
	candies			0.05	
13	eggplants	25	0.20		
	tomatoes	42	0.41		
	cabbages	13.1 kg	0.47		
	candies			0.05	
	candles			0.01	
14	eggplants	64	0.60		
	tomatoes	79	0.55		
	candies			0.05	
15	eggplants	418	2.60		
	tomatoes	6	0.05		
	cabbages	6 kg	0.30		
	apron			0.15	
	candies			0.05	
16	eggplants	110	0.65		
	pumpkins	4 (3.75 kg)	0.20		
	cabbages	3 kg	0.15		
	tomatoes	70 (7.5 kg)	0.60		
	polished rice	1.8 L		0.27	
	postage for mailing photo film			0.10	
17	eggplants	140	0.82		
	tomatoes	25 (3 kg)	0.20		
	pumpkins	4 (5.6 kg)	0.30		
	cabbages	15 kg	0.60		
18	eggplants	90	0.55		
	tomatoes	56 (6 kg)	0.35		
	pumpkins	6 (5.6 kg)	0.30		
	cabbages	8.6 kg	0.40		
	(school) alumni association party fee			1.50	
	processed sugar			0.10	
	candies			0.05	
19	eggplants	111	0.75		
	tomatoes	42 (3.75 kg)	0.25		
	pumpkins	6 (6.4 kg)	0.33		
	cabbages	5.25 kg	0.25		
	book			2.00	
20	eggplants	170	0.88		
	tomatoes	90 (9 kg)	0.52		
	pumpkins	3 (5.25 kg)	0.30		
	cabbages	6.75 kg	0.30		
	salt	1.8 L		0.10	
	fish			0.10	
	candies			0.05	

Date	Item	Quantity	Income (in yen)	Expenditures (in yen)	Difference (in yen)
21	eggplants	80	0.45		
	tomatoes	124 (8.6 kg)	0.55		
	pumpkins	4 (4.9 kg)	0.25		
	cabbages	9.4 kg	0.40		
	chemicals			0.65	
22	eggplants	85	0.45		
	pumpkins	3 (4.5 kg)	0.25		
	watermelon	1 (3.4 kg)	0.20		
	cabbages	9.4 kg	0.40		
23	eggplants	80	0.40		
	tomatoes	37 (3.75 kg)	0.25		
	pumpkins	2 (3 kg)	0.20		
	cabbages	8.6 kg	0.35		
24	eggplants	135	0.90		
	tomatoes	20 (1.9 kg)	0.10		
	salt			0.20	
	candies			0.05	
25	eggplants	84	0.50		
	tomatoes	50 (4.5 kg)	0.30		
	watermelons	9.75 kg	0.45		
	cabbages	6.75 kg	0.20		
	polished rice	1.8 L		0.25	
26	eggplants	70	0.35		
	watermelons	31 (13.5 kg)	0.62		
	cabbages	6.4 kg	0.30		
	cotton cloth			0.20	
	mochi			0.05	
27	eggplants	120	0.60		
	tomatoes	40	0.20		
	watermelons	4 (20.6 kg)	1.05		
	cabbages	11.25 kg	0.40		
28	eggplants	125	0.63		
	tomatoes	15 (1.1 kg)	0.05		
	pumpkins	3 (6 kg)	0.25		
	cabbages	9 kg	0.35		
	rolled oats	1 bag		2.40	
29	eggplants	110	0.53		
	tomatoes	15 (1.1 kg)	0.07		
	watermelons	4 (20.25 kg)	0.95		
	cabbages	9.75 kg	0.35		
	conical hat (*karakasa*)			0.60	
	fish-shaped candies			0.15	
30	eggplants	144	0.75		
	tomatoes	15 (1.1 kg)	0.06		
	watermelons	1 (4.9 kg)	0.25		
	onions	14.25 kg	0.60		

Date	Item	Quantity	Income (in yen)	Expenditures (in yen)	Difference (in yen)
31	eggplants	105	0.50		
	tomatoes	18 (2.25 kg)	0.10		
	watermelons	4 (11.25 kg)	0.53		
	onions	5.25 kg	0.27		
	whale			0.10	
	tsukudani rice topping			0.50	
	Subtotal (August 1935)		**55.37**	**23.19**	**32.18**
Sept. 1	eggplants	100	0.40		
	watermelons	1 (9 kg)	0.30		
	onions	3.75 kg	0.20		
	candies			0.05	
2	eggplants	198	0.80		
	tomatoes	20 (1.3 kg)	0.07		
	watermelons	1 (5.6 kg)	0.20		
	cabbages	5.25 kg	0.25		
	rice			0.35	
3	eggplants	122	0.60		
	watermelons	2 (9.4 kg)	0.35		
	summer cabbage seeds			1.40	
	postcards			0.15	
4	eggplants	190	0.73		
	watermelons	2 (9.75 kg)	0.35		
	kerosene			0.12	
5	eggplants	200	0.95		
	watermelons	15 kg	0.55		
	rice polishing fee			0.15	
6	eggplants	160	0.90		
	watermelons	2 (7.5 kg)	0.28		
7	eggplants	57	0.35		
	pumpkins	27 (3 kg)	0.10		
	watermelons	2 (8.25 kg)	0.25		
	fish			0.10	
	sweets			0.05	
8	eggplants	110	0.65		
	watermelons	12 kg	0.45		
	train fare			0.18	
9	eggplants	65	0.55		
10	eggplants	93	0.45		
	pumpkins	5 (22.5 kg)	0.72		
11	eggplants	150	0.85		
	watermelons	2 (12 kg)	0.42		
	cabbages	3.75 kg	0.20		
	cotton			0.30	
12	eggplants	25	0.20		
	watermelons	5.25 kg	0.65		
	cabbages	6.4 kg	0.30		
	coarse brown sugar			0.10	
	pears			0.10	
	sweets			0.05	

Date	Item	Quantity	Income (in yen)	Expenditures (in yen)	Difference (in yen)
13	eggplants	20	0.10		
15	eggplants	40	0.15		
	rice	7.2 L		1.00	
16	eggplants	123	0.60		
	watermelons	3 (13.1 kg)	0.50		
18	eggplants	60	0.35		
	matches			0.05	
	shijigo (fish)			0.10	
	mochi			0.05	
19	eggplants	20	0.10		
	white sugar			0.05	
	soy sauce			0.06	
	pears			0.10	
	fish			0.10	
20	hospital fee			16.75	
	fertilizer (anchovy powder)	5.6 kg		0.70	
22	eggplants	36	0.20		
	candies			0.02	
23	train fare			1.05	
	canned whale and sweets			0.40	
	book			1.60	
24	eggplants	190	0.95		
	young rabbits (sold)	3	0.45		
	postage stamps			0.03	
25	tobacco (cigarettes or powder)			0.10	
26	eggplants	75	0.40		
	young rabbits (sold)	2	0.40		
	"get well soon" cash for Fusae		0.50		
	paper			0.05	
	sweets			0.05	
27	hospital fee			2.76	
28	eggplants	200	0.77		
	daikon		0.10		
	soy sauce	1.8 L		0.15	
29	eggplants	180	0.80		
	daikon		0.15		
30	salted trout			0.40	
	Subtotal (September 1935)		**19.59**	**28.67**	**-9.08**
Oct. 2	tooth powder			0.10	
	kerosene			0.10	
	sweets			0.10	
3	rice	7.2 L		1.20	
	white sugar			0.10	
5	fish			0.10	
	work gloves			0.12	
	brown sugar			0.08	
7	postage stamps			0.10	
9	eggplants	60	0.20		
10	eggplants	20	0.10		

Date	Item	Quantity	Income (in yen)	Expenditures (in yen)	Difference (in yen)
11	*shijigo* (fish)			0.10	
	brown sugar			0.05	
13	candies, apples			0.05	
14	eggplants	18	0.10		
15	salt			0.10	
	salted salmon			0.50	
	brown sugar			0.05	
16	candies for the kids			0.05	
17	*shijigo* (fish)			0.10	
	white sugar			0.05	
18	rolled oats	1 bag		2.30	
	sweets			0.05	
19	eggplants	170	0.50		
	taros	10.8 L	0.75		
	cabbages	3 kg	0.15		
	soy sauce			0.10	
	sweets			0.05	
	salt			0.15	
20	eggplants	190	0.50		
	taros	9.9 L	0.60		
	cabbages	3.75 kg	0.20		
	pears			0.05	
21	eggplants	190	0.40		
	taros	14.4 L	0.80		
22	cabbages	7.5 kg	0.30		
	taros	12.6 L	0.72		
	rice	60 kg		11.15	
	candies			0.05	
	soap			0.05	
23	eggplants (tiny)	9 L	0.40		
	taros	12.6 L	0.72		
	napa cabbages	4.9 kg	0.15		
	candies for the kids			0.03	
24	eggplants	9 L	0.50		
	taros	13.5 L	0.81		
	daikon	8	0.12		
	brown sugar			0.05	
25	eggplants	9 L	0.50		
	daikon	6	0.10		
	taros	12.6 L	0.70		
	brown sugar			0.10	
	white sugar			0.10	
26	green onions	6 bunches	0.10		
	daikon		0.10		
	taros	14.4 L	0.70		
27	cabbages	6.75 kg	0.25		
	daikon	10	0.15		
	taros	14.4 L	0.70		
	shijigo (fish)			0.10	
	pears	7.5 kg		0.60	

Date	Item	Quantity	Income (in yen)	Expenditures (in yen)	Difference (in yen)
28	postage stamps			0.06	
	sweets			0.04	
29	eggplants	3.6 L	0.20		
	daikon	7	0.10		
	cabbages	7.5 kg	0.30		
	taros	12.6 L	0.65		
	scale			0.33	
30	napa cabbages	5.6 kg	0.25		
	daikon	10	0.15		
	taros	12.6 L	0.70		
	udon			0.22	
	soy sauce			0.15	
	dried anchovies			0.05	
	sweets			0.03	
31	bulb daikon (radishes)	4	0.10		
	napa cabbages	5.6 kg	0.20		
	cabbages	4.5 kg	0.20		
	taros	9 L	0.50		
	geta (sandals)	2 pairs		0.10	
	Subtotal (October 1935)		**14.67**	**18.96**	**−4.29**
Nov. 1	radishes	2	0.10		
	napa cabbages	6 kg	0.25		
	taros	14.4 L	0.85		
	salt			0.10	
	salted trout			0.10	
	stamps			0.06	
2	daikon	10	0.10		
	napa cabbages	6.4 kg	0.20		
	taros	12.6 L	0.70		
	straw			5.25	
	fish			0.10	
3	radishes	4	0.08		
	napa cabbages	5.6 kg	0.20		
	cabbages	3.75 kg	0.14		
	taros	10.8 L	0.60		
	sweets			0.10	
4	daikon	10	0.15		
	napa cabbages	6 kg	0.20		
	cabbages	2.6 kg	0.15		
	taros	12.6 L	0.75		
	granulated sugar			0.10	
5	radishes	8	0.15		
	napa cabbages	6.75 kg	0.25		
	taros	10.8 L	0.70		
6	daikon	5	0.05		
	napa cabbages	6.75 kg	0.25		
	cabbages	4.9 kg	0.20		
	taros	10.8 L	0.65		
	spending money for my wife			0.25	

Date	Item	Quantity	Income (in yen)	Expenditures (in yen)	Difference (in yen)
7	*tawara* of rice	2 (120 kg)		22.00	
	hulling/polishing			0.10	
	rice bran			0.40	
	kerosene			0.15	
8	green onions		0.20		
	napa cabbages	6.75 kg	0.25		
	taros	10.8 L	0.65		
9	green onions	12 bundles	0.20		
	napa cabbages	7.7 kg	0.25		
	taros	7.2 L	0.35		
	fish			0.10	
	sweets			0.05	
10	napa cabbages	15.75 kg	0.40		
	taros	9 L	0.50		
	writing paper			0.10	
11	chicken eggs	2		0.05	
13	rubber boots (for adult)			2.20	
	rubber boots (for child)			0.50	
14	rabbit pelt tanning fee	2 pelts	0.50		
16	napa cabbages	18.75 kg	0.50		
	taros	4.5 L	0.25		
17	Chinese yams	5.6 kg	0.50		
	napa cabbages	18.75 kg	0.50		
	salt			0.30	
	candies			0.05	
18	Chinese yams	4.5 kg	0.40		
	napa cabbages	75 kg	2.00		
	soy sauce			0.05	
	sweets			0.05	
19	cabbages	5.25 kg	0.25		
	napa cabbages	26.25 kg	0.70		
	young rabbit	1	0.50		
	loose-leaf paper			0.15	
	candies and sweets			0.05	
20	Chinese yams	5.25 kg	0.40		
	napa cabbages	22.5 kg	0.60		
21	Chinese yams	3.75 kg	0.35		
	napa cabbages	22.5 kg	0.55		
22	participation fee for event			4.00	
	train fare			1.00	
	a new diary for 1936			0.35	
	loose-leaf writing paper			0.35	
	folding umbrella			0.80	
	meals			0.60	
	Golden Bat cigarettes (for friend)	5 packs		0.35	
	magazine			0.10	

Date	Item	Quantity	Income (in yen)	Expenditures (in yen)	Difference (in yen)
23	Chinese yams	3.75 kg	0.35		
	napa cabbages	22.5 kg	0.60		
	soy sauce	1.8 L		0.15	
	tofu	1 block		0.06	
	toothbrushes			0.15	
	loose-leaf paper			0.10	
24	napa cabbages	63.75 kg	1.56		
	refined sugar			0.10	
26	napa cabbages	28.1 kg	0.73		
	red flannel cloth			0.23	
27	napa cabbages	24.4 kg	0.65		
28	napa cabbages	16.9 kg	0.42		
	cabbages	6 kg	0.30		
	postage stamps			0.05	
	candies			0.04	
	a small pot			0.10	
29	kerosene	720 ml		0.15	
	brown sugar			0.05	
	Subtotal (November 1935)		**22.13**	**41.09**	**-18.96**
Dec. 1	cabbages	11.25 kg	0.40		
	green onions	6 bunches	0.10		
	salted salmon			0.50	
	sweets			0.05	
2	tofu			0.06	
	refined sugar			0.05	
	sweets			0.15	
3	*dokata* labor wages		1.40		
	having old cotton redone	5.6 kg		0.60	
	sweets			0.05	
4	green onions	7.5 kg	0.30		
5	brown sugar			0.05	
7	salmon			0.10	
	sweets			0.05	
8	soy sauce			0.20	
	konnyaku			0.05	
9	wedding gift (*hide*)			0.50	
10	kerosene			0.10	
11	peanuts (*nankin*/*kanto* beans)			0.05	
12	charcoal (from Hokkaidō)	37.5 kg		1.90	
13	salt	5.8 L		0.40	
16	*hatahata*	2 barrels		1.75	
17	green onions	22 bundles	0.40		
	burdock roots	5.6 kg	0.30		
	cabbages	5.25 kg	0.28		
	dried alum	1 bag		0.30	
	laundry detergent	2 bags		0.20	
	headache medicine			0.20	
	chinomichi gusuri			0.30	

Date	Item	Quantity	Income (in yen)	Expenditures (in yen)	Difference (in yen)
18	postage on small package			0.10	
	scarfs (*erimaki*) for the kids' kimonos			0.15	
19	postage stamps			0.10	
	brown sugar			0.10	
20	*dokata* labor wages		4.62		
	wooden barrel	1		1.00	
	loose-leaf paper			0.45	
	candies			0.05	
	rolled oats	2 bags		4.60	
22	brown sugar			0.10	
23	green onions	1.9 kg	1.10		
	kerosene	1.8 L		0.30	
26	green onions	32 bundles	0.60		
	burdock roots	5.6 kg	0.30		
	mandarin oranges (*mikan*)			0.05	
27	green onions	30 bundles	0.55		
	burdock roots	5.6 kg	0.30		
	cabbages	4.1 kg	0.19		
	sweets and candies			0.05	
29	green onions	36 bundles	0.65		
	burdock roots	5.6 kg	0.30		
	cabbages	5.6 kg	0.30		
30	green onions	36 bundles	0.70		
	burdock roots	5.6 kg	0.35		
	konnyaku			0.05	
	brown sugar			0.15	
	sheets of cotton	2		0.40	
	cotton cloth			0.20	
	hair oil			0.05	
	candies			0.05	
	Subtotal (December 1935)		**13.14**	**15.56**	**–2.42**
Jan. 2	postage stamps			0.10	
	sweets			0.10	
3	candies for the kids			0.03	
4	chicken eggs	5		0.15	
	candies			0.02	
6	brown sugar			0.20	
8	magazine			0.25	
10	soybeans	1.8 L		1.70	
	candies			0.03	
11	postage stamps			0.06	
	brown sugar			0.05	
12	rabbit pelt tanning fee		0.50		
14	toothbrushes (*hamigaki yōji*)			0.10	
	Lion toothbrushing powder			0.10	
	candies			0.05	
16	haircuts for the kids			0.20	
	candies			0.03	

Date	Item	Quantity	Income (in yen)	Expenditures (in yen)	Difference (in yen)
18	matches			0.05	
	brown sugar			0.05	
19	kids' *geta*	2 pairs		0.30	
	saké lees (*sakagasu*)			0.15	
	kelp (*konbu*)			0.10	
	whale			0.25	
	medicine for "snow burn"			0.10	
	futoko candy			0.15	
	apples			0.05	
	sweets			0.10	
	shagushi			0.10	
	metal screen (*wadashi*)			0.10	
20	burdock roots	37.5 kg	2.00		
	green onions	68 bundles	1.20		
21	green onions	50 bundles	1.00		
	burdock roots	3.75 kg	0.20		
	cabbages	5.25 kg	0.25		
22	cabbages	72.2 kg	0.80		
	green onions	31 bundles	0.70		
	spending money for my wife			1.40	
	tofu	1 block		0.05	
	soybean curds (*okara*)			0.10	
	candies			0.05	
23	bucket	1		0.30	
	cod and squid			0.30	
	soy sauce			0.21	
	udon			0.24	
28	fee for mailing package			0.40	
	social expenses			0.10	
29	train fare			0.14	
	postage stamps			0.06	
	green onion seeds	541.2 ml		2.40	
	seed peas	721.6 ml		0.52	
	burdock seeds	360.8 ml		0.40	
	tomato seeds	72.2 ml		0.50	
	cabbage seeds	36.1 ml		0.35	
	daikon seeds	180.4 ml		0.50	
	postal order fee			0.10	
30	tofu	1 block		0.06	
	kerosene	1.8 L		0.30	
31	*konnyaku*	1 block		0.02	
	Subtotal (January 1936)		**6.65**	**13.17**	**–6.52**
Feb. 2	brown sugar			0.10	
3	salt			0.10	
4	clock repair			0.80	
	train fare			0.14	
5	brown sugar			0.20	
6	apples			0.05	

Date	Item	Quantity	Income (in yen)	Expenditures (in yen)	Difference (in yen)
7	*udon*			0.23	
	fish (*shinjo*, halibut, flounder)			0.25	
	soy sauce	360.1 ml		0.06	
	tofu	1 block		0.06	
8	*tabi*			0.35	
	sweets			0.05	
9	postcards			0.15	
	apples and mikan			0.05	
11	fish			0.50	
	tofu			0.06	
	dried roasted sardines			0.05	
	udon			0.33	
	soy sauce			0.15	
	konnyaku			0.05	
	saké			0.20	
	envelopes			0.10	
12	postal fee			0.10	
15	rabbit pelt tanning fee		0.25		
	twine			0.05	
	candies			0.03	
16	cabbages	16.1 kg	0.85		
	cotton cloth			2.63	
	tofu dregs (*kasu*)			0.06	
	kelp (*konbu*)			0.05	
	candies			0.03	
19	candies			0.05	
20	postage stamps			0.02	
21	cabbages	7.5 kg	0.38		
	tofu dregs (*okara*)			0.10	
22	fish			0.20	
	tofu			0.06	
	soy sauce			0.06	
	sugar			0.08	
23	fertilizer			17.05	
	saké lees (*sakagasu*)			0.10	
	kelp (*konbu*)			0.15	
24	train fare			1.00	
	gift for a friend			0.65	
	sweets			0.05	
25	ink			0.35	
26	lumber processing fee			0.80	
28	magazine(s)			1.10	
	postal fee			0.10	
	candies and sweets			0.05	
	Subtotal (February 1936)		**1.48**	**28.95**	**–27.47**
Mar. 1	candies			0.02	
	fee for telephone call			0.04	
2	postage stamps and postcards			0.36	
3	candies			0.02	

Date	Item	Quantity	Income (in yen)	Expenditures (in yen)	Difference (in yen)
4	kerosene	1.8 L		0.30	
	mandarin oranges (*mikan*)			0.03	
	charcoal	30 kg		1.60	
5	dried alum (*yaki myōban*)			0.13	
	octopus tentacle			0.15	
6	tofu dregs (*okara*)			0.06	
	postage stamps			0.03	
7	tofu dregs (*okara*)			0.09	
	tofu			0.24	
	udon			0.28	
	dried roasted anchovies			0.08	
11	candies			0.03	
	Subtotal (March 1–12, 1936)		**0.00**	**3.46**	**-3.46**
	TOTAL		**199.88**	**234.03**	**-34.15**

Appendix 4: Expected and Actual Income from Labor and Produce Sales by Calendar Year (in Yen)

Actual produce sales figures, and expected figures, by year (in yen)	1930	Expected figure for 1931	1931	Expected figure for 1932	1932	Expected figure for 1933	1933	Expected figure for 1934	1934	Expected figure for 1935	1935	Expected figure for 1936
eggplant seedlings	3.55	4.00	2.93	3.00	3.40	5.00	x	x	x	x	x	x
peas (*endō mame*)	3.42	5.00	2.33	5.00	2.70	5.00	4.40	5.00	3.80	4.00	3.42	6.00
cucumbers	18.95	15.00	3.17	10.00	4.80	6.00	6.00	5.00	6.17	5.00	4.61	2.00
eggplants	81.65	60.00	31.55	60.00	40.00	60.00	40.80	60.00	36.20	50.00	53.01	40.00
burdock roots	2.86	3.00	3.55	5.00	0.80	1.00	1.20	2.00	0.60	5.00	3.75	4.00
Chinese yams	1.11	2.00	1.40	1.00	1.40	2.00	2.70	3.00	4.20	5.00	2.00	4.00
napa cabbages	3.00	10.00	12.70	20.00	19.50	20.00	3.00	20.00	16.80	15.00	11.67	10.00
green onions	21.00	20.00	10.35	20.00	4.50	10.00	16.00	15.00	13.00	15.00	7.70	40.00
carrots	2.30	2.00	x	x	x	x	0.90	x	x	x	x	x
green edamame	1.85	1.00	2.33	3.00	1.42	1.00	x	x	x	x	x	x
oriental melon (*amauri*)	3.44	3.00	0.85	0.80	x	x	x	x	x	x	x	x
watermelons, *shiro-uri*, tomatoes, bean sprouts, flowers	8.00	10.00	x	x	x	x	x	x	x	x	x	x
cowpeas (*sasage*)	x	x	1.72	x	0.55	x	x	x	x	x	x	x
summer cabbages	12.00	20.00	9.40	20.00	32.40	35.00	21.85	30.00	39.70	3.00	38.31	40.00
watermelons	x	x	2.56	2.50	0.35	x	2.50	5.00	x	x	9.07	3.00
pumpkins	x	x	0.60	0.70	1.22	3.00	3.50	3.00	3.00	2.00	3.35	1.00
tomatoes	x	x	4.33	10.00	2.75	10.00	7.50	10.00	13.40	12.00	15.40	10.00
taros	4.18	5.00	5.60	1.00	2.20	6.00	5.70	5.00	8.40	25.00	13.95	25.00
autumn cabbages	x	x	6.00	8.00	2.82	5.00	7.00	5.00	3.70	3.00	5.69	5.00
autumn daikon	x	x	x	x	2.00	2.00	3.00	2.00	1.20	1.00	1.70	3.00
eggplant and tomato seedlings	x	x	x	x	x	x	3.30	5.00	8.45	4.00	4.12	10.00
summer daikon	3.70	5.00	4.00	10.00	8.70	9.00	7.00	10.00	11.00	10.00	8.14	12.00
cash payments for day labor	x	x	2.60	10.00	9.65	10.00	15.34	15.00	25.00	10.00	10.00	15.00
rice payments for day labor	x	x	180.40 L	180.40L	180.40 L	180.40L	216.48 L	180.40L	243.54 L	216.48L	247.15 L	234.52 L
miscellaneous	x	10.00	x	x	x	x	x	x	15.00	10.00	10.40	15.00
totals (cash)	171.01	175.00	107.97	190.00	141.16	190.00	151.69	200.00	209.62	179.00	206.29	245.00
totals (rice)	x	x	180.40 L	180.40L	180.40 L	180.40L	216.48 L	180.40L	243.54 L	216.48L	247.15 L	234.52 L

Appendix 5: Expected and Actual Expenditures by Calendar Year (in Yen)

	1930	Expected figure for 1931	1931	Expected figure for 1932	1932	Expected figure for 1933	1933	Expected figure for 1934	1934	Expected figure for 1935	1935	Expected figure for 1936
property rent	20.00	20.00	20.00	20.00	20.00	20.00	20.00	20.00	20.00	20.00	20.01	20.00
fertilizer	12.30	15.00	11.00	15.00	15.46	15.00	13.30	15.00	10.30	15.00	13.40	15.00
seeds	7.52	10.00	8.85	10.00	6.94	10.00	8.86	8.00	8.64	10.00	12.93	10.00
rice	(a)	60.00	45.20	70.00	50.20	70.00	62.05	80.00	79.23	80.00	70.15	85.00
medicine and agricultural chemicals	7.00	40.90	11.00	10.00	18.62	15.00	17.85	15.00	4.26	15.00	33.14	20.00
miscellaneous expenses	35.00	5.00	40.30	40.00	42.80	40.00	44.00	40.00	55.64	45.00	86.84	56.00
books	12.71	3.00	2.70	5.00	0.42	5.00	4.70	5.00	3.00	5.00	4.80	5.00
communications	3.46	2.00	3.00	3.50	3.05	3.00	x	x	3.67	4.00	5.78	5.00
kerosene	1.14	2.00	1.70	2.00	1.40	2.00	1.85	2.00	2.00	2.00	1.32	2.00
haircuts	1.50	2.00	1.60	2.00	1.40	2.00	0.40	(b)	x	x	x	x
soy sauce	x	x	1.00	1.50	0.88	1.00	0.80	(c)	x	x	x	x
taxes	x	x	x	x	1.20	1.50	1.10	1.20	1.45	1.50	1.52	2.00
totals	100.63	159.90 (d)	146.35	179.00	162.37	184.50	174.91	186.20	188.19	197.50 (e)	249.89 (f)	220.00

(a) received from main house.
(b) bought clippers.
(c) added to miscellaneous expenses.
(d) This figure was 167.00 (a miscalculation) in the original document.
(e) This figure was 205.70 (a miscalculation) in the original document.
(f) This figure was 249.88 (a miscalculation) in the original document.

Appendix 6: March 1935 Letter from Shibusawa Keizō to Yoshida Saburō

I carefully read your March 18 letter. Please allow me to reply.[1] First, I would like to apologize for being too busy to spend much time with you when you were in Tokyo. Regarding your book, although it isn't yet widely available, all who have read the book were deeply impressed. I am very happy about it.

As for the matter of the daily journal, which we discussed while you were in Tokyo, it is going to be quite an undertaking, but I'm sure that you can complete it, and I am looking forward to seeing the finished product. I will address each of the issues you raised below.

With regard to whether you should record details about food preparation for three meals daily or not, it will be good if you could in fact do this as best you can. You don't have to write the details daily, especially if there are no major changes; explaining every few days, or once each week, would be good, I think. But if there is any sort of new dish, please elaborate. This also goes for other foods, aside from your daily meals. It's very important to explain whether food was roasted, boiled/simmered, or eaten raw, or prepared with vinegar, salt, sugar, soy sauce, or some other flavoring or seasoning. Furthermore, noting the preparation of the side dishes and where the ingredients came from—that is, whether they were collected, raised, caught, or purchased—is also probably necessary.

Recording daily income and expenditures: this is extremely important. As it is a private matter for most people, researchers can't easily compel them to share such information. But if you provide such details yourself, your daily record will be an unparalleled resource from an academic standpoint; its value will increase, and I for one will be very pleased.

Regarding wet paddy rice production:

1. You are able to create a record of your natal household's daily work details but not of its income and expenditures. It will be fine if you only record the daily work details.
2. You won't be able to keep track of your natal household's daily food details. It will be fine if you don't do that.
3. Rice paddy work and the autumn harvest: please write about these matters in as much detail as possible.

About the photographs, please send the exposed negatives to me as soon as you can. I will have them processed and send the prints to you. I think it will be necessary to make a photographic record of your farming work—holes for planting seeds and such. And I believe it would be very interesting if you could select some representative plants and photograph them at ten days and at two weeks, perhaps, or document the daily growth of the same plants, depending on what you grow.[2]

Regarding the thermometer, it would be best to place it outside somewhere. And you could check it at nine or ten o'clock in the morning, or at seven, or at noon. It's just important to make sure you check it at set times every day and record the temperatures; you don't necessarily have to get the actual highs and lows. The thermometer should probably be in a place where the sunlight doesn't hit it directly.

As for recording the weather, I would like for you to note any changes. It would be good to have a record of especially strong winds, but not necessarily of an unchanging daily breeze.

I'm sure that writing this daily record, in addition to your regular daily work, is going to require a lot of extra effort. However, I believe that once it's complete it's going to be an extremely valuable resource for our country—perhaps the very first of its kind.

My apologies for being so late in responding to you; I have been extremely busy. Please give my best regards to your wife and your daughters, and to your elder brother and other family members whom I recently met as well.

Yours in haste,

Shibusawa Keizō

To: Yoshida Saburō
March 27, 1935

Appendix 7: April 1937 Letter from Shibusawa Keizō to Yoshida Saburō

It has been a long time; I hope you are well.[1] Your book is now in the printing stage, so it won't be long now before it is ready.[2]

It's a bit sudden, but there is a matter that I would like to discuss with you. We're building a Japanese Society of Ethnology[3] research facility. The details of the project are as follows.

The property—about 33,000 square meters—is located in Hōya Village, Musashi.[4] It is close to Takahashi Buntarō's home.[5] He and I have acquired the land and donated it to the society together.

The building will cover about 1,157 square meters and have an approximately 400 square meter specimen room. You can think of it as the new home of the Attic Museum's folk crafts division (*mingubu*).

On the 33,000 square meter property, I would like to start a botanical folk garden (*minzoku shokubutsu-en*).

We're going to need a guard/gardener/caretaker for the facility, and I am thinking of you.

Inside the building, there will be a living area with a six-tatami mat room, a four and a half-tatami mat dining room/kitchen, and a guard office—plenty of living space for your family. Of course, it will be rent-free.

For the foreseeable future, your salary would be 25 yen per month. You might like to buy rice and such from people in your home area.

Regarding firewood, the 33,000 square meter plot of land will provide enough to last for quite some time.

In sum, I would like to ask you to come and lend us your skills and expertise in support of our project. No matter how much I think about it, I can't see anyone else but you as the person for this position.

I'd like to ask you to come to Tokyo at the end of this month—perhaps on the 29th or 30th—or in early May, at your convenience, so that we can discuss this matter in detail. I will cover the travel cost myself. I have a trip scheduled to begin around May 13th, so it would be best if you could come before then.

At any rate, let's discuss the details when we meet in person. In the meantime, give your wife and daughters my best regards.

Shibusawa Keizō

To: Yoshida Saburō
April 22, 1937

Appendix 8: Name Index

People Identified by First Name Only

People Identified by Only a Nickname

People Identified by Pseudonyms

Notes

Translator's Introduction

1 Partner, *The Mayor of Aihara,* 3.

2 Dower et al., *The Brittle Decade,* 12, 21.

3 Fujioka, "The Development of Department Stores in Japan."

4 Francks, *Technology and Agricultural Development,* 22.

5 Dower et al., *The Brittle Decade,* 14.

6 This was probably mainly driven by growth in per worker labor productivity, which rose by a factor of 2.5 between 1880 and 1935: Hayami and Yamada, *Agricultural Development of Japan,* 254.

7 Kerry Smith, "Building the Model Village"; Kimura and Minohara, "Introduction"; Shōji, *Kindai Nihon no Nōgyō,* 116.

8 Garon, *State and Labor*; Kerry Smith, *A Time of Crisis,* chap. 1.

9 Nishida, "Dimensions of Change in Twentieth-Century Rural Japan," 21–22.

10 Francks, *Technology and Agricultural Development,* 22.

11 Yamaguchi Kazuo, "Development of Agricultural and Marine Industries," 421–433.

12 Chō, "Kindai Nihon ni Okeru Nōson Kajō."

13 Hayami and Yamada, *Agricultural Development of Japan,* 65–66; Yamaguchi Kazuo, "Development of Agricultural and Marine Industries," 439.

14 Hayami and Yamada, *Agricultural Development of Japan,* 254.

15 Nishida, "Dimensions of Change in Twentieth-Century Rural Japan"; Kerry Smith, *A Time of Crisis,* 113–132.

16 Nishida, "Dimensions of Change in Twentieth-Century Rural Japan," 17. For a detailed discussion, see Waswo, "In Search of Equity," and *Japanese Landlords.*

17 Tama, "Changes in the Number of Farm Households."

18 Thomas C. Smith, *Agrarian Origins,* 5.

19 Beardsley, Hall, and Ward, *Village Japan,* 139–142; Fukutake, *Rural Society,* 3–7; Yamaguchi Takehide, *Nōmin Undō Nyūmon,* 25–26.

20 Nagatsuka, *Tsuchi;* Kobayashi Takiji, *Fuzai Jinushi.* Kobayashi's 1933 murder at the hands of Tokyo police officers serves as testament to the accuracy and power of his and other like-minded writers of the era.

21 Nozoe and Ueda, *Kosaku Nōmin no Shōgen,* 5–6; Mori, *Kindai Nōmin Undō to Shihai Taisei,* 12.

22 There are 366 entries because 1936 was a leap year.

23 Other important characteristics of the manuscript are discussed below.

24 Yoshida, *Oga Kanpūsanroku Nōmin Shuki.*

25 I thank Dr. Nakamichi Takurō, Yoshida's son-in-law, for the graduate student/professor analogy.

26 Yoshida, *Mono Iu Hyakushō,* 63.

27 Taguchi, *Kindai Akita-ken Nōgyō-shi no Kenkyū,* 386–390. It is also worth noting that about 65 percent of rice farming households, and 41 percent of vegetable farming households, in the same county were tenant farmers in 1935 (389).

28 Craig, *Middlemen of Modernity,* 9; Kerry Smith, *A Time of Crisis,* 20–24.

29 Ōnishi, *Nihon Rōnōden,* 125–166.

30 Craig, *Middlemen,* 76, 90–95.

31 Craig, *Middlemen.*

32 Iizuka, *Meiji-ki no Chihō Seido to Meibōka.*

33 Craig, *Middlemen,* 16; Ōshima, *Rōnō, Nakai Taichirō to Nōmintachi no Kindai.*

34 See Dec. 24 entry.

35 Yoshida, *Mono Iu Hyakushō,* 57; *Oga no Koboredane,* 270–271. For a discussion of *dekasegi* and its centrality in regional Japan in the twentieth century, see Bailey, *Ordinary People, Extraordinary Lives.*

36 A major source is the monumental 72-volume *Nihon Nōsho Zenshū* (Nōsangyoson Bunka Kyōkai), released first in 1993, which contains records from Akita households.

37 See, e.g., Takagi, *Kinsei Nōmin Nikki no Kenkyū.*

38 See, e.g., Thomas C. Smith, *Agrarian Origins,* and Kerry Smith, *A Time of Crisis.*

39 An early record giving balance to this androcentric field is the Edo-era memoir of Ōgimachi Machiko (*In the Shelter of the Pine*). Born into an aristocratic family in the 1670s, she became a concubine of a samurai connected to the Tokugawa Shogunate. The fascinating story of her life with him in Edo is seen as historically accurate, but it reflects an anti-peasant bias, and of course is not much concerned with them. A more recent and relevant example is Partner's 2023 presentation of the nineteenth-century diary of the wife of a Wakayama samurai family (*Koume's World*). Even more relevant is Kazuko Smith's engaging translation of the detailed 1910 diary of the wife of a Kyoto merchant (Nakano Makiko, *Makiko's Diary*), which nicely complements Yoshida's 1938 record, as an urban counterpart of the same general time period. Finally, an autobiographical memoir of a woman's life as a servant and, later, a geisha in a hot springs resort in Nagano Prefecture from the 1930s to the mid-1940s resonates very well with Yoshida's chronicle (Masuda, *Autobiography of a Geisha*). This frank, first-person, account of a very hard life is viscerally gripping and deeply endearing, but sheds no light on village life or on farming.

40 Teeuwen et al., *Lust, Commerce, and Corruption.*

41 Partner, *The Mayor of Aihara,* 63.

42 For the original language, see Miyamoto, *Wasurareta Nihonjin.*

43 Saga, *Memories of Silk and Straw.*

44 Partner, *Toshie.*

45 Nishida, "Dimensions of Change in Twentieth-Century Rural Japan."

46 Similarly, Yoshida's chronicle stands out among other prewar Japanese farmer diaries, discussed by Govindasamy, in "Smell of the Soil" and "Not Without Blood, Sweat and Tears," for its uniformity, breadth, and detail, and also for its focus on vegetable farming. Furthermore, these diaries were not originally written for publication, and their authors were wealthier and of higher social status than Yoshida. No parts of these appear to be available in English.

47 Gotō, "Kakeikiroku yori mita Shōwasenzenki ni Okeru Nōka no Shōhi Kōzō"; Imai, "Senzen no Chihō Nōson"; Kobayashi, "Relationship Between Attic Museum and the Dainippon Youth Group"; Masui, Sawada, and Tsuda, "Changes in Grassland Vegetation"; Ogawa, "Oga Hantō Wakimoto Mura no Magi."

48 Yanagita and Miki, *Yukiguni no Minzoku;* Okada, "Yukiguni no Minzoku"; Fujii, "Yanagita Kunio and the Culture Film."

49 See Mar. 25, Apr. 9–10, Apr. 15, May 1, Jun. 4, Jun. 21, Nov. 18, Dec. 12, and Jan. 3 entries.

50 Partner, *The Mayor of Aihara,* 4.

51 Kwakiutl ethnographer and photographer George Hunt (1854–1933) is another rare case. See Jacknis, "George Hunt, Kwakiutl Photographer."

52 See Wood, "Photography in the 1930s 'Native' Ethnography of Yoshida Saburō."

53 See esp. Apr. 10, Dec. 24, and Mar. 11 entries.

54 Craig, *Middlemen,* 68–69.

55 Craig, *Middlemen,* 104–108.

56 See Mar. 21, Apr. 1, May 5, Jun. 24, Sept. 28, Oct. 22, Nov. 7, Dec. 24, and Feb. 28 entries.

57 For more, see Lewis, *Becoming Apart* (esp. pp. 12–13).

58 See Jul. 1, Dec. 3, Dec. 15, Jan. 31, and Feb. 13 entries.

59 Arimoto, "Participatory Rural Development in 1930s Japan"; Kerry Smith, *A Time of Crisis* (and see Dec. 3 and Mar. 11 entries).

60 This is called *e* by Yoshida (see Jun. 2 entry) but has been identified as *yui* or *i* by Embree (*Suye Mura,* 99). See also Shinpo, *Three Decades in Shiwa,* and Thomas C. Smith, *Agrarian Origins.*

61 Beardsley, Hall, and Ward, *Village Japan,* 260; Fukutake, *Rural Society,* 99–100.

62 Wiswell was widowed by Embree's untimely death in 1950. Also see Tanaka, *Wasurerareta Jinruigakusha.*

63 See esp. Mar. 13, May 10, May 13, Jun. 2, and Mar. 4 entries.

64 See Mar. 13, 21–22, Apr. 2, 6, 11, May 30, Jun. 17, Aug. 6, Sept. 5, 10, Nov. 13, Dec. 10, 20, 28, 31, Jan. 3, 6, 24, Feb. 7–8, 10, and Mar. 8 entries, and Appendix 1.

65 To make matters worse, Akita had bad harvests in 1931, 1932, 1934 (see Mar. 16 entry), and 1935, which made life harder for poor farmers. and helped foment unrest; 1935 saw 471 tenant farmer revolts in Akita Prefecture. For more, see Nozoe and Ueda, *Kosaku Nōmin no Shōgen,* 5.

66 See Appendices 3–5.

67 This is partly due to the immense historical and cultural importance of rice, and to its association with Japanese identity. For a greater discussion, see Ohnuki-Tierney, *Rice as Self.*

68 Suzuki, *Meiji, Taishō ni Okeru Nōminsō Bunkai,* 183–215. See also Love, "Mountain Vegetables and the Politics of Local Flavor in Japan," and Assmann, "Reinventing Culinary Heritage in Northern Japan," on recent vegetable-centered community initiatives in areas near Akita Prefecture.

69 Hayami and Yamada, *The Agricultural Development of Japan,* 27.

70 Regarding the latter issue, see especially the Dec. 26 entry.

71 Good early postwar examples include Beardsley, Hall, and Ward, *Village Japan,* and Dore, *Land Reform.*

72 Yoshida, *Mono Iu Hyakushō,* 3.

73 The exact ages at which Yoshida started and finished school are unclear, due apparently to a record-processing error and perhaps also to the old practice of calculating a person's age as one year upon birth.

74 His mother had given birth to four more children by that time, but the last-born did not survive.

75 Kobayashi, "Relationship Between Attic Museum and the Dainippon Youth Group." Other, unrelated, magazines called *Aozora* appear to have existed at the time.

76 Boyle, *Modern Japan,* 156; Garon, *State and Labor,* 3–4; Reischauer and Craig, *Japan,* 234–244.

77 Or *magi,* in local speech.

78 For a greater discussion and other examples, see Aruga, *Nihon Kazoku Seido to Kosaku Seido*; Beardsley, Hall, and Ward, *Village Japan,* 169–172; Fukutake, *Rural Society,* 58–66; Shimpo, *Three Decades in Shiwa,* xiii, 113–114.

79 Yoshida, *Mono Iu Hyakushō,* 1–2; Ogawa, "Oga Hantō Wakimoto Mura no Magi."

80 See Apr. 10 entry.

81 Yoshida, *Mono Iu Hyakushō,* 50. But when his eldest daughter, Fusae, entered elementary school in 1936, Yoshida covered her shoulder bag with critical political commentary (see Fig. 0.55). As a child, she was unaware of the meaning of the writing, but it attracted the attention of nearly everyone who saw it. "I was so embarrassed by all that attention," she said many years later (unrecorded and undated interview).

82 See Dec. 24 entry.

83 Kuwayama, *Native Anthropology.*

84 Some of Ōnishi's writings indicate that he made his writing suggestion to Yoshida in 1933 and that Yoshida's manuscript reached him in 1934, but Yoshida's version of the timeline is used here, as it is probably correct.

85 For a discussion of this, and of the Attic Museum and its origins, see Christy, *A Discipline on Foot,* 137–151.

86 Ishiguro (1884–1960) was associated with the Attic Museum and was married to one of Shibusawa's many cousins. He would later help Shibusawa found the Japanese Society of Ethnology (see Appendix 7).

87 Miyamoto (1911–1979) was a Tokyo-born Attic Museum researcher who did much of the association's film documentary work, and with whom Yoshida got along well, according to Yoshida's daughter, Fusae, who recalled Miyamoto as having treated her and her sister very kindly.

88 See Feb. 7 entry and Appendix 7.

89 Ogawa (1914–2001) was an Okayama-born Attic Museum researcher from Kanagawa Prefecture who was heavily involved in editorial work during his earlier years with Shibusawa's organization (https://ifm.minpaku.ac.jp/hoya/detail_collector/?id=86). Ogawa was probably the chief, if not the sole, editor of Yoshida's 1935 and 1938 Attic Museum books.

90 Yoshida, *Mono Iu Hyakushō,* 86.

91 It is unclear exactly when, and how, these were provided to Yoshida, but the camera and all accessories and film appear to have been handed to him when the assignment was given, as Yoshida began taking photos almost immediately. (Shibusawa probably also provided Yoshida with a thermometer.)

92 Notably, Yoshida began writing daily entries five days before he wrote to Shibusawa about his concerns. Furthermore, he had already completed more than two weeks' worth of entries by the time Shibusawa's response

(Appendix 6) arrived, helping to explain some of the differences between the ways in which Yoshida wrote at the beginning and how he wrote after the initial period of adjustment.

93 I also found a copy of Yoshida's 1935 book, which I photocopied.

94 See Jan. 22, Feb. 3, and Feb. 7 entries.

95 This decision met with the approval of Yoshida's daughter, Fusae.

96 I created a name index for better organization and clarification, at my wife's suggestion. With this, Yoshida's eldest daughter, Fusae, helped me identify people and correct the pronunciations, as popular readings of Chinese characters have changed over time.

97 I also hope he would have appreciated the error corrections and clarifications.

98 Yoshida, *Mono Iu Hyakushō,* 93–94. But the option may have continued to linger in the back of his mind; it appears that he told Shibusawa about the idea at one point and was advised against going.

99 They left just in the nick of time. Soon after their departure, police officers arrived in Wakimoto, looking for people who had made their blacklist for having spoken at a labor rally, for having attended certain gatherings, or simply for having been associated with blacklisted people. Officers apparently visited Yoshida's little house, only to find it abandoned (Yoshida, *Mono Iu Hyakushō,* 95–96).

100 Yoshida later picked up extra salaried work teaching agricultural methods at two different schools. He also converted a section of Shibusawa's decorative gardens to a vegetable garden, which Shibusawa subsequently farmed himself. His wife learned, for example, how to cook curry rice from Miyamoto Tsuneichi's wife (Yoshida Fusae, unrecorded undated interview).

Author's Preface

1 Shibusawa Keizō.

2 Ōnishi Goichi.

3 See Appendix 2.

4 See Appendices 3–5.

Chapter 1: March 1935

1 A broth flavored with miso (fermented soybean paste) and dried fish (usually sardine), often containing vegetables and perhaps tofu. See Figs. 0.25 and 0.26, and later entries, on production and other usages.

2 Giant white radishes of continental Asian origin. "Pickled daikon" (*daikonzuke*) generally refers to plain salt-preserved daikon, or other vegetables if not daikon.

3 Regarding fish and vegetables, "salted" means cured, or salt-preserved, throughout this book.

4 Japanese sandfish (*Arctoscopus japonicus*)—caught off the coast of the Oga Peninsula and commonly eaten in Akita during the late autumn.

5 Saburō's closest neighbor was a man with a mental disorder who lived alone in a shack, on relatives' support. Yoshida refers to him as the *tonari no kyōjin* ("madman next door" or "crazy neighbor"). Elsewhere (*Mono Iu Hyakushō,* 81), he explains that the man had left the village, married, and established a household near Tokyo years earlier but had at one point become "a madman" and subsequently returned to Ōkura. Saburō rented property from the man's relatives from 1930 to 1937.

6 A hotbed with a wooden enclosure and cover, warmed from below by decaying organic matter, used for raising seedlings.

7 Young agricultural workers, generally under twenty-five, who would either commute daily to, or reside at, their employing household (Sutō, *Wakaze*).

8 Saburō's father died on May 9 or June 9, 1927 (accounts vary). The ninth day of every month, therefore, was noted as his father's *tsuki meinichi.* In the 1930s, old lunar calendar dates tended to have more meaning than Gregorian calendar dates (see also Embree, *Suye Mura,* 38; Partner, *Toshie,* 31).

9 Local pronunciation for *Okano mochi:* large, round "rice cakes" made from pounded glutinous rice formed into ovals, eaten on February 9 in Saburō's village in honor of the folk god Okano (Kōshin) (Yoshida, *Oga Kanpūsanroku Nōmin Shuki,* 41–42). These may also be offered on trays to the gods Daikoku and Ebisu at certain times of the year. According to Yanagita Kunio (Miki and Yanagita, *Yukiguni no Minzoku,* 40), they were eaten only by (male) household heads in the region north of Akita City, but in Saburō's village this does not seem to have been the case.

10 *Undaria pinnatifida*, a commonly eaten seaweed.

11 This was made in the village by cooking oats together with soybeans, and then mixing in bits of *gomagi* (*Viburnum sieboldii*) leaves and letting the mixture partially ferment—causing a mold to grow—and subsequently adding salt and water and allowing it to ferment more. The liquid that came off the mixture could be used as a substitute for soy sauce, and the rest could be mixed with *amazaké* (sweet saké or malted rice) and eaten with cooked rice (Yoshida, *Oga Kanpūsanroku Nōmin Shuki,* 13).

12 Human excrement used as fertilizer, generally collected at night (historically).

13 Protective covers, generally made of straw or cloth, for the lower legs that wrap around the shins and calves. Kanezō would have made them with straw (see Mar. 6 entry).

14 Saburō always refers to his brother's wife as "my elder sister" in his manuscript, in accordance with cultural norms, but this translation uses "my brother's wife" for clarity.

15 *Ami no shiokara.* Saburō normally uses the local term *isaja* instead of the standard *ami* (krill) but written as he and his fellow villagers probably pronounced the word: *ishaja.* Commonly eaten today, *ami no shiokara* is prepared by pressing krill in salt for about six months and then cooking it (Nihon no Shokuseikatsu Zenshū Akita Henshū Iinkai, 85–86).

16 By "whale," Saburō generally means small pieces of subcutaneous blubber with skin. References to miso soup with whale in it might indicate *kayaki,* an Akita term for a soup with very little broth.

17 Kōshin-en was a communal, self-sufficient farming initiative located in the foothills of Mt. Kanpū that was owned, funded, and managed by Yoshida Kōtarō, to whom Saburō refers here as "master." Yoshida Kōtarō was the vice-mayor of the village, and held several other local government positions during the early postwar years (Oga City, *Ogashishi,* 309–315).

18 The head of a household located near Saburō's natal home, possibly a relative.

19 *Honson* 本村: the main part of Wakimoto Village, lying to the south of Yoshida's hamlet, near the coastline and beneath the former castle site.

20 Saburō made one hotbox first, and then two more in April.

21 A common unit of measurement for volume that differs according to situation. Saburō uses *ka* to refer to the amount that he could carry at one time or in one trip—most likely divided into two containers, attached to the ends of a pole, with the pole across the shoulders.

22 The highland area, in the foothills of Mt. Kanpū.

23 The local expression *wappaga,* denoting the amount of work that could be done in one day, is used here.

24 Saburō sometimes facetiously refers to his elder brother in this way.

25 *Koshukai* 戸主会.

26 Daikon that has been hanged to dry and then preserved in a mix of rice bran and salt, and perhaps also sugar—usually served in slices (see Dec. 14 entry).

27 One rin was equal to one-tenth of one sen (one one-hundredth of one yen). One *bu* would have been one-tenth of one *rin.*

28 From Meiji times, one *hiro* was legally defined as six *shaku,* with one *shaku* being equal to 30.3 cm, and five *shaku* therefore equaling about 1.5 m.

29 In referring to his elder brother as "master," and his natal household's sweets (*wagashi*) shop as a "little shop" (*ko-mise*), Saburō reveals a glimpse into his complex feelings toward his elder brother, who was unable to engage in much hard labor due to an illness contracted in early adulthood—a problem that their father had solved by apprenticing his eldest son to a *wagashi* maker and subsequently building a shop on the family property. The shop is mentioned many times in this book. What is not mentioned is that (according to his daughter, Fusae) Saburō frequently helped himself to tasty items in the shop when it was unattended, much to the consternation of his elder brother's wife.

30 A period of about one week that coincides with the spring equinox, considered by Buddhist sects in Japan as a time for people to visit and clean the family graves and pray for the souls of the ancestors and perhaps other relatives, and in some cases to have sutras chanted. Certain foods are often prepared and eaten. In the past, more than today, it was a time for rest and participation in Buddhist rituals. There is also autumn *higan* (see Sept. 12 entry).

31 Small smooth dumplings made from rice flour, often eaten on skewers.

32 Yoshida Sangobē was the head of a wealthy village household, and probably a relative of Saburō's, through his grandfather.

33 Saburō refers to his paternal grandfather's natal household as the "big main house" because his own natal household (the main house) was originally established as a branch household (*bunke*) by his father. The big main house stood next to the main house (see Fig. 0.21).

34 Saburō often spent small sums on candy (*ame*). He may have at times bought malt-honey (*mizu-ame*): a substitute for refined sugar for cooking, as the latter was becoming increasingly expensive.

35 Nirvana Day (*nehan*) marks the day of the Buddha's corporeal death and attainment of nirvana.

36 Saburō uses the term *kome-kuzu:* a mix of powder, small bits, and broken pieces of rice.

37 "Later in the morning" means "after breakfast," as it was common to work some before breakfast.

38 *Dioscorea opposita*, also known as *tororoimo.*

39 A firm but soft jelly-like food made from *konnyaku* potatoes (*Amorphophallus konjac*)—the roots of devil's tongue plants—that can be cut into various shapes and cooked.

40 Miso with crushed poppy seeds mixed in.

41 Siddhārtha Gautama, the Buddha.

42 The main temple of the village.

43 It appears that the family in fact started eating *takuan* on March 16.

44 An entire daikon (*ippon*) that has been pickled (*-zuke*) whole.

45 Saburō's elder brother.

46 A barrier made of wood or bamboo poles and sticks and thatching usually set up around a house during wintertime to protect it from snow and strong wind.

47 Although Saburō's eldest daughter, Fusae, once recalled that her mother had had no elder brother, a labeled photograph in one of her father's albums reveals the existence of an elder brother of Fusae's mother, and his family. Why this brother appears to have maintained a household separate from that of his birth, even though he was the eldest son (*chōnan*), remains unclear.

48 Sawayama (1891–1965) authored a paper or pamphlet on tanning rabbit pelts that was published in 1934: probably why he was invited to speak in Akita. An expert on tanning, Sawayama wrote reports on the subject after the war as well.

49 A work surface with curved edges, the name coming from its resemblance to a block of *kamaboko* (a smooth, firm "cake" made from whitefish, which is curved across the top).

50 A curved blade with handles on each end.

51 *Hyakumanben nenbutsu:* a Buddhist ritual ceremony that varies by temple and region. Temple members would gather and sit on the floor and string a set number of beads onto long ropes while chanting sutras for the souls of their deceased ancestors, family members, or neighbors. Afterward special prayers might be said, and there would likely be a feast.

52 Yoshida is lamenting the low turnout as well as the temple's failure to provide rope, both probably reflections of the poor economy (as attending requires the preparation of certain dishes).

53 Local term for *itako,* a female spirit medium (see Ivy, *Discourses of the Vanishing,* esp. 169–188).

54 A séance, for the spirit of a deceased person—a child in this case. If successful, the spirit would inhabit the body of the *itako* for a short time and perhaps be able to communicate with his or her relatives. *Itako* generally received either cash or rice for their services. (Yoshida also uses the term *nanakurayohi*—probably how most elders of his village pronounced the word.)

55 Yoshida means that many older village women attended, but not necessarily all village women.

56 One *kamasu* was an amount of something that could be wrapped in a standard-size (about 1.7 m^2) straw mat (*mushiro*), folded over once, or held in a *mushiro*-like bag (see Aug. 6 entry).

57 Yoshida probably received permission from the landowner, and apparently was allowed to take some of the stumps for his own use.

58 "A field should be plowed so that its rows are 2 *shaku* 5 *sun* apart."

59 It is a common practice in Akita to attach the non-standard diminutive *-kko* to the end of nouns—as in *ame* (candy) plus *-kko* to make *amekko.*

60 Yoshida's ambivalent feelings toward his elder brother are evident. His words also belie his attitude toward the value of hard work, and his overall focus on farm labor in his chronicle. (Note that he does, however, continue to write about his brother and mother in the rest of the book, even when they did not engage in such work.)

61 The village of Tennō was about 10 km from Wakimoto, to the southeast, along the coast. Today it is part of Katagami City.

62 This was probably a public cooperative entity formed by area businessmen for collectively marketing local products.

63 This was likely the same as the Nōkai Agricultural Warehouse.

64 Yoshida Chūzō's first name is written in several different ways in the original book, but the version in Appendix 8 (the name index) is probably correct, as no record of a person with a name written in the other forms appears in Saburō's 1977 follow-up study of his home village. Yoshida Kinjirō and Yoshida Kasen were cousins of Saburō's. The former was the head of the "big main house" next door to Saburō's natal household, and a member of the village governing council during the early postwar years (Oga City, *Ogashishi,* 311).

65 Tayazawa, about 1 km west of Ōkura, was one of Wakimoto's six major hamlets. Saburō's mother hailed from Tayazawa (see Jul. 31 entry).

66 Minor disputes of this sort were extremely common at the time, with no fewer than 6,824 major disputes across Japan in 1935 (Dore, *Land Reform,* 72). In many cases, regarding disputes of the sort Yoshida describes, other local landlords, politicians, or temple priests might have stepped in to mediate. In other cases, a judge would consider the matter in court and make a ruling. This may be what happened here since a fee of 10 yen (a fair sum) is mentioned.

67 One *tawara* (see Mar. 26 entry).

68 Yoshida's elder daughter (b. 1930).

69 Yoshida's younger daughter (b. 1932).

70 Short straw sandals that leave the heels free to touch the ground (see Feb. 22 entry).

71 A straw barrel-shaped container for rice grains (see Figs. 0.14 and 0.15). Today, one standard *tawara* (one *hyō*) of rice weights 60 kg.

72 As an adult, Tsugio served in the local government in different capacities, and was eventually elected to a position on the Oga City assembly (*gikai*).

73 A simple rice gruel: rice boiled in water until it is soft.

74 Today soy sauce is usually added to such ground sesame seeds rather than miso, but this varies by region. Yoshida and his family members probably put a bit of this on top of their rice as they ate.

75 A local way of saying *shirauo* (*Salangidae*): small whitish fish, several centimeters long.

76 *Funa* (*Carassius*).

77 *Cucumis melo var. conomon* (*shima-uri*): a soft, fleshy type of melon.

78 See photos of Yoshida's tanning tools in Mar. 21 entry.

79 Minami Iso Village, of which Hirasawa would have been a hamlet, was a coastal settlement about 12 km to the southwest of Saburō's hamlet of Ōkura. Minami Iso remained an independent municipality until it merged with Funagawa Minato Town in 1942, which in turn merged with Oga City in 1954 (along with Saburō's home village of Wakimoto).

80 *Nereke* was also made with rice flour at times.

81 *Shottsuru* is a strong and pungent sauce characteristic of Akita that is distilled from fermented *hatahata* (Japanese sandfish), similar to Italian *ciolatura di alici* or Thai *naam plaa.*

Chapter 2: April 1935

1 The Nōkai, the predecessor of today's Agricultural Cooperative Association.

2 The elementary school to the south, in the main part of the village (*honson*).

3 Yoshida's hamlet of Ōkura was technically a part of the larger hamlet of Tominaga (named after the small mountain at the center of the village), one of the six major hamlets of the village of Wakimoto; others included Tarusawa and Urata (Urada, as probably pronounced locally). These place names exist today, but Wakimoto is no longer classified as a village.

4 The Shakai Taishūtō, a center-left political party active from 1932 to 1940 (see also Garon, *State and Labor,* 193–194).

5 Generally known as Musantō, a short-lived leftist party that existed from the mid-1930s.

6 Local terms for *itoyo*—three-spined stickleback (*Gasterosteus aculeatus*).

7 *Dekasegi* refers to away-from-home seasonal labor in which people—especially farmers—living in peripheral areas of Japan have engaged, mostly in urban areas and during the agricultural downtimes, to supplement household income, or simply to survive (Bailey, *Ordinary People,* 146–154; Robert J. Smith, *Kurusu,* 83). Although recognized as a largely twentieth-century phenomenon, *dekasegi* was common throughout the Edo Period (1603–1868),

and still exists today (Hayami, *Historical Demography,* 138–153; Nozoe, *Dokyumento Dekasegi*). It is recorded that 18,042 people left Akita Prefecture—548 from Wakimoto Village—for such work in 1935 (Oga City, *Ogashishi,* 253). Notably, in December of the same year a group of *dekasegi* laborers met and formed the Akita Prefecture Dekasegi Labor Union in the city of Noshiro (about 30 km north of Wakimoto) to demand unemployment insurance benefits and better work standards, sparking the rapid establishment of other labor unions across the prefecture (Aoki, *Nihon Nōmin Undōshi,* 530–531). By January of 1936 one had been established in Wakimoto (see Jan. 28 entry).

8 A brand of cooking oil generally made from plants and more recently soybeans (literally, "white squeezed oil"). In the original 1938 book, an editorial error resulted in Shirashimeyu being written as Shirasameyu ("white shark oil").

9 Written as *mototsuna* 本綱, or "main rope," *modotsu* being local speech.

10 *Sekku,* traditionally celebrated at set times each year with special foods and activities—the spring *sekku* typically being marked by the display of *hina* dolls for Girls' Day. In many places such annual events are celebrated up to one month late, reflecting the lingering influence of the old lunar calendar. This is probably why the festival was celebrated in April in 1930s Wakimoto.

11 Saburō uses both the standard *sudare* and the local *shidare* to refer to a type of screen or blind that is generally made of reeds (see Feb. 28 entry), but *sudare* is used throughout this translation.

12 *Bikki* is a common term for frog in many parts of Japan. Saburō seems to refer to a "bearded frog," but exactly what this was is unclear. Elsewhere (*Oga Kanpūsanroku Nōmin Shuki,* 165) he notes that *bikki* also meant "baby" in his village.

13 Maeda was a paddy area to the east of Saburō's natal household (the main house).

14 This was actually *tachikiri* (*tazukiri* is local speech), probably called so because it involved cutting (*kiri*) the paddies with a tool resembling a large, bladed weapon (*tachi*) (see Fig. 0.22).

15 This is, literally, making a *naeshiro* (or *nawashiro*)—a small temporary paddy for raising seedlings, later to be transplanted into a larger paddy where they will be grown to maturity (*nashiro* is a local pronunciation).

16 Local pronunciation for the standard *komazari*—a long tool for dredging and scooping (see Yanagita and Miki, *Yukiguni no Minzoku,* 63).

17 Also *sanbonguwa,* but locally known as *sanbon makka.*

18 A type of daikon said to date back to the 1820s and still grown today.

19 An annual custom, the "hoe *mochi*" probably having been offered to the spirits of the hoe, of farming, of the paddies, or of the household, before being eaten.

20 Sahitori refers to leading a horse through a paddy with a long pole—the pole being attached to the lead rope and held at about a 90° angle to the horse. Considered to be difficult and arduous even for adults, it was especially hard for children, to whom the work was often assigned (Yoshida, *Oga Kanpūsanroku Nōmin Shuki,* 104) (see Fig. 0.28).

21 Kanezō and Saburō's brother's wife probably took care of this.

22 "Horse-drawn plow *mochi*" and "New Year's eyeball *mochi,*" respectively: handful-sized *mochi* that were stuck on bunches of rice plant straw (*wara*) before being eaten (Yoshida, *Oga Kanpūsanroku Nōmin Shuki,* 42).

23 This was probably a subtype of small fish commonly known as *gori,* including gobies and other species.

24 *Kashi* in Japanese—not necessarily sweet snacks, but "sweets" is used throughout this translation.

25 Straw soaked in draft animal excrement that has been allowed to begin decomposing, also known as *marogi-goe* (see Fig. 0.10).

26 The paddy area below Saburō's farm, known today as Kogawasaki or Kogawazaki.

27 See the translator's introduction.

28 *Kigitsuke* (March 9 according to the old calendar) was celebrated by making small, round *mochi* and slipping them onto reed skewers, and fixing these on the house and at all entranceways and windows, etc. (Yoshida, *Oga Kanpūsanroku Nōmin Shuki,* 60; *Oga Kanpūsanroku Hōgen Minzokushi,* 103).

29 Local pronunciation for the standard *itoyo* (three-spined stickleback).

30 The man's shack was located beyond Saburō's sloping fields, about 100 m from his house (Yoshida Fusae, unrecorded interview, Mar. 29, 2022).

31 Although Yoshida reveals this man's name in the original book, a pseudonym is used here.

32 It is possible that the man's delusions were influenced by Charles Lindbergh's (1927) and Amelia Earhart's (1932) Atlantic Ocean crossings, which were widely reported across Japan.

33 The Naimushō: created in 1873, abolished in 1947. The ministry was heavily involved in rural and agrarian issues during the first half of the twentieth century (see Craig, *Middlemen of Modernity*).

34 This was probably either a wooden hamlet *dōsojin* or a Buddhist deity.

35 This was apparently not done.

36 *Mugi-maki:* a simple cake made by pouring batter (flour, eggs, sugar, and water) into a skillet while constantly rolling it over on itself as it cooks, and then cutting the resulting roll into sections.

37 Nagane was within a kilometer of Yoshida's hamlet, to the southeast, along the train tracks.

38 Powdered remains of squeezed and crushed soybeans.

39 *Sōmokubai,* made by burning plant matter, sticks, and other pieces of wood.

40 A roughly two-week seasonal period that comes four times annually.

41 *Meshi* refers to cooked rice, and *tororo* to grated Chinese yam (*nagaimo* or *tororoimo*).

42 *Nereke:* a thick, *mochi*-like substance made from *soba* (buckwheat) flour, or perhaps with finely ground rice powder (see Mar. 31 entry).

43 One of Saburō's younger sisters married into a family by the name of Takakuwa that lived near Wakimoto Station, on the Oga train line, which had opened in 1914. In 1930 he bought the wood for building his little house in the hills from this family's lumber business for 20 yen.

44 Bishamon is a paddy area below and west of Saburō's house and farm, just beyond Kogawasaki (Kogasaki) (see Apr. 10 entry).

45 Saburō jokingly uses the term *oyakata:* a tongue-in-cheek reference to his elder brother.

46 "Rest" for Saburō and his fellow villagers often actually meant time for performing tasks that were less physically demanding, such as making things with straw, sewing, cooking, etc.

47 *Wappaga* generally refers to a full day's work, but here it seems to mean permission or a work order.

48 This style—a two-handed method—seems to be essentially the same as *furite-uchi.*

49 This appears to have been done during Saburō's time as well.

50 *Temadori* is written as *musansha* 無産者 and *oyage* is written as *yūsansha* 有産者. *Temadori* and *oyage* are local words. Another for the latter was *kamado.*

51 Probably (local pronunciation for) *kamaboko:* cured surimi, or "fish cake."

52 This is written as "head-shaving fee" (*atama shiri dai*).

53 He appears to have forgotten to follow up on this.

54 *Sudare* and *komo* can serve as blinds. The former are screens made of bamboo or hard reeds, and the latter are made of lighter straws that are sometimes pounded. Both are made by simply tying the components together, *mushiro* mats, often mentioned by Saburō, are comprised of woven straws.

55 *Itoyo:* three-spined stickleback (*Gasterosteus aculeatus*—see Apr. 2 entry).

56 The local word *hadageru* is used.

57 *Shirashimeyu* (white squeezed oil): mistakenly printed as "white shark oil" in the original.

58 Saburō uses Akita dialect: *udzuge-da* ("the plowing is done").

59 Saburō probably planted these in a hotbox.

60 These he appears to have planted in a field.

61 Literally, plowing (*kō*) with a horse (*ma* or *ba*).

62 *Sanbon:* "three trees" in this case. The official name for the area then was, and today is, Noda. In an Attic Museum film from a September 1934 visit to Wakimoto, two large pine trees, standing tall above the paddies, are visible. However, one appears to have no needles on its branches. This tree must have been felled by April of the following year, leaving one.

63 Sadly, Ichisuke succumbed to tetanus in 1939, at the age of eighteen.

64 *Jinomi,* or *marutobimushi* (*Sminthurus viridis annulatus*).

65 A local term for an unidentified edible wild plant. *Bonji* is an Akita word for a Buddhist priest, and *koromo* refers to clothing.

66 The term *shirako* is widely recognized.

Chapter 3: May 1935

1 *Shida mochi:* glutinous rice cakes made from *shiina* powder (powdered rice grains that failed to fully mature before being harvested).

2 *Fugo:* a carrying device made from woven straw, either rounded like a basket or box-shaped.

3 Shinden ("new paddies"): this was probably an area that had been developed as paddy land sometime after the end of the Edo Period in 1868.

4 Saburō uses the local term *shiguri* instead of the standard *mabiki.*

5 Probably a local term for *sugegasa:* hat-like conical "umbrellas" worn on the head.

6 Bordeaux mixture—a blend of copper sulfate and quicklime used as a fungicide and bactericide—was invented in France's Bordeaux region in the late nineteenth century and had become a common element of farm life in Japan by the 1930s. It has preventive properties but is said to be ineffective on established fungi.

7 *Umamokko:* a carrying device comprised of wooden poles and sling-like baskets that would go over the back of a horse so that one basket could hang on each side—analogous to a carrying pole held across a person's shoulders, with one container hanging from each end (see May 24 entry, and Yanagita and Miki, *Yukiguni no Minzoku,* 184).

8 Saburō often adds such comments, as if to emphasize his attention to detail, his dedication to efficient and productive farming, and perhaps also his progressiveness. There is also frequently a hint of admonishment, directed at his fellow villagers, in such remarks.

9 The local word *dofura* for pumpkin is used. The term was also applied to people who had taken on a bloated appearance due to disease (Yoshida, *Oga Kanpūsanroku Hōgen Minzokushi,* 224).

10 This would be *wasedaizu* in standard Japanese, but the pronunciation *bonmame* is given. *Bonmame* were probably edamame. Saburō's records indicate that he grew edamame annually from 1930 to 1932, but his later soybean crops are not identified as edamame.

11 *Mochi* made from regular cooked white table rice (*meshi*), and not from special *mochi* rice.

12 This contradicts Saburō's 1935 field map, which indicates that he had at least 30 *tsubo* of tomato plants.

13 The terms *umayagoe* or *kyūhi* are used, but with the pronunciation *mayagoe.*

14 *Tsuchiyose:* pulling soil up onto the rows in a field with a hoe to make the layer of soil over the plants' roots deeper. Saburō (and his wife) did this regularly.

15 South Sakhalin, an integral part of Japan and home to some 300,000 people at the time; see Dusinberre, *Hard Times,* 108–110.

16 Saburō's eldest daughter, Fusae, probably fell while running with the ball.

17 Yoshida calls this post-planting hoeing *chūkō* 中耕: literally, middle-plowing.

18 *Merikenko* (American flour): probably not imported but called so because the term had been used to differentiate between wheat flour imported from the United States and domestic *udon* flour during the final decades of the nineteenth century.

19 *Adenophora triphylla var. japonica,* also known as *tsurigane ninjin.*

20 *Fallopia japonica,* commonly known as *itadori,* but here the local/regional term *sashidori* is used.

21 *Innyo* and *deshika* are also edible wild plants.

22 Local/regional terms for ladybugs (*tentōmushi*).

23 In other words, Saburō used proportions for making 4 *to* (72.2 L) of Bordeaux mixture. He also at times made six-*to*-style Bordeaux mixture (proportions for making 6 *to* (108.2 L) of the solution).

24 This term is often used to refer to a cluster of seeds planted in one spot; in most cases, only one plant will be allowed to grow to maturity.

25 The nuance here is clear: Saburō believes that his careful, methodological way of planting seeds is superior to the dominant, careless, method (*bara-maki*).

26 A *seirō:* a bamboo steamer to put inside a pot.

27 A ring-shaped device that could help one aim when pounding or stomping.

28 *Sekihan:* rice cooked with red *azuki* beans.

29 Written as *uma no mame* with the unusual pronunciation of *mu no mame.*

30 This would normally be *mizukuchi* or *minakuchi,* but is given the local pronunciation *minagudzu.*

31 The Wakimoto Nōkai facility near the train station.

32 *Kinpi* 金肥: fertilizer purchased with cash, whether organic or chemical.

33 A device for carrying a load—a *mokko* for one would be a pole carried across the shoulders with a basket or other container hanging from each end, and a *mokko* for two would most likely consist of two poles and a large hammock-like sling, so that they might carry a large load together (see May 5 entry).

34 Yoshida Kinjirō: Saburō's cousin and head of his natal household's "main house."

35 *Nigorozaké* was probably thick, white saké made from strained fermented mash.

36 In this case, the *kobumaki* was herring wrapped with kelp and cooked in soy sauce.

37 An unidentified fish—possibly red gurnard (*Chelidonichthys spinosus*).

Chapter 4: June 1935

1 Saburō uses the local expression *hakayutta* (standard: *hakadotta*—"made good progress").

2 Saburō uses this local term throughout his one-year record to refer to such a dish, made with either dried daikon or pieces of burdock root and typically known today as *kinpira*. However, 80 years later his eldest daughter Fusae recalled the correct word as having been *denbu*.

3 Saburō uses the local verb *magushi* (*kata o magushi ni iku*) instead of the standard *mawashi* (to roll, turn, or spin). This is to roll a wood frame device (*kata*) through wet paddies to make perfectly spaced lines in the mud for even, straight planting (see Fig. 0.30).

4 Saburō's wording indicates that these were stones for large decorative lanterns, and that he blamed the boy's employer for the incident.

5 Written as 初植 in the text; the reverse form 植初 (*uezome*), is a pre-rice-planting ritual.

6 One *kari* is a unit of land area needed to grow one cluster of rice seedlings to maturity in a paddy.

7 This is "e" as in "egg." It refers to a labor-exchange arrangement between households, which, as Yoshida's record attests, was extremely important in the villages at the time. It was analogous to *yui* or *i* as described by Embree (*Suye Mura*, 99–103).

8 The eldest son of Saburō's cousin Yoshida Kinjirō, head of the big main house (see May 29 entry).

9 Hōchi was a major prewar newspaper; it was eventually absorbed by Yomiuri Shimbun.

10 *Ana* (hole) plus *jashi* (to stab or thrust, from the verb *sasu*).

11 *Kabuhajire:* loose or missing seedling bunches—*hajire* is *hazure* (having been removed, or missing).

12 Records point to vast numbers of deer living on the Oga Peninsula in the past, and to large-scale culling due to resulting crop damage—as many as 27,000 in a single year at one point prior to the modern era (Oga City, *Ogashishi*, 157). By 1935 they had vanished due to overhunting. An attempt was made to repopulate the peninsula in November of 1937 by introducing a small number of deer from elsewhere, but complaints about crop damage grew as the animals multiplied, and once again they were eradicated from the peninsula (Oga City, *Ogashishi*, 158).

13 The original text reads, "*Nga baraida mono no nae—dare toru mono da te?*" and, "'*Nga baraida mono*' is like '*kisama no yō na mono*'."

14 Saburō laments the breakdown of the formerly strong cooperative nature of farming village life, and increasing focus on households and individuals, and on bottom-line profits. See also Dore, *Shinohata*, 2–43, and Thomas C. Smith, *Agrarian Origins*, 180–181.

15 This may reflect an error in the original book.

16 *Katchiki:* a local term for green (or dried) grass and weeds, etc. This matter would be put into the nursery paddies in May or June, once all seedlings had been removed, and would serve as fertilizer for the following year (Yoshida, *Oga Kanpūsanroku Hōgen Minzokushi*, 83).

17 This is the common term; the scientific name is *Weigela hortensis*.

18 This is written as "head-shaving" (*atama-suri*).

19 A post-planting rest period for farmers, sometimes involving a party or feast, and possibly a trip.

20 The word "basket" (*kago*) is used, but this may have been a flat sieve (*zaru*).

21 This is the past tense of *shudo o mawasu:* to spread "neck-soil" in a circular motion.

22 Probably Japanese sand lance (*Ammodytes personatus*).

23 Tsuchizaki is currently the port area of Akita City, but was a separate municipality in the 1930s.

24 Rubber-soled cloth footgear with separate big toe space, typically worn by carpenters.

25 The Shinōkai (支農会) was a predecessor of the Nōkai (農会), which eventually merged with other associations to become the Nōgyōkai (農業会) in 1943 (which was abolished in 1947).

26 *Futon* can refer to mattresses stuffed with cotton or other soft material that are normally laid on tatami floors, and to thick covers (usually filled with cotton) for sleeping (i.e., Japanese *futon* serve as mattresses or as covers, and do not have wooden frames).

27 Funagawa is now the main port and administrative center of Oga City. It lies about 6 km to the southwest of Yoshida's hamlet of Ōkura (see map in Appendix 2).

28 The Ishikawa Association (Ishikawa-kai) was dedicated to protecting the legacy and spreading the philosophy of Akita's great *rōnō* Ishikawa Rikinosuke (see Translator's Introduction). The association was founded in 1920 by one of Ishikawa's grandsons and other disciples in Toyokawa, Ishikawa's home village: approximately 15 km east-southeast of Wakimoto (Ishikawa Rikinosuke-ō Tanjō 150-nen Kinen Jigyō Jikkō Iinkai, 56).

29 A relative of Saburō's wife—probably an uncle—and a member of the village governing council during the early postwar years (Oga City, *Ogashishi,* 311).

Chapter 5: July 1935

1 Lt. Col. Tominaga Kyōji (1892–1960) was a leader of the powerful, imperialistic, right-wing Kōdō-ha faction of the army. In February of 1936 some of its officers launched a coup d'état that failed and significantly reduced the faction's influence in the military, but at the time Yoshida heard Colonel Tominaga speak the Kōdō-ha was dominant. A *teishintai* is a battalion or other group organized for a purpose for which its members are resolved to sacrifice their lives.

2 Momokawa was several kilometers north of Yoshida's hamlet of Ōkura.

3 It was monthly festival time, and also the day of the annual Tanabata star festival.

4 Tarusawa was a hamlet located about two kilometers to the north, just south of Momokawa. Tarusawa Elementary School eventually became Wakimoto No. 2 Elementary School, and subsequently merged with Wakimoto No. 1 Elementary School (Saburō's alma mater—now simply Wakimoto Elementary School—in the main part of the village). The remaining buildings are now used for storing and preserving local historical and archaeological records and items, and as a community center.

5 Saburō has not yet mentioned harvesting any cowpeas (*sasage*), but he appears to have picked at least enough to begin consuming them.

6 *Koshimaki:* a thick sash worn around the waist, usually with a kimono or a summer *yukata*.

7 This was probably for the children.

8 *Amegashi* (*ame no kashi*): traditional candy made with sugar extracted from rice or potatoes, etc.

9 *Koejaro* in local speech (Yanagita and Miki, *Yukiguni no Minzoku,* 185); see Fig. 0.24.

10 A traditional loincloth (*fundoshi*) of the Etchū (Toyama Prefecture) style, with a long front piece.

11 See Mar. 22 entry.

12 One *pa* was equal to one bundle, probably what one man could carry in one load.

13 Saburō clearly feels that his way is superior.

14 Kinzaburō was a cousin of Saburō's.

15 These were probably the bulbs of golden-rayed lilies (*Lilium auratum*).

16 These may have been an experimental strain of napa cabbage, possibly a hybrid.

17 This was the *senkyo shukusei dōmei* 選挙粛正同盟, perhaps part of the election cleanup movement (*senkyo shukusei undō*), which by 1935 had probably come to serve at least in part to foster nationalistic sentiments.

18 Futto was a village about three kilometers away, toward the northeast, near the shore of Lake Hachirō. Now, like Wakimoto, it is part of Oga City.

19 Again, Saburō makes it clear that he considers his way superior to the typical local method.

Chapter 6: August 1935

1 These (*misokagi/misokaki* cucumbers) seem to have been allowed to lie in miso for some time, different from *momi* cucumbers, which were generally eaten soon after being flavored with miso and vinegar—see Jul. 17 entry.

2 Saburō made this insecticide and used it on his crops a number of times during August.

3 This probably refers to so-called *min-min* cicadas (*Hyalessa maculaticollis*): a possible editorial error.

4 Kita Iso was a settlement on Oga's northern shore, about 16 km northwest of Ōkura. Saburō uses the community's old name; it had officially become Kitaura Town more than three decades earlier.

5 This was probably either a straw skimmer hat or a Panama hat.

6 This may have been organized by the local branch of the Agriculture Support Association (Shinōkai), partly for pleasure but mainly for observing other farmers' operations and techniques.

7 Possibly spicy pickled cucumbers (using *shishitō* peppers), but most likely cucumber *sushizuke:* cucumber preserved together with vinegared sushi rice—probably a gift.

8 *Natsuna:* a green, leafy *Brassica rapa* subspecies, similar to bok choy.

9 Bon: The annual Buddhist celebration, when the spirits of the ancestors (*hotoke sama*) are believed to briefly return to the world of the living. It is generally observed during July in the Kantō region, but in August in northeastern Japan, where it is also called *okuri-bon*.

10 Aokubi Hōryō: a type of daikon originally from Aichi Prefecture with a green (*ao*) top, or "neck" (*kubi*) (http://www.city.ama.aichi.jp/bussiness/sangyo/nougyou/1003864/1003872.html).

11 *Kusakezuri:* chopping up (*kezuri*) weeds or other plant matter (*kusa*) for disposal or composting.

12 Local pronunciation for *samukaze,* or *samui kaze* (cold wind).

13 A reference to the existence of one of more pear (*nashi*) trees (*ki*) in the area.

14 Obara Kuniyoshi (1887–1977) was a national leader in Japan's New Education Movement. A dedicated educator, Obara founded Tamagawa Gakuen in 1929, and later served as president of Tamagawa University.

15 The slopes and foothills of Mt. Kanpū, where households were allowed to cut hay in certain areas (see Aug. 20 entry and Fig. 0.34).

16 *Komazari:* a fine-toothed wooden rake (*komajari* in local speech); see below, and Apr. 6 entry.

17 *Midoko tabake:* tied up in three places (*mitsu no tokoro*).

18 Strands of rice plant straw (*wara*) tied together (*tsunagi*) in three places (*sanba*) or four places (*yonba*) to make a single rope for bundling.

19 *Yujige:* a local way of saying *yuzuke*—cooked rice with hot water poured over it. Also called *yumeshi* (see, e.g., Mar. 29 and Jun. 10 entries).

Chapter 7: September 1935

1 The Seinen Gakkō (Young Men's School) was a nationwide educational program for boys, established by the government in 1935. Nearly 17,000 branches existed as of the first year, with almost 2,000,000 boys enrolled (https://www.mext.go.jp/b_menu/hakusho/html/others/detail/1317715.htm).

2 This was probably a very small nearby hamlet; no record seems to exist of it or of the shrine today.

3 The use of the suffix *kun* reveals Saburō's familiarity with the person and indicates that the two were probably close in age.

4 Also, *karasu mochi:* small, round *mochi* made from grinding rice into powder and adding water and kneading (Yoshida, *Nōmin Shuki,* 60; *Hōgen Minzokushi,* 334; *Oga no Koboredane,* 98–101). See also Apr. 11 entry.

5 This coincided with autumn *higan* (see Mar. 17 entry) and roughly with the autumnal equinox.

6 This is analogous to *yoriai,* as described by Miyamoto Tsuneichi (*Wasurerareta Nihonjin,* 36–58; *The Forgotten Japanese,* 21–29).

7 Probably *sujihaze* (*Acentrogobius virgatulus*), a member of the goby family of fish, caught in Lake Hachirō (see multiple mentions of *gunji,* another goby often eaten by the family).

8 The diagnosis was encephalitis (Yoshida, *Oga no Koboredane,* 413).

9 Saburō probably included travel, food, and other associated costs in this figure.

10 Okada (1902–1975) was a politician who represented the Socialist Party in the National Diet (https://kotobank.jp/word/岡田%20宗司-1641054).

11 Yaeko (1932–2014): Saburō's younger daughter.

12 Probably essentially the same as *konoha-sarai* (see Apr. 6 entry).

13 Train, food, and other associated costs were likely included.

14 A tool varying in size and shape made from a flat board, with holes drilled in it and small wooden sticks or pegs shoved into the holes (Yoshida, *Hōgen Minzokushi,* 168).

15 The main house of Saburō's "main house" (i.e., natal household), located next door (see Fig. 0.21). The second son mentioned here was probably the younger brother of Yoshida Kin'etsu (see Jun. 3 entry).

16 A ceremony marking the presentation of a grandchild (*mago*) to the public (Yoshida, *Nōmin Shuki,* 159).

17 Local term for fish otherwise known as *haze,* members of the goby family (*Acanthogobius flavimanus*) (Sugiyama, *Akita no Tansuigyo,* 107–108). Although still found in Lake Hachirō and around its mouth today, these were far more plentiful before the lake was largely converted to farmland through a Dutch-style polder dam reclamation process commencing in the 1950s (Wood, *Ogata Mura*).

18 The original text reads "Michinoku" or "Mutsu" ("Riku-oku") 陸奥, but this is probably an error. Riku'u 陸羽 No. 132 is a type of domesticated rice that emerged in Akita in the early 1920s (https://www.jataff.jp/senjin2/40.html).

Chapter 8: October 1935

1 This exceeded Saburō's income for the 366-day period covered by his 1938 book by about 80 yen, and amounted to approximately one-third of the national 1935 average annual income for farming families (Wood, "Trial and Error," 259–260). Also see Appendix 4.

2 A *bakurō* was a person who traded in horses and/or tended to them in general, providing preventive care, treating maladies, and performing traditional rituals to protect them from harm. No explanation is given for the switching of the two horses.

3 The term "village" (*mura*) is used here, but it may have been a reference to a different hamlet of Wakimoto, as "village" and "hamlet" were used interchangeably at times. See Nov. 14 entry.

4 *Tsunagi* cords were made of rice plant straw (*wara*).

5 Grated Chinese yam (*nagaimo/tororoimo*) on rice, probably flavored with soy sauce, or possibly more like *tororo jiru* (see Apr. 14 entry).

6 This was an impressive achievement, considering that the national 1935 rice crop is considered to have not been a bountiful one, and that the local farmers had been concerned about the effects of rain and low temperatures on their rice crops only several months earlier (see Jul. 18 entry).

7 This section has been rewritten slightly for clarity, based on careful inspection of photographs in Yoshida's personal collection.

8 This expression (*ine o keshi*) means to do *keshi* (stacking) to the rice plants (*ine*): in this case, re-stacking bundles for drying.

9 The standard pronunciation is used here, instead of the local style (*shijigo*).

10 Probably for a time slot at the communal work shed or for a good fee elsewhere.

11 *Bon-odori* dances are normally performed across Japan for the summer Bon Festival, when the spirits of the ancestors are said to briefly return to the land of the living. Many associated songs contain *saku-odori* dance lyrics, which express hope for a bountiful harvest or for a good catch of fish. Today there is generally no separation between *bon-odori* and *saku-odori,* with the latter having faded into the background. However, there may have been separate dances in the 1930s (see http://www.akita-minzoku-geino.jp/archives/4331).

12 Probably a member of the stargazer family (*Uranoscopidae*) of perciform fish (*Xenocephalus*).

13 *Waraniyo:* a large, circular, stack of rice plant bunches, covered with a rooflike top made of the same (see Fig. 0.43, and Yanagita and Miki, *Yukiguni no Minzoku,* 113). According to Saburō's eldest daughter, Fusae, her father would build these in the space between the house and the *wagashi* shop on the property of the main house (unrecorded interview, Mar. 8, 2022).

14 Saburō uses the local verb *tsubokeru* (to make *tsuboke* clusters of plants—see Figs. 0.39 and 0.41–0.42) here and below (Yoshida, *Hōgen Minzokushi,* 206; see also Oct. 19 entry).

15 *Ogi* is local pronunciation for *oki*—to skip a space or day or week, for example, while progressing along a course of action, or while following a schedule.

16 Java waterdropwort (*Oenanthe javanica*), also known as Japanese parsley and Chinese celery. In Akita, *seri* grew naturally among rice paddies in the past, but today it is generally commercially produced. It is one of seven herbs traditionally consumed annually on January 7 in rice gruel (*nana kusa gayu*), and is a necessary ingredient in *kiritanpo nabe,* a well-known trademark dish of Akita.

Chapter 9: November 1935

1 *Migo* 身子 (also *waramigo* and *warashibe*) are the soft inner stems of rice plants, to which the rice grains themselves are attached. These were valuable in the 1930s for making strong and thin ropes, and for crafting straw coats and the like (Yoshida, *Hōgen Minzokushi,* 309).

2 See Apr. 11 entry.

3 This is probably a reference to a 1589 event. The castle was abandoned in 1590. A more detailed explanation of this is given in Yoshida, *Oga no Koboredane,* 98–101.

4 *Ashi:* foot, feet, leg, or legs.

5 Shioya (1895–1972) was an Akita native who traveled widely inside Akita Prefecture as well as beyond its borders to entertain the public—especially children—with retellings of fables and folktales, and the teachings of the Bible (Akita Sakigake Shinpōsha, *Akita Jinmei Daijiten,* 292).

6 This is written as "young cod" (*kotara*), but it is likely that the fish was not always cod. Therefore, the more general "scrod" is used in the translation.

7 These are identified as *chūkan nawa,* pronounced *tsūken nawa.*

8 Pieces of daikon that have been cut and chopped with a cleaver (*nata*), salted, squeezed, and pickled with sweet saké (*amazake*) and/or malted rice. Yoshida generally writes the word as he and his fellow villagers pronounced it (standard: *natazuke*).

9 This was a séance (also *nanakurayose,* which is probably standard). See Mar. 22 entry.

10 A spirit medium. The standard form is *itako* but Saburō writes both *idako* and *edako* (see Mar. 22 entry).

11 Momokawa, beyond Tarusawa to the north, was a hamlet of Wakimoto and had not been an independent village since it merged with the other five hamlets of Wakimoto to form the village in 1889. Saburō often uses "hamlet" (*buraku/shūraku*) and "village" (*mura*) interchangeably (see, e.g., Oct. 11 entry).

12 A type of seaweed, probably *Hypnea flexicaulis. Egeshi* is a regional term.

13 Standard term for pickled vegetables (*gakko* in Akita dialect).

14 A laborer, usually doing civil engineering–type earth-moving work or carpentry for set wages (see Bernstein, *Haruko's World,* 78–81).

15 This is probably a reference to the area called Sakura, where the family owned or rented some paddies. Kari appears as a place name in parts of Wakimoto, but here one *kari* probably means the unit of land area needed to grow one cluster of rice seedlings to maturity in a paddy.

16 Partially fermented soybeans, usually put on top of a bowl of rice and eaten.

17 Ishikawa Rikinosuke (1845–1915): the famous *rōnō* of Akita (see Translator's Introduction, and Jun. 28 and Dec. 24 entries).

18 Most of the area that was Kawabe County at the time is now part of Akita City.

19 This would have been part of the Akita Farmers' Festival.

20 Conventional metal folding umbrellas commonly used today were called *kōmori gasa* (bat umbrellas) in Japan since at least the 1860s, and still in the 1930s, because they were thought to resemble bat wings.

21 She may have done the *futon* work at the main house or at her natal home, as both had rice paddies.

22 Loose, durable cotton pants (standard: *monpe*).

23 A séance (also, *nanagurayohi*). See Mar. 22 and Nov. 13 entries.

Chapter 10: December 1935

1 *Eitoku Maru* was a fishing vessel; see Dec. 22 entry and Appendix 1.

2 The Nōsangyoson Keizaikōsei Undō was a government initiative launched in 1932 to improve the living conditions for the nation's farmers, particularly those strongly affected by the Great Depression. (See Arimoto, "Participatory Rural Development in 1930s Japan"; Kerry Smith, *A Time of Crisis,* 206–214; Mar. 11 entry.) It offered debt repayment extensions, financial assistance for fertilizer purchases, and up to 50 million yen for moving to colonial Manchuria or Inner Mongolia.

3 It is unclear why Saburō was paid for two people.

4 Saburō's brother probably took care of the task.

5 *Mujinkō* or *tanomoshikō:* these had all but vanished from Japan's villages by the 1950s. See Beardsley, Hall, and Ward, *Village Japan,* 260, and Fukutake, *Rural Society in Japan,* 99–100.

6 Upon reading this passage, Yoshida's eldest daughter, Fusae, concluded that the groom was probably one of her mother's cousins, as her mother's unmarried younger brothers had all left the village by that time, and because the household consisted of a large, extended family within which consanguineous marriage was common (unrecorded interview, Sept. 14, 2021).

7 Gifts given to neighboring households on special ceremonial occasions, such as weddings and funerals, have been referred to as *hide* in rural parts of Akita. Tomiki, *Nihon no Minzoku: Akita,* 129.

8 *Shakudori* 酌人: pourers of saké for the ceremony, and perhaps servers for the banquet.

9 *Sahainin* (standard): local leaders or managers who would often serve as representatives of powerful elites, such as landlords, acting on their behalf and authority. This *sahainin*'s role as master of ceremonies may have been associated with a national program aimed at cutting wedding ceremony costs. Marutani, "Akita Kennai ni Okeru Kekkonshiki no Kaizen Undō," 77.

10 A Noh play that features a long-married, loving couple. Excerpts have been frequently sung at weddings.

11 Yoshida's elder daughter, Fusae, believed that it was probably a fruit salad containing pears (unrecorded interview, Sept. 14, 2021).

12 Although candy sea bream is mentioned above, this appears to be a reference to other similar types of colorful confectionaries made of glutinous rice flour and filled with sweet bean paste that are generally served on such occasions (Yoshida Fusae, unrecorded interview, Sept. 14, 2021).

13 *Kishi:* probably blackhead sea bream (*kurodai*).

14 This is likely a reference to the Amur stickleback (*Pungitius sinensis*), a species of small (3–5 cm) fish with sharp spines on its back, known as *tomiyo* across most of Japan. Sugiyama, *Akita no Tansuigyo,* 26.

15 Specifically, the female reproductive organ. An alternative term was *dappe.* Words for the male organ included *keppe, henoko,* and *gamo.* Yoshida, *Nōmin Shuki,* 165.

16 Saburō would collect his pay on a later date (see Dec. 20 entry).

17 Saburō means that there were about 80 households in the hamlet of Ōkura; there were many more in the entire village of Wakimoto.

18 Saburō uses the term *kosakunin:* small-scale (tenant) farmers (see Apr. 10 and Nov. 18 entries).

19 The Akita Ondo is a famous song of Akita that is said to be at least several centuries old. Hachimori, an independent village from 1889 until 1954, is a small coastal community approximately 50 km north of Yoshida's hamlet, close to the Aomori Prefecture border.

20 Kon Sakunosuke (1885–1957), a native of the town of Moriyoshi (now Kitaakita City) in Akita Prefecture, was a proponent of farming lifestyle improvement and farmer education, and a politician. The Rikken Yōseikai 立憲養正会 was a rightist Nichiren Buddhist sect–associated political party (Drea, *The 1942 Japanese General Election,* 42). Kon ran on the party's ticket multiple times between 1930 and 1942.

21 *Chinomichi gusuri* 血の道薬 was probably a Chinese (*kanpō*) remedy for menstrual problems, for Saburō's wife. The headache medicine was probably also for her.

22 *Deshiko mochi* were made annually on November 23 to pay respect to the goddess Odeshi, who had 23 children, for whom she cared well, feeding each fairly and equally using long poles like chopsticks because she could not reach them all otherwise. Odeshi is said to have starved (sacrificed) herself to adequately feed her children and raise them (Yoshida, *Nōmin Shuki,* 80–81).

23 *Fumidawara: tawara* for wearing like boots in winter (see Mar. 7 entry).

24 Although dogs were not commonly eaten, their pelts were worn for warmth in northern areas, including Akita Prefecture. Yanagita and Miki, *Yukiguni no Minzoku,* 129.

25 Saburō's daughter Fusae recalled this relative of her father's as having been an "unusual person" (*kawarimono*), hence the odd nickname, which was often used for servants (unrecorded interview, Mar. 8, 2022). She has also applied the *kawarimono* label to her father.

26 Grated potato—strained, squeezed, and thickened with other starch and cooked, ultimately having a consistency similar to that of conventional *mochi.* Nihon no Shokuseikatsu Zenshū Akita Henshū Iinkai, 36–37.

27 The Mitsui Kindness Repayment Association: a philanthropy venture launched by the Mitsui Corporation in 1932, when it and other zaibatsu conglomerates were struggling to improve their public images. Samuels, *Machiavelli's Children,* 155; Yasuoka, *Japanese Zaibatsu.*

28 Kosaku-kan 小作官 were local public officials charged with mediating between tenant farmers and landlords, and with helping to facilitate the legal arbitration process, during the time of great tenant farmer unrest in the late Taishō and early Shōwa periods. Yagi, "Saikin ni Okeru Kosakusōgi no Dōkō." However, these were often accused of siding with the landlords, as the unflattering depiction of a Kosaku-kan in Akita-born Kobayashi Takiji's 1929 novel, *The Absentee Landlord,* suggests. Kobayashi, *Fuzai Jinushi,* 2010. Saitama-born Nishida Kintarō (1891–1969), mentioned here, differed from most such officials of the time in that he tended to side with the farmers: a stance that is said to have earned him the reputation of a humanist, and the unofficial nickname "father of the farmers" (Nishida Kintarō Tsuisōshū Henshū Iinkai). A connection existed between Nishida and Ishiguro Tadaatsu, who was related by marriage to Shibusawa Keizō. Nishida was heavily influenced by Ishiguro. Washiya, "Nishida-sensei wa 'Kokoro' aru Hito no Naka ni Mada Ikiteiru!" 33. Saburō praised Nishida's kindness and humanism and suggested that Nishida might have been hand-picked by Ishiguro as the Kosaku-kan for Akita, which was known at the time as a hotbed of farmer unrest, with more than its fair share of heated tenant-landlord disputes. Yoshida, "Shōten Shitemo Kanashii." Yoshida also became close to Nishida; in 2019 Yoshida's daughter Fusae recalled that Nishida often visited her father at his postwar home. See also the Jan. 27 entry.

29 The original text reads "Hiratsuka County," but this was probably an editorial error, as no such county ever existed in, or anywhere near, Akita Prefecture. Akita's Hiraka County, located about 80 km away from Yoshida's village, is part of a major rice- and apple-producing region of the prefecture.

30 Their visit probably related to the Mitsui Hōonkai's support for the Ishikawa Association (see Jun. 28 entry). The Hōonkai published an encyclopedia of Ishikawa Rikinosuke's agricultural philosophy in 1938.

31 This and the following *bishaku* were written as *hiyaku* and *biyaku,* respectively, in the 1938 book: typographical errors (see Apr. 1 entry).

32 Local words used here include *kui kui* ("stinky, stinky") and *dara* ("night soil").

33 *Konogazuke* was a local term for *takuan.* Yoshida, *Nōmin Shuki,* 12.

34 Monzen is about 14 km west-southwest of Yoshida's hamlet of Ōkura.

35 For this celebration, red bean (*azuki*) rice (*sekihan*) was formed into *mochi*-like cakes and served on special plates made from rice plant straw, on trays, with pickled vegetables (*tsukemono*) and *hatahata,* and eaten. Yoshida, *Nōmin Shuki,* 78. Ebisu is the god of prosperity, commerce, and fishing. In the past, many such New Year worship celebrations for different gods were observed, and this continues in parts of Japan today (e.g., http://furusato.fmii.co.jp/ichinohe/2013/12/20/神様の年取り行事/).

36 According to the original text, she boiled *shifuro* water. This may have been bath (*furo*) water.

37 This was mixed with sugar and consumed, generally by licking it off a flat surface.

38 The Hada god (Hada no Kami) was recognized as the deity of looms and weaving by the people of Wakimoto, a reflection of the great past importance of making clothing at home from hemp fiber grown in local fields. Villagers paid appreciation to Hada no Kami annually on February 6 by making *kohin* and offering it to the god, and of course also by eating it. Yoshida, *Nōmin Shuki,* 79.

39 This would have been mixed with some sugar and either shaped into bite-sized pieces by adding some water or simply licked off a plate or other flat surface.

Chapter 11: January 1936

1 Degogu is local pronunciation for Daikoku, the god of wealth, who is traditionally believed to have received a bride on December 9 (http://www.creative-tsuruoka.jp/project/utigozen/fuyu-otoshiya.html). In Yoshida's time, this day was generally celebrated as the New Year festival day of Daikoku. In Ōkura, daikon were offered to Daikoku. Yoshida, *Nōmin Shuki,* 79.

2 Tall, solid *geta* (wooden sandals) fitted with metal skates. The name derived from their similarity to a warship (*gunkan*). *Gane* is a reference to the metal skate (*kane*).

3 This was a local pronunciation for Jūōdō 十王堂, a hall (temple) for the ten kings who preside over the gates of hell in Buddhist lore; there may have been such a place of worship in Wakimoto in the past.

4 One *tsubo* is equal to 3.306 m^2.

5 *Yabatsu* is Akita slang for the commonly used *yabai* (bad, dangerous) (http://common3.pref.akita.lg.jp/koholib/search/html/027/027_040.html).

6 Yoshida uses the expression *sanbyaku daigen* 三百代言.

7 One *hiro* 尋 is approximately equal to 1.8 meters (but Yoshida defined it as 1.5 meters (see Jan. 12 entry)).

8 At times, Yoshida writes *umamono* (horse fodder) instead of *umagusa* (hay). This might indicate that something other than proper hay was cut.

9 This is commonly known today as canola oil.

10 This would have been either about 13.5 m or 16.2 m, depending on how Saburō defined one *hiro* in this instance (see Mar. 16 entry).

11 This would have been either 15 m or 18 m, or somewhere in between.

12 Villagers involved in mountain work would take the day off (i.e., refrain from doing such work on that day) and pound the type of *mochi* mentioned here and offer one dozen to the mountain god on a tray. It was said that if a person were to venture into the mountains to work that day, he would be injured by an arrow shot by the mountain god: a risk no local person was willing to take. Yoshida, *Nōmin Shuki,* 80.

13 A similar procedure is described by Embree, *Suye Mura,* 88.

14 This would have been the hearth (*irori*) inside the house.

15 Mutō Tetsujō (1896–1956) was a meticulous and influential archaeological, ethnological, and historical researcher of his home prefecture of Akita. He was also a prolific writer. At least two of his books were published by Shibusawa Keizō's Attic Museum. Yoshida was reading Mutō's 1935 book.

16 The local term *doyashita* is used here.

17 See Mutō, *Chōchū Sōboku Minzokugaku-teki Shiryō,* 223.

18 *Hamigaki yōji:* small wooden sticks with frayed ends with which to brush (*migaku*) one's teeth (*ha*).

19 See Mutō, *Chōchū Sōboku Minzokugaku-teki Shiryō,* 192–193 (and below).

20 Yoshida uses the verb *shibiru:* probably a regional term (i.e., the *mochi* becomes freeze-dried).

21 The grains have firm centers, as they are not thoroughly cooked.

22 A general discriminatory attitude toward such people is apparent.

23 Literally, trees that don't allow birds to land on them.

24 Attaching "*kko*" to the end of nouns is a common regional speech pattern in Akita (*tori* means "bird"). See Mar. 23 entry.

25 "Snow burn": *yukiyage* in the local vernacular.

26 Pieces of candy that look like human faces. Perhaps *Kintarō-ame.*

27 Probably one or more long-handled wooden ladles; there is evidence of this terminology for such items in Yamagata Prefecture (http://koichikato.world.coocan.jp/minpou/minpou2005/muranokurasitodogu2/newpage8.html).

28 Local pronunciation for *yuzuri* (*Daphniphyllum macropodum*): used along with pine branches for decorating for the New Year celebrations.

29 One of the *maki* (*dōzoku* lineage groups) of Yoshida's hamlet.

30 Probably the household or the natal household of Takase Takezō (see Jun. 21 entry).

31 Mysterious New Year's Eve visitors (masked and specially garbed young men of the village) who would traditionally search the households of Oga's villages for naughty children and lazy young mothers (*yome*) and frighten them to ensure good behavior throughout the coming year. Namahage are generally placated with saké, *mochi,* and promises by the head of each household. See Yamamoto, *The Namahage,* and Feb. 3 and Feb. 7 entries.

32 In the southeastern (Yuzawa) part of Akita Prefecture, young men who lit and tended cremation fires were known as *dami wakaze.* Kanezō may have served in this capacity. However, he may have simply represented the household that employed him as a guest at the funeral, as he was not there all day.

33 Customary New Year decorations. See Feb. 7 entry explanations and photographs. Also see Yanagita and Miki, *Yukiguni no Minzoku,* 37, for comparison.

34 *Tara* tree (*Aralia elata*) sprouts: often collected and eaten in northeastern Japan in the springtime.

35 *Nanakusa* refers to *nanakusa-gayu,* a simple rice gruel containing seven specific plants that is typically eaten on the seventh of January across Japan today, for good health. *Jōnimochi* appears to be a reference to *(o)zōni* soup (typically consumed during the New Year holiday season today) with *mochi* in it. *Nanakusa Jōnimochi* was probably a combination of the two. See the following day's lunch menu.

36 This man had become a widower nearly six months earlier (see Aug. 3 entry).

37 The *haha no kai* 母の会: a prewar precursor of the PTA.

38 Wazaki Haru (1885–1952), a native of Akita City and leader of the local women's suffrage movement, was the first woman of the prefecture to win a seat in the National Diet, and a member of the first group of 39 women elected to the legislature, in 1946 (Akita Sakigake Shinpōsha, *Akita Jinmei Daijiten,* 579–580; Pflugfelder, *Seiji to Daidokoro*). In the original text, the common female name suffix *-ko* was added to her name in error.

Chapter 12: February 1936

1 A customary New Year decoration for welcoming the spirits of the ancestors or perhaps local or harvest deities, featuring pine branches and other elements (today, usually bamboo). These are not generally to be simply discarded but rather to be incinerated or, as in 1930s Wakimoto, tied up in a tree (see Figs. 12.12 and 12.31).

2 *Wakagi mukai:* split pieces of wood, usually brought down from forested mountains or hillsides, displayed in the house as a blessing to the household for the new year and as protection from wood scarcity or calamity.

3 *Koshōgatsu:* the "Little New Year," the first full moon festival, generally coinciding with January 1 according to the old lunar calendar.

4 *Nihonka ni suru:* to make one long object (e.g., a sheaf of rice plant straws) into two such objects (i.e., to divide one long object into two parts).

5 This was the predecessor of today's Akita Prefectural Library. Ono, a former school principal, served as head of the library from 1934 to 1946, when the library was making efforts to increase interest in reading and education

in local areas by cooperating with schools and Seinen-dan (Young Men's Association) chapters, etc. (Nihon Toshokan Kyōkai, *Kindai Nihon Toshokan no Ayumi,* 94).

6 The family probably simply ate leftover miso soup and rice from breakfast with the usual side dishes.

7 The Namahage tradition of the Oga Peninsula is a UNESCO Intangible Cultural Heritage of unknown origin. In the past, young local men dressed as Namahage would burst into houses at will and frighten young children and their mothers, leaving only after having been properly placated. Namahage are seen as quasi-deities, emissaries of gods, and bringers of good fortune in the coming year. The tradition, one of many similar traditions scattered along Japan's western coastline, has thrived and ebbed over the centuries, passing through periods of intentional suppression, but is relatively healthy today. Oga's Namahage are now representative of Akita Prefecture itself. Nevertheless, the tradition has changed significantly. For example, the Namahage do not now visit all houses in Oga communities that continue the tradition; they are always accompanied by a "point man" who secures permission in advance. Moreover, they normally do not visit houses with no children. See Matanle, Rausch, and the Shrinking Regions Research Group, *Japan's Shrinking Regions in the 21st Century,* 341–360; Nihonkai-iki Bunka Kenkyūsho, *Namahage;* Yamamoto, *The Namahage;* Yoshida, *Nōmin Shuki,* 47–57; and see Jan. 22 and Feb. 3 entries.

8 These were probably rock greenlings (*Hexagrammos lagocephalus*), common fish along the shorelines of the region (https://www.fishbase.se/Summary/SpeciesSummary.php?id=4033&lang=japanese).

9 Yoshida writes *hatahata no shishizuke* (*shishi* being the local way of saying *sushi*). Generally known today as *hatahata-zushi,* this is made by packing *hatahata* in a container with cooked rice and saké lees and other ingredients and allowing it to ferment (see Aug. 13 entry).

10 *Kyū jūsan-ka-getsu kichijitsu* 旧十三ヶ月吉日; "former" refers to the old lunar calendar.

11 This is probably a reference to a type of ladle, or one used for this specific purpose, as *minja* refers to flowing water (see Nov. 24 entry), and the suffix *shi* often forms a noun. However, the meaning remains uncertain (Yoshida's eldest daughter, Fusae, was unable to recall the word in Sept. 2019).

12 See Yoshida, *Nōmin Shuki,* 47.

13 In Yoshida's village, on the evening of the sixteenth day of the New Year (according to the old calendar) a large fire (*hido*) would be lit in front of the shrine, and in this fire *mochi* would be roasted. Villagers would take these home with them and share the *mochi* with all family members. It was said that eating this *mochi* would help protect the people from illness throughout the year. Yoshida, *Hōgen Minzokushi,* 263.

14 This is an apparent reference to an Imperial Army division.

15 Although Yoshida writes Seinenkai, this was probably the same as the Seinendan.

16 This Akita City school, founded in 1900, is now officially known as the Akita Prefectural Akita Kita High School. Although it is public, admittance was open only to women until 2008, when it became a coeducational institution. Wazaki Haru was a graduate of this school (see Jan. 31 entry).

17 Written as *gotomori* here: a possible editorial error. See Apr. 6 entry.

18 Covers for the lower legs. See Mar. 6 and 14 entries.

19 See May 29 and Jan. 21 entries.

20 This probably reflects local usage, *fundoshi* being the standard term for a loincloth (see Jul. 22 entry). *Ōhaba fudoshi* translates as "big wide loincloth."

21 Mata means "groin" or "crotch." *Ippe* (in this case) is unclear, but it may have been used because of its similarity to the male first name Ippei.

22 *Mochi* with sugar on it.

23 This is a play on words. *Dobin* is also a term for a small serving pot (containing saké in this case), with a spout and a curved handle attached in two places, and *mata ippe* can mean "another cupful" (*mo ippai*).

24 Local pronunciation for the standard *senaka-ate,* a device worn over the shoulders to provide cushion when carrying a heavy load on one's back.

25 One *rin* was equal to one-tenth of one *sen* (see Mar. 16 entry).

26 Although the term *mikko nawa* sounds like "a rope made with three ropes," Saburō's explanation here indicates a different meaning.

27 The bath was heated by burning wood in a compartment beneath it, accessible from outside the house.

28 Half sandals that cover only the balls of the feet (*ashi* ("foot") + *naka* ("half")). See Mar. 26 entry.

29 Whereas *ashinaka* typically fit only beneath the toes and the balls of the feet, Yoshida seems to have made them somewhat longer.

30 One *kamasu* of each the materials mentioned here probably weighed between 30 and 40 kilograms. See Mar. 22 and Aug. 6 entries.

31 Hōnen was a popular brand name at the time. The meal Yoshida bought was probably left over from oil extraction, as Hōnen was a major producer of cooking oil.

32 This is a reference to the Sanriku Coastline of Aomori, Iwate, and Miyagi Prefectures: possibly a brand name. The sardine bits were probably leftovers from processing for making fish stock, etc.

33 The *shigana* seaweed mentioned here may have been *Petalonia binghamiae* or *Scytosiphon lomentaria*. Both were referred to as *shigana* or *sugana* in the general region, according to a 2002 *Japanese Journal of Phycology* report (50(1):37–40) (http://sourui.org/publications/sorui/list/Sourui_PDF/Sourui-50-01.pdf).

34 Normally, the name Tarō or a variant (e.g., Kintarō) would be given only to a firstborn son.

35 Yoshikin, a family-operated store, did exist as of 2012. A liquor and general goods shop that doubled as a newspaper distributor, this was built into the house, fronting the street, next door to where Saburō's brother's sweets shop once stood. The most notable item on Yoshikin's shelves was a saké brand called Imajin 今人, named after John Lennon's 1971 song. Made by an Akita brewery but with Mt. Kanpū spring water from the Wakimoto vicinity, this label had been founded by Yoshida Kinji, the late husband of the woman who ran Yoshikin at the time. She was the *yome-san,* having married into the family, but her husband Kinji had been the eldest son of Kin'etsu, eldest son of Saburō's cousin, Kinjirō. Unfortunately, Yoshikin's doors had been permanently closed by the end of the decade.

Chapter 13: March 1936

1 The original text reads "agriculture" (*nōgyō*) but this may have been an error. Yoshida probably meant (and may have written) "farming village" (*nōson*)—a reference to a national movement to improve general living conditions in farming villages, which was strong at the time.

2 Yoshida's daughter Fusae identified this man as one of her mother's younger brothers (unrecorded interview, Sept. 14, 2021).

3 *Fukei zadankai* ("father–elder brother conference"): this would be a parent-teacher (or PTA), conference today. However, in the 1930s the male household head (generally the father, or the elder brother in the absence of the father) was the official guardian (or, "master") of the household's members—especially the younger ones and the women.

4 This would have been her second son, Tsugio, who was the same age as Saburō's eldest daughter, Fusae. See Fig. 0.49.

5 Generally made of straw, these were wrapped around the lower legs, ankles, or possibly as low as the heels, for warmth in winter. Yoshida identifies the location for *agutomaki* (*agutomagi* in local dialect) as the "heel" (*kagato,* local pronunciation for *kakato*) in *Hōgen Minzokushi,* 24.

6 *Tawara* for wearing like boots when walking in deep snow.

7 This was done in accordance with a farming village improvement project: probably the 1932 Economic Revitalization Campaign / Economic Rehabilitation Movement. See Kerry Smith, *A Time of Crisis,* 207–214; Arimoto, "Participatory Rural Development in 1930s Japan"; Dec. 3 entry. Amano Gen'ichi was mayor of Wakimoto from Sept. 1934 to May 1945. Oga City, *Ogashishi,* 210. Amano family members were prominent in village politics for many decades, from as early as the 1890s.

8 Although not as radical as postwar reforms, there were many prewar paddy land reorganization programs and attempts (Kerry Smith, *A Time of Crisis*).

9 A newspaper article from the time reports on the mayor's determination to succeed in merging the three municipalities, to create a new town with a population of about 13,000. Akita Sakigake Shinpōsha, "Oga sanchōson o gappei." The article mentions the mayor's plan to merge schools, and progress regarding water irrigation, as well as the opening of nearly two square kilometers of new paddy land. It also points to raised taxes and hope for tourism development: a progressive concept at the time. Despite his determination, however, the mayor was unsuccessful; Wakimoto, Funakoshi, and Futto each remained independent, eventually joining Oga City in 1954, 1955, and 2005, respectively.

10 See Fig. 0.04. Saburō's snow-covered *waraniyo* straw stack is visible, just beneath his house.

11 Onagawa was located approximately 3 kilometers to the southwest.

12 See Dec. 12 entry for more on this incident.

Appendix 1: *Miko Tales*

1 Perhaps actually the Wakimoto Village hamlet of Tayazawa, from which she hailed.

2 Yoshida uses the term *miko* (shrine maiden), but the woman was an *itako* (see Mar. 22 and Nov. 13 entries).

3 The *miko* uses the male personal pronoun *ore* here and later as well (when speaking for Yoshida Kenji).

4 Here, and just below, the *miko* uses the expression *makoto ni* (truly/really) but with hard consonants (in the local dialect) and stretches the vowels for emphasis: *māāgōōdo ni.*

5 A Taoist Kōshin folk belief deity, revered and worshipped by many in the past. Kōshin worship stones are often seen alongside roads and paths in regional Japan today (see Feb. 10 entry for a photograph (Fig. 12.13) of one in Wakimoto in 1935). Also see Embree, *A Japanese Village,* 185–186, and Nakamaki, *Japanese Religions at Home and Abroad,* 17.

6 Today, Yabase is part of Akita City, but in the 1930s it belonged to South Akita County. The county is now much smaller, containing only the Hachirōgata reclaimed land area and several communities on its east side.

7 Although this expression (*Kyō-no-machi*) appears to refer to Kyoto, it could be a reference to Tokyo.

Appendix 6: March 1935 Letter from Shibusawa Keizō to Yoshida Saburō

1 I thank Yoshida Sōkō, Saburō's grandson, for providing scans of this letter and its envelope, and for kindly typing the letter, which greatly simplified the translation process.

2 This is the only request of Shibusawa's that Yoshida appears to have not fulfilled. However, there is no evidence that he did not try.

Appendix 7: April 1937 Letter from Shibusawa Keizō to Yoshida Saburō

1 I thank Yoshida Sōkō, Saburō's grandson, for providing scans of this letter and its envelope, and for kindly typing the letter, which greatly simplified the translation process.

2 Publication of Yoshida's book was delayed; it would not appear for another 13 months.

3 The society (Nihon Minzoku Gakkai)—predecessor to the Japanese Society of Cultural Anthropology (JASCA)—had been founded by Shibusawa and others, including Ishiguro Tadaatsu and Aruga Kizaemon, the previous November, with Shibusawa serving as the first president.

4 Hōya is now part of Nishitōkyō City. The land mentioned here is located at 1 Higashichō, along Kaede Dōri St.

5 Takahashi Buntarō (1903–1948) was a wealthy Hōya resident and an Attic Museum researcher. Major works of his included several ethnographic studies of mountain life (see, e.g., Takahashi, *Yamabito no Jinsei*) and a detailed investigation of his native village of Hōya (1935). See also Christy, *A Discipline on Foot,* and Kimura, "Takahashi Buntarō," 200.

Bibliography

Akita Sakigake Shinpōsha. "Oga sanchōson o gappei: Higashi Oga Machi jitsugen e." *Akita Sakigake Shimbun,* February 29, 1936.

Akita Sakigake Shinpōsha, ed. *Akita Jinmei Daijiten.* Akita: Akita Sakigake Shinpōsha, 2000.

Aoki, Keiichirō, ed. *Nihon Nōmin Undōshi.* Tokyo: Nihon Hyōron Shinsha, 1962.

Arimoto, Yutaka. "Participatory Rural Development in 1930s Japan: The Economic Rehabilitation Movement." *The Developing Economies* 50 (2012): 170–192.

Aruga, Kizaemon. *Nihon Kazoku Seido to Kosaku Seido.* Tokyo: Miraisha, 2000.

Assmann, Stephanie. "Reinventing Culinary Heritage in Northern Japan: Slow Food and Traditional Vegetables." In *Japanese Foodways: Past & Present,* edited by Eric C. Rath and Stephanie Assmann, 243–274. Urbana: University of Illinois Press, 2010.

Bailey, Jackson H. *Ordinary People, Extraordinary Lives: Political and Economic Change in a Tōhoku Village.* Honolulu: University of Hawai'i Press, 1991.

Beardsley, Richard, John W. Hall, and Robert E. Ward. *Village Japan.* Chicago: University of Chicago Press, 1959.

Bernstein, Gail Lee. *Haruko's World: A Japanese Farm Woman and Her Community.* Stanford, CA: Stanford University Press, 1983.

———. *Isami's House: Three Centuries of a Japanese Family.* Berkeley: University of California Press, 2005.

Boyle, John Hunter. *Modern Japan: The American Nexus.* Fort Worth, TX: Harcourt Brace Jovanovich, 1993.

Chō, Tan. "Kindai Nihon ni Okeru Nōson Kajō Ryushutsu to Toshi Rōdōryoku no Keisei." *Gendai Shakai Bunka Kenkyū* 36 (2007): 157–172.

Christy, Alan. *A Discipline on Foot: Inventing Japanese Native Ethnography, 1920–1945.* Lanham, MD: Rowman & Littlefield Publishers, Inc., 2012.

Craig, Christopher. *Middlemen of Modernity: Local Elites and Agricultural Development in Modern Japan.* Honolulu: University of Hawai'i Press, 2022.

Dore, Ronald. P. *Land Reform in Japan.* London: Oxford University Press, 1959.

———. *Shinohata: A Portrait of a Japanese Village.* New York: Pantheon, 1978.

Dower, John W., Anne N. Morse, Jacqueline M. Atkins, and Frederic A. Sharf. *The Brittle Decade: Visualizing Japan in the 1930s.* Boston: Museum of Fine Arts, 2012.

Drea, Edward J. *The 1942 Japanese General Election: Political Mobilization in Wartime Japan.* Lawrence: University of Kansas Center for East Asian Studies, 1979.

Dusinberre, Martin. *Hard Times in the Hometown: A History of Community Survival in Modern Japan.* Honolulu: University of Hawai'i Press, 2012.

Embree, John F. *A Japanese Village: Suye Mura.* London: Keegan Paul, Trench, Trubner and Co., Ltd., 1946 (orig. 1939, University of Chicago Press).

Francks, Penelope. *Technology and Agricultural Development in Pre-war Japan.* New Haven, CT: Yale University Press, 1984.

Fujii, Jinshi. "Yanagita Kunio and the Culture Film: Discovering Everydayness and Creating/Imagining a National Community, 1935–1945." *Arts* 9, no. 54 (2020).

Fujioka, Rika. "The Development of Department Stores in Japan: 1900s–1930s." *Japanese Research in Business History* 31 (2014): 11–27.

Fukutake, Tadashi. *Rural Society in Japan.* Tokyo: University of Tokyo Press, 1980.

Garon, Sheldon. *The State and Labor in Modern Japan*. Berkeley: University of California Press, 1987.

Gotō, Kazuko. "Kakeikiroku yori mita Shōwasenzenki ni Okeru Nōka no Shōhi Kōzō: Tōhoku Chihō ni Okeru Jisakuhinnōsō no Jirei Toshite." *Iwate Daigaku Kyōikubunka Kenkyū Nenpō* 40, no. 1 (1980): 61–75.

Govindasamy, Balatchandirane. "Not Without Blood, Sweat and Tears: Peasant Farming in Prewar Japan." *China Report* 54, no. 2 (2018): 194–212.

———. "Smell of the Soil: Peasant Diaries of Prewar Japan." *China Report* 51, no. 2 (2015): 1–15.

Hayami, Akira. *The Historical Demography of Pre-modern Japan*. Tokyo: Iwanami Shoten, 2001.

Hayami, Yujiro, and Saburo Yamada. *The Agricultural Development of Japan: A Century's Perspective*. Tokyo; University of Tokyo Press, 1991.

Iizuka, Kazuyuki. *Meiji-ki no Chihō Seido to Meibōka*. Tokyo: Yoshikawa Kōbunko, 2017.

Imai, Masayuki. "Senzen no Chihō Nōson Seinen o Torimaku Shisōteki Shakaiteki Jōyko ni Tsuite." *Kokusai Jōmin Bunka Kenkyū Sōsho* 13 (2019): 37–55.

Ishikawa Rikinosuke-ō Tanjō 150-nen Kinen Jigyō Jikkō Iinkai, ed. *Ishikawa Rikinosuke-ō Tanjō 150-nen Kinenshi*. Akita: Ishikawa Rikinosuke-ō Tanjō 150-nen Kinen Jigyō Jikkō Iinkai, 1995.

Ivy, Marilyn. *Discourses of the Vanishing: Modernity, Phantasm, Japan*. Chicago: University of Chicago Press, 1995.

Jacknis, Ira. "George Hunt, Kwakiutl Photographer." In *Anthropology & Photography*, edited by Elizabeth Edwards, 143–151. New Haven, CT: Yale University Press, 1992.

Kimura, Hiroki. "Takahashi Buntarō." In *Taishō Shōwa Kurashi no Hakubutsukan: Minzokugaku no Chichi, Shibusawa Keizō to Achikku Myūzeamu*, edited by Masaki Kondō, 98–101. Tokyo: Kawade Shobō Shinsha, 2001.

Kimura, Masato, and Tosh Minohara. "Introduction." In *Tumultuous Decade: Empire, Society, and Diplomacy in 1930s Japan*, edited by Masato Kimura and Tosh Minohara, xiii–xxii. Toronto: University of Toronto Press, 2013.

Kobayashi, Kōichirō. "The Relationship Between Attic Museum and the Dainippon Youth Group: A Case Study of the Attic Coterie of Ōnishi Goichi." *Kanagawa University Institute for the Study of Folk Culture*, 2020 (working paper).

Kobayashi, Takiji. *Fuzai Jinushi*. Tokyo: Iwanami Shoten, 2010.

Kuwayama, Takami. *Native Anthropology*. Melbourne: Trans Pacific Press, 2004.

Lewis, Michael. *Becoming Apart: National Power and Local Politics in Toyama, 1868–1945*. Cambridge, MA: Harvard University Asia Center, 2000.

Love, Bridget. "Mountain Vegetables and the Politics of Local Flavor in Japan." In *Japanese Foodways: Past & Present*, edited by Eric C. Rath and Stephanie Assmann, 221–242. Urbana: University of Illinois Press, 2010.

Marutani, Hitomi. "Akita Kennai ni Okeru Kekkonshiki no Kaizen Undō: Kōminkan Kekkonshiki o Chūshin ni." *Annual Report of the Akita Prefectural Museum*, 41 (March 2016): 73–86. https://www.akihaku.jp/publication/report/41/aktpmrep41_073-086.pdf.

Masuda, Sayo. *Autobiography of a Geisha*. Translated by G. G. Rowley. New York: Columbia University Press, 2003.

Masui, Taiki, Yoshihiro Sawada, and Satoshi Tsuda. "Changes in Grassland Vegetation in Mt. Kanpu, Oga Peninsula of the Akita Prefecture." *Keikan Engei Kenkyū* 19 (2017): 1–11.

Matanle, Peter, Anthony S. Rausch, and the Shrinking Regions Research Group. *Japan's Shrinking Regions in the 21st Century*. Amherst, NY: Cambria, 2011.

Miyamoto, Tsuneichi. *The Forgotten Japanese: Encounters with Rural Life and Folklore*. Translated by Jeffrey S. Irish. Berkeley, CA: Stone Bridge Press, 2010.

———. *Wasurareta Nihonjin*. Tokyo: Iwanami Shoten, 2008.

Mori, Takemaro. *Kindai Nōmin Undō to Shihai Taisei*. Tokyo: Kashiwashobō, 1985.

Mutō, Tetsujō. *Ugo Kakunodate Chihō ni Okeru Chōchū Sōboku Minzokugaku-teki Shiryō* (Ethnological Findings Concerning Birds and Insects, and Plants and Trees, of the Kakunodate Region). Tokyo: Attic Museum, 1935 (羽後角館地方に於ける鳥虫草木の民俗学的資料).

Nagatsuka, Takashi. *Tsuchi.* Tokyo: Shinchōsha, 2021.

Nakamaki, Hirochika. *Japanese Religions at Home and Abroad: Anthropological Perspectives.* New York: Routledge, 2003.

Nakano, Makiko. *Makiko's Diary: A Merchant Wife in 1910 Kyoto.* Translated by Kazuko Smith. Stanford, CA: Stanford University Press, 1995.

Nihonkai-iki Bunka Kenkyūsho, ed. *Namahage: Sono Men to Shūzoku.* Akita: Akita Bunka Shuppan, 2004.

Nihon no Shokuseikatsu Zenshū Akita Henshū Iinkai. *Akita no Shokuji.* Tokyo: Nōson Gyoson Bunka Kyōkai, 1986.

Nihon Toshokan Kyōkai. *Kindai Nihon Toshokan no Ayumi: Nihon Toshokan Kyōkai Sōritsu Hyakunen Kinen (Chihō-hen).* Tokyo: Nihon Toshokan Kyōkai. 1992.

Nishida Kintarō Tsuisōshū Henshū Iinkai. *Nishida Kintarō Tsuisōshū.* Akita: Akita Bunka Shuppansha, 1972.

Nishida, Yoshiaki. "Dimensions of Change in Twentieth-Century Rural Japan." In *Farmers and Village Life in Twentieth-Century Japan,* edited by Ann Waswo and Nishida Yoshiaki, 7–37. London: Routledge Curzon, 2003.

Nōsangyoson Bunka Kenkyūkai. *Nihon Nōshozenshū.* Tokyo: Nōsangyoson Bunka Kenkyūkai, 1993–.

Nozoe, Kenji. *Dokyumento Dekasegi.* Tokyo: Shakaishisōsha, 1994.

Nozoe, Kenji, and Yōichi Ueda. *Kosaku Nōmin no Shōgen.* Akita: Akita Shobō, 1977.

Oga City, ed. *Ogashishi.* Oga, Akita: Oga City, 1974.

Ogawa, Tōru. "Oga Hantō Wakimoto Mura no Magi to Oyagu ni Tsuite." *Minzokugaku Kenyū* 12, no. 2 (1947): 121–129.

Ōgimachi, Machiko. *In the Shelter of the Pine: A Memoir of Yanagisawa Yoshiyasu and Tokugawa Japan.* Translated by G. G. Rowley. New York: Columbia University Press, 2021.

Ohnuki-Tierney, Emiko. *Rice as Self: Japanese Identities through Time.* Princeton, NJ: Princeton University Press, 1993.

Okada, Ikumi. "Yukiguni no Minzoku: Eigajin ga Kiritotta Akita." *Akita Sakigake Shinpō,* February 12, 2023, 1.

Ōnishi, Goichi. *Nippon Rōnōden.* Tokyo: Nōbunkyō, 1985.

Ōshima, Sachiko. *Rōnō, Nakai Taichirō to Nōmintachi no Kindai.* Kyoto: Shibunkaku Shuppan, 2013.

Partner, Simon. *Koume's World: The Life and Work of a Samurai Woman Before and After the Meiji Restoration.* New York: Columbia University Press, 2023.

———. *The Mayor of Aihara: A Japanese Villager and His Community, 1865–1925.* Berkeley: University of California Press, 2009.

———. *Toshie: A Story of Village Life in Twentieth-Century Japan.* Berkeley: University of California Press, 2004.

Pflugfelder, Gregory M. *Seiji to Daidokoro: Akita-ken Joshi Sanseiken Undōshi.* Tokyo: Domesu Shuppan, 1986.

Reischauer, Edwin O., and Albert M. Craig. *Japan: Tradition and Transformation.* Boston: Houghton Mifflin Company, 1989.

Saga, Junichi. *Memories of Silk and Straw: A Self-Portrait of Small-Town Japan.* Translated by Garry O. Evans. Tokyo: Kodansha International, 1987.

Samuels, Richard J. *Machiavelli's Children: Leaders and Their Legacies in Italy and Japan.* Ithaca, NY: Cornell University Press, 2005.

Shimpo, Mitsuru. *Three Decades in Shiwa: Economic Development and Social Change in a Japanese Farming Community.* Vancouver: University of British Columbia Press, 1976.

Shōji, Shunsaku. *Kindai Nihon no Nōgyō: Nōsei no Genten o Saguru.* Tokyo: Yoshikawa Kōbunkan, 2003.

Smith, Kerry. "Building the Model Village: Rural Revitalization and the Great Depression." In *Farmers and Village Life in Twentieth-Century Japan,* edited by Anne Waswo and Nishida Yoshiaki, 126–155. London: Routledge Curzon, 2003.

———. *A Time of Crisis: Japan, the Great Depression, and Rural Revitalization.* Cambridge, MA: Harvard University Asia Center, 2001.

Smith, Robert J. *Kurusu: The Price of Progress in a Japanese Village, 1951–1975.* Stanford, CA: Stanford University Press, 1978.

Smith, Robert J., and Ella Lury Wiswell. *The Women of Suye Mura.* Chicago: University of Chicago Press, 1982.

Smith, Thomas. C. *The Agrarian Origins of Modern Japan.* Stanford, CA: Stanford University Press, 1959.

Sugiyama, Hideki. *Akita no Tansuigyo.* Akita: Akita Sakigake Shinpō, 1985.

Sutō, Isao. *Wakaze: Dewaguni no Nōgyō o Sasaeta Wakamonotachi.* Akita: Mumyōsha Shuppan, 2015.

Suzuki, Tatsurō. *Meiji, Taishō ni Okeru Nōminsō Bunkai: Suzuki Tatsurō-cho Sakushū.* Akita: Akita Kindai-shi Kenkyūkai, 2016.

Taguchi, Shōichirō. *Kindai Akita-ken Nōgyō-shi no Kenkyū.* Akita: Mishima Shobō, 1984.

Takagi, Shunsuke. *Kinsei Nōmin Nikki no Kenkyū.* Tokyo: Hanawashobō, 2013.

Takahashi, Buntarō. *Musashi Hōya Mura Kyōdo Shiryō.* Tokyo: Attic Museum, 1935.

———. *Yamabito no Jinsei: Shinpi to Fushigi no Minzoku o Tazunete.* Tokyo: Kawade Shobō Shinsha, 2019.

Tama, Shinnosuke. "Changes in the Number of Farm Households and its Regional Character from 1908–1940 in Japan." *Nōgyō Keizai Kenkyū* 86, no. 1 (2014): 1–11.

Tanaka, Kazuhiko. *Wasurerareta Jinruigakusha: Enburī Fusai ga Mita "Nihon no Mura."* Fukuoka: Bōyōsha, 2017.

Teeuwen, Mark, and Kate W. Nakai, eds. *Lust, Commerce, and Corruption: An Account of What I Have Seen and Heard, by an Edo Samurai.* Translated by Mark Teeuwen, Kate W. Nakai, Fumiko Miyazaki, Anne Walthall, and John Breen. New York: Columbia University Press, 2017.

Tomiki, Ryuzō. *Nihon no Minzoku: Akita.* Tokyo: Daiichihōki, 1973.

Washiya, Kaei. "Nishida-sensei wa 'Kokoro' Aru Hito no Naka ni Mada Ikiteiru!" In *Nishida Kintarō Tsuisōshū,* edited by Nishida Kintarō Tsuisōshū Henshū Iinkai, 29–34. Akita: Akita Bunka Shuppansha, 1972.

Waswo, Ann. "In Search of Equity: Japanese Tenant Unions in the 1920s." In *Farmers and Village Life in Twentieth-Century Japan,* edited by Ann Waswo and Nishida Yoshiaki, 79–125. London: Routledge Curzon, 2003.

———. *Japanese Landlords: The Decline of a Rural Elite.* Berkeley: University of California Press, 1977.

Wood, Donald C. *Ogata Mura: Sowing Dissent and Reclaiming Identity in a Japanese Farming Village.* New York: Berghahn Books, 2012.

———. "Photography in the 1930s 'Native' Ethnography of Yoshida Saburō: The Eye of the Seikatsu-sha and the Legacy of Shibusawa Keizō's Attic Museum." *Japanese Review of Cultural Anthropology* 18, no. 1 (2017): 37–68.

———. "Trial and Error, Study and Sweat: Yoshida Saburō's Smallholding in Northeastern Japan, 1935." *Research in Economic Anthropology* 34 (2014): 235–266.

Yagi, Yoshinosuke. "Saikin ni Okeru Kosakusōgi no Dōkō to Kosakurippō." *Keizai Ronsō* 42, no. 5 (1936): 887–906.

Yamaguchi, Kazuo. "Chapter Nine: Development of Agricultural and Marine Industries." In *Japanese Society in the Meiji Era,* vol. IV, edited by Shibusawa Keizō, translated by Aora H. Culbertson and Kimura Michiko, 401–472. Tokyo: Ōbunsha, 1958.

Yamaguchi, Takehide. *Nōmin Undō Nyūmon.* Tokyo: Gōdō Shuppansha, 1960.

Yamamoto, Yoshiko. *The Namahage: A Festival in the Northeast of Japan.* Philadelphia: Institute for the Study of Human Issues, 1978.

Yanagita, Kunio and Miki Shigeru. *Yukiguni no Minzoku.* Tokyo: Daiichi Hōki, 1977.

Yasuoka, Shigeaki. *Japanese Zaibatsu.* Campden FB, 2002. (https://www.campdenfb.com/article/japanese-zaibatsu) (last accessed Feb. 22, 2022).

Yoshida, Saburō. *Mono Iu Hyakushō.* Tokyo: Keiyūsha, 1963.

———. *Oga Kanpūsanroku Hōgen Minzokushi.* Akita: Akita Bunka Shuppan, 1971.

———. *Oga Kanpūsanroku Nōmin Nichiroku.* Tokyo: Attic Museum, 1938.

———. *Oga Kanpūsanroku Nōmin Shuki.* Tokyo: Attic Museum, 1935.

———. *Oga no Koboredane.* Akita: Akita Bunka Shuppansha, 1977.

———. "Shōten Shitemo Kanashii." In *Nishida Kintarō Tsuisōshū,* edited by Nishida Kintarō Tsuisōshū Henshū Iinkai, 61–63. Akita: Akita Bunka Shuppansha, 1972.

INDEX

Page numbers in bold refer to illustrations.

About the Translator

Donald C. Wood is an associate professor in Akita University's college of medicine, in Akita Prefecture, Japan. After spending two years in Akita on the JET Program in the 1990s, he earned a master's degree in anthropology at Texas A&M University and a doctoral degree in cultural anthropology at the University of Tokyo. He has lived in Akita, which he truly loves, since 2001.